D1765625

ON REFERENCE

Landslide risk assessment

E. M. Lee and D. K. C. Jones

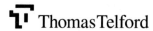 ThomasTelford

Published by Thomas Telford Publishing, Thomas Telford Ltd, 1 Heron Quay, London E14 4JD.
URL: http://www.thomastelford.com

Distributors for Thomas Telford books are
USA: ASCE Press, 1801 Alexander Bell Drive, Reston, VA 20191-4400, USA
Japan: Maruzen Co. Ltd, Book Department, 3–10 Nihonbashi 2-chome, Chuo-ku, Tokyo 103
Australia: DA Books and Journals, 648 Whitehorse Road, Mitcham 3132, Victoria

First published 2004

Also available from Thomas Telford Books
Investigation and management of soft rock cliffs. E M Lee and A R Clark. ISBN 07277 2985 3
Landslides in research, theory and practice. Edited by E Bromhead, N Dixon and M Ibsen. ISBN 07277 2872 5
Coastal Defence – ICE design and practice guide. ISBN 07277 3005 3

A catalogue record for this book is available from the British Library

ISBN: 978 0 7277 3171 5
© Thomas Telford Limited 2004

Typeset by Academic + Technical, Bristol

Contents

Preface and acknowledgements

In recent years it has become fashionable to talk about risk in a variety of contexts and to actively undertake risk assessments. Closer inspection reveals, however, that in many instances the notions of risk that are employed differ greatly or are unclear, and that the risk assessments themselves are often vague and of limited scope. This situation is not helped by the fact that until recently the literature has itself been characterised by diverse views on the nature of risk and the ideal sequence of steps required in the undertaking of a risk assessment. Hence the idea of producing a book to clarify the situation with reference to landslide risk assessment, a subject of considerable interest to both authors.

Originally envisaged as a book of examples to actively assist field scientists in developing landslide risk assessments, the final version has evolved to include more detailed consideration of the nature of landslide hazard and risk, as well as their relevance in a global context. Nevertheless, the worked examples gleaned from personal experience and the literature are considered a key feature of the book and it is hoped that the blend of consultancy experience and academic background has produced a text that all readers will find both stimulating and useful. As Confucius said: 'Man has three ways of learning: firstly, by meditation, this is the noblest; secondly, by imitation, this is the easiest; and thirdly, by experience, this is the most bitter.'

It is important to stress that this is not a book about landsliding but about how landsliding can affect human society. Whilst an understanding of landsliding is crucial in the development of a landslide risk assessment, it is only the first step in the process. Landslide risk assessment is concerned with establishing the likelihood and extent to which future slope failures could adversely impact society. Two landslides can be identical in all physical respects yet pose very different levels of risk, emphasising that the focus should not be on the *what* but on the *so what!*

Other important underlying issues that are worth stressing at the outset are:

- *Uncertainty* is an inevitable part of the risk assessment process because of incomplete knowledge of both the probability of future events and their consequences. All risk assessments need to be supported by a clear statement of the uncertainties in order to inform all the parties of what is known and unknown, and the weight of evidence for what is only partly understood.
- *Precision*. Statements as to the probability of landsliding and the value of adverse consequences can only be *estimates*. The temptation for increasing precision in the risk assessment process needs to be tempered by a degree of pragmatism that reflects the reality of the situation and the limitations of available information. Numbers expressed to many decimal places can provide a false impression of detailed consideration, accuracy and precision. The use of numbers also conceals the fact that the potential for error is great because of the assumptions made and the computations involved. Estimates will generally need to be 'fit for purpose' rather than the product of a lengthy academic research programme. The quality of a landslide risk assessment is related to the extent to which the hazards are recognised, understood and explained – aspects not necessarily related to the extent to which they are quantified.
- *Expert Judgement*. The limited availability of information dictates that many risk assessments will rely on expert judgement. Indeed, Fookes (1997) noted that the art of geological or geotechnical assessment is 'the ability to make rational decisions in the face of imperfect knowledge'. Because of the reliance on judgement, it is important that effort be directed towards ensuring that judgements can be justified through adequate documentation, allowing any reviewer to trace the reasoning behind particular estimates, scores or rankings. Ideally the risk assessment process should involve a group of experts, rather than single individuals, as this facilitates the pooling of knowledge and experience, as well as limiting bias.
- *Defensibility*. The world is becoming less tolerant of the losses caused by landslides, especially those associated with the failure of man-made or man-modified slopes. Increasingly, engineers get blamed for their actions or inactions. In an increasingly litigious world there will be a need for practitioners to demonstrate that they have acted in a professional manner appropriate to the circumstances.
- *Trust*. It is important for practitioners to appreciate that their judgements about risk may provoke considerable disagreement and controversy,

especially if the judgements have implications for property values or the development potential of a site. Acceptance of the risk decisions by affected parties is critical to the successful implementation of landslide risk management strategies. Such acceptance is dependent on the establishment of trust and this, in turn, is dependent on openness, involvement and good communications.

The completion of the book has brought sharply into focus the debt of gratitude owed to many others. Both authors would like to acknowledge the huge influence of Professor Peter Fookes in nurturing the involvement of geomorphologists in the 'real world' of consultancy. They would also like to thank all those colleagues who have, over the years, provided them with stimulation when working on landslides, most especially Professor Denys Brunsden, Dr John Doornkamp, Dr Jim Griffiths, Professor John Hutchinson, Dr Jim Hall, Dr Roger Moore, Dr Fred Baynes (remember Bakuriani), Rick Guthrie, Mike Sweeney (*Shoa*), David Shilston, Saul Pollos, John Charman, Dr Alan Clark and Maggie Sellwood (forever frozen in Whitby). They are also grateful to Jane Pugh and Mina Moskeri of the London School of Economics for producing 34 of the diagrams (EML did the easy ones), and to their wives Claire and Judith for putting up with the seemingly endless disruption to normal life.

MARK LEE (emarklee@compuserve.com)

DAVID JONES (d.k.jones@lse.ac.uk)

May 2004

Scientific notation

Numbers are presented in a variety of ways in this book. In additional to the conventional forms, scientific notation is used. Numbers in scientific notation always consist of an integer to the left of the decimal point and the remainder of the number after the decimal point, with the entire number expressed as a multiple of 10; alternatively, the power of 10 can be expressed as the number of decimal places to the right (positive exponent) or left (negative exponent) of the first significant digit:

Number	Scientific notation	
0.0001	1.0×10^{-4}	1.0 E-4
0.001	1.0×10^{-3}	1.0 E-3
0.01	1.0×10^{-2}	1.0 E-2
0.1	1.0×10^{-1}	1.0 E-1
1	1.0×10^{0}	1.0 E0
10	1.0×10^{1}	1.0 E1
100	1.0×10^{2}	1.0 E2

Beware of the liberal use of decimal places, as they can convey a sense of precision that may not be warranted by the judgements and assumptions made in assessment processes. However, we have generally avoided rounding figures up or down in order to help the reader see where the answers have come from.

1

Background to landslide risk assessment

Introduction

To many readers a book on landslide risk assessment should begin with a consideration of landslides, their nature, form, generation, distribution and potential to cause damage. However, there are good grounds for arguing that this would be a mistake, especially as the main purpose of this book is to explore how the relatively new but rapidly expanding field of risk assessment can be applied in the context of landsliding. Thus, although the nature of landsliding will have to be examined at some point, the most logical place to start is actually with a brief consideration of risk. This is important, for risk is now recognised to be 'a universal concept, inherent in every aspect of life' (Chicken and Posner, 1998), an element of which increasingly pervades all decision-making.

Emphasis on risk and its assessment has grown dramatically in recent decades, raising the subject from relative obscurity as the preserve of the financial sector and certain hazardous industries and high-tech activities, to achieve such prominence that some claim it to be one of the most powerful concepts in modern society; a transformation that is sometimes called the *risk revolution*. Such dramatic growth has produced a huge and confusing – some would say confused – literature that is highly compartmentalised because contributions focus on a wide variety of different subjects, are written from a range of backgrounds or perspectives and tend to use differing terminologies, so that the results often appear contradictory. As a consequence, one of the main purposes of this introductory chapter is to clarify terminology and provide a framework for landslide risk assessment.

As regards landsliding, itself, the relevant aspects are discussed in Chapter 2. For the moment, the following statement provides a sufficient basis to proceed:

It has to be recognised that the terms 'landslides', 'mass movements' or 'slope movements' are convenient umbrella terms which cover a multitude of gravity-dominated processes that displace relatively dry earth materials downslope to lower ground by one or more of the three main mechanisms: falling, flowing (turbulent motion of material with a water content of less than 21%) and sliding (material displaced as a coherent body over a basal discontinuity or shear plane/surface). In reality these three basic mechanisms combine with geologic and topographic conditions to produce a bewildering spectrum of slope failure phenomena, many of which do not conform to stereotype views of landsliding, thereby resulting in serious underestimation of the extent and significance of landslide hazard (Jones, 1995a).

Risk

There is, as yet, no single agreed set of definitions of risk, although there are signs of some convergence of views in the literature. At a generic level, risk can be defined as: *the potential for adverse consequences, loss, harm or detriment* or *the chance of loss.*

Risk is a human concept. It is also a human-centred concept and can only be applied in those instances where *humans and the things that humans value* could be adversely impacted at a foreseeable future date. Growth in knowledge dictates that the threats are many and ever increasing in variety and scale, due mainly to developments in science and technology. Similarly, adverse consequences can be exceptionally diverse (e.g. death, injury, destruction, damage, disruption, reduced efficiency, loss of amenity etc.). As the various elements are not valued similarly by different individuals, groups or even societies, it is necessary to view risk as not merely a human concept but also as a *cultural construct.* In other words, every individual has a slightly different construction of the risk they face in terms of the likelihood and scale of threat and the potential for harm, based on background, past experience, beliefs, etc. (see Section 'Risk evaluation'; pages 24–29). Thus, what may constitute acceptable/tolerable levels of risk vary across the world at any one point in time and also change with time; a good example of the latter is the way in which views as to what constitute acceptable levels of health risk have changed dramatically in Europe over the last hundred years as expectations of health and safety have risen.

On a more scientific level, risk has been defined as: 'A combination of the probability, or frequency, of occurrence of a defined hazard and the magnitude of the consequences of occurrence' (Royal Society, 1992). This definition is useful because it identifies the importance of an agent in generating risk (e.g. landsliding) and the significance of consequences in the determination of risk. However, it makes the assumption that consequences of a hazard have to be detrimental; an assumption which does not hold true for the majority of hazards, including landsliding, where hazard events can and do result in benefits as well as losses. Similar misleading definitions are widespread, as is illustrated by MAFF Guidance FCDPAG4 (MAFF, 2000), where risk is defined as: 'a combination of both the likelihood and consequences of an event'. It cannot be stressed enough that risk is concerned with adverse outcomes. Nobody refers to *the risk of good health* or *the risk of happiness* or *the risk of secure foundations*.

It also has to be emphasised that risk is not concerned with the cause of potentially adverse outcomes; this is *hazard* (see later and Chapter 2). However, due to the flexibility of the English language, there has been some blurring of the distinction between hazard and risk so that the term risk is frequently used in the context of the probability (or likelihood) of failure in technological systems (e.g. nuclear plant meltdown, aircraft accidents etc.), failure of engineering structures (e.g. dam failure) or the occurrence of extreme natural hazard events (e.g. meteor/comet impacts, tsunami generated by the collapse of volcanic cones on islands etc.), where significant levels of loss are assumed but not specified; that is there are high levels of probability that great adverse consequences will be generated by such events. However, references to *increased risk of rain* or *growing avalanche risk* are misleading, for they refer to the increased likelihood of the occurrence of phenomena rather than an increase in the likelihood and scale of adverse consequences.

The real arena of risk is when the potential for adverse outcomes can be compared with the benefits to be gained from actions, activities, locations and events. It is these comparisons, through the processes of risk–benefit analysis, cost–benefit analysis and project appraisal that facilitate informed decisions about future developments. The applications of cost–benefit analysis (CBA) and project appraisal in the context of landsliding are discussed in Chapters 6 and 7. The term risk–benefit analysis is employed in the context of risk taking activity (e.g. financial markets) and in those instances where the comparison of risks and benefits from a development, such as a nuclear facility, is exceedingly complex and involves a combination of technical appraisal and assessment of social tolerance (see Chapter 7).

Following on from the above, rather better definitions of risk are:

- *The likelihood of specified adverse consequences arising from an event, circumstance or action.*
- *The likelihood of differing levels of potential detriment arising from an event, circumstance or action.*
- *An amalgam of the likelihood and magnitude of potential adverse consequences arising from an event, circumstance or action.*

These definitions emphasise that risk is concerned with yet to be realised harms and losses or, to quote the sociologist Giddens (1999), 'risk resides in the future'. Herein lies a fundamental problem, for despite advances in science, humans still have limited knowledge and thus imperfect appreciation of recognised or known risks, while remaining unaware of yet to be identified or even encountered risks that will/may emerge in the future. As a consequence, it is usual to limit consideration of risk within reasonable spatial and temporal limits which, in turn, requires that the following postscript be added to the above definitions:

> . . . *within a stated period and area.*

It is now generally recognised that risk is ubiquitous and that no human action or activity can be considered to be risk free. Risk can be increased, decreased, transferred but rarely eradicated. For example, the eradication of the small-pox virus has not removed entirely the risk, as has been highlighted by recent concerns regarding terrorism. Prevailing approaches to risk management, therefore, seek to keep risk within tolerable levels and to minimise risk where it is economically feasible, commercially necessary or politically desirable. It is, however, possible to talk of *the removal of risk* in specific contexts through the process of risk avoidance. Thus in the case of flood hazard, the development of flood control schemes (reduction of hazard), and relocation of assets away from hazardous areas to *safe* locations (avoidance of risk) may be viable options, but flooding will still pose a risk in other contexts and the relocation may well increase other risks. Thus the once fashionable aspirations for a *zero-risk* society are now seen to be pure fantasy.

But risk management cannot be achieved without some understanding of the levels of risk associated with various activities, localities or phenomena. As a consequence there is the requirement for risk assessment, a process which seeks to establish a measure of the likelihood and scale of potentially adverse consequences within a particular context.

Hazard and vulnerability

Hazard is another human/cultural concept. It is the label applied to objects, organisms, phenomena, events and situations which emphasises *the potential to affect adversely humans and the things that humans value.* Put simply, hazard is a perceived danger, peril, threat or source of harm/loss. Perhaps the best generic definitions of hazard are:

- 'situation that in particular circumstances could lead to harm' (Royal Society, 1992).

- '... the potential for adverse consequences of some primary event, sequence of events or combination of circumstances' (British Standards Institution, 1991).
- 'Threats to humans and what they value: life, well-being, material goods and environment' (Perry, 1981).

For hazard to exist, situations have to arise or circumstances occur where human value systems can be adversely impacted. Put another way, where humans and the things that humans value could be detrimentally affected. Thus landslides on remote uninhabited mountains are not hazards. Even if humans and the things humans value occur in the same area as the landslides, the landslides must be capable of causing costs from a human perspective to be classified as hazard. Clearly size (magnitude) of hazard (landslide) is an important determinant but it is not the sole determinant, for of at least equal importance is the notion of vulnerability.

Vulnerability is also a human concept and can be defined as *the potential to suffer harm, loss or detriment from a human perspective*. It is the reverse of robustness or durability and in some contexts other words may be used, such as fragility.

The recognition of vulnerability results in a simple generalised relationship:

Hazard Event × Vulnerability = Adverse Consequences

which can be expressed as

$$H \times (E \times V) = C$$

where H = specific hazard event, E = total value of all threatened items valued by humans, known as *elements at risk* and including population, artefacts, infrastructure, economic activity, services, amenity etc., V = vulnerability or the proportion of E reduced by the hazard event, C = adverse consequences of hazard event. However, if the impact of a particular type of hazard (i.e. landsliding) is considered in a generic sense, then it immediately becomes obvious that both *elements at risk* and *vulnerability* vary over time. Slight changes in the timing of a hazard event (landslide) can result in very different levels of detriment depending on exactly what is present to be impacted. As a result, some consider that vulnerability can be regarded as consisting of two distinct aspects:

- the level of potential damage, disruption or degree of loss experienced by a particular asset or activity subjected to a hazard event (i.e. a landslide) of a given intensity. For example, the different vulnerability of a timber-framed building to slow ground movement, compared with a rigid concrete structure;

5

- the proportion of time that an asset or person is exposed to the hazard. For example, the contrasting exposure of a person walking underneath an overhanging rock, compared with a stationary beach hut.

In the second case, however, it is not vulnerability that changes in the short-term but the value of elements at risk. A person is equally vulnerable to a falling rock irrespective of whether or not s/he is actually present when the rock falls. What is of crucial importance is *presence* or *absence* at the time when the event occurs. If the person is present within a threatened area, then the potential for loss is greater because the value of elements at risk has been raised by the addition of a highly valued and vulnerable element (i.e. temporal vulnerability). This variation is known as *exposure*. Theoretically, exposure (E^*) is a measure of the ever-changing value (E) and vulnerability (V) of elements at risk within an area, so that

$$(E \times V) = E^*$$

and the general relationship becomes

$$H \times E^* = C$$

or

$$\text{Adverse Consequences} = \text{Hazard} \times \text{Exposure}$$

However, when it comes to developing risk assessments, the problem of temporal vulnerability has to be tackled in a different way. In these cases, the values and vulnerabilities of every different category of element at risk are retained, but each is multiplied by a figure representing the proportion likely to be present at the time of impact by the hazard. This different measure of exposure (E_x) results in the simple equation becoming

$$H \times (E \times V \times E_x) = C$$

where H = specific hazard, E = maximum potential value of all elements at risk, V = vulnerability or the proportion of E that could be reduced by the hazard event, E_x = the proportion of each category of elements at risk present at the time of event. As a consequence, the generic relationship becomes

$$C = H \times \sum (E \times V \times E_x)$$

where E, V and E_x have to be computed for each and every relevant category of element at risk.

Hazards, such as landslides, are therefore the agents that can cause loss and thereby contribute to the generation of risk. As a consequence, their magnitude–frequency characteristics are important in helping to determine risk, but they are not, in themselves, risk. It is incorrect, therefore, to consider the probabilities of differing magnitudes of hazard as estimations of risk; they are merely measures of hazardousness. Thus the 100-year flood, the recurrence interval of a Magnitude 8 earthquake in a particular region or the probability of a major cliff fall are all essential ingredients in estimating risk, but are not actual measures of risk. It has to be seen as regrettable that the wholly respectable field of hazard assessment and prediction should recently have become swamped by and largely subsumed within what some authors (e.g. Sapolsky, 1990) have viewed as a 'collective mania with risk'.

A re-emphasis of the significance of the term *hazard* also requires some clarification of associated terms. Hazards have traditionally been subdivided into *natural* and *human-made* categories, later subdivided into *natural, technological* and *societal* categories with *natural hazard* defined as: 'Those elements of the physical environment harmful to humans and caused by forces extraneous to human society' (after Burton and Kates, 1964).

However, it is now recognised that the growing extent and complexity of human activity has had a dramatic influence on the operation of environmental systems, thereby blurring this simple division, with an increasing proportion of hazards now interpreted as *hybrid hazards* (Fig. 1.1) reflecting their complex origins, including the well-known *na-tech* and *quasi-natural* groups. It is also true that the term *natural hazard* has increasingly fallen out of favour to be replaced by biogeophysical hazard or geophysical hazard (*geohazard* for short).

As landsliding can be a natural phenomenon as well as human-accentuated and human-induced, it has to be included within both geohazard and hybrid hazard categories (quasi-natural and na-tech) under the general label of environmental hazard. The term *environmental hazard* remains highly contentious, with some authors arguing that it refers to hazards that

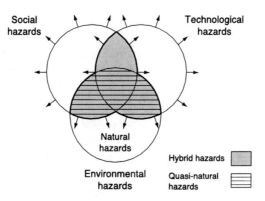

Fig. 1.1. The hazard spectrum (based on Jones, 1993)

7

degrade the quality of the environment (Royal Society, 1992) and others who follow the definition of Kates (1978): 'the threat potential posed to humans or nature by events originating in, or transmitted by, the natural or built environment'. The latter is to be preferred.

Finally, it is relevant at this point to introduce the term *disaster* and to clarify its usage. Disaster is yet another human/cultural construct and a word much beloved by the media. There is an extensive literature on the various aspects of disaster, but the term essentially means a level of adverse consequences sufficiently severe that either outside assistance is required to facilitate the recovery process or that the detrimental effects are long-lasting and debilitating. Disaster is, therefore, a term that refers to relative rather than absolute level of impact. It has to be used in context, which means that there is no absolute scale of disaster, despite the fact that the term is usually associated with sheer numbers and concentrations of casualties and damage. Size of event (e.g. landslide) is not a basis for using the term; indeed, many studies go so far as to state that disasters are more to do with people, societies and economies than with hazard events. It is the significance of vulnerability and exposure in facilitating disaster that has resulted in growing criticism of the term *natural disaster* and led to recommendations that its use should be discontinued. The relationship between disaster, calamity and catastrophe is also open to debate and it is suggested here that use of all three terms should be avoided if at all possible.

From hazard to risk

As has been shown above, risk is concerned with the likelihood and scale of adverse consequences that are the product of the interaction of hazard and vulnerability. Thus both the magnitude–frequency characteristics of hazard and variations in vulnerability and exposure over space and time contribute to risk (Fig. 1.2).

The simple relationships described earlier can, therefore, be turned into basic statements of risk, so long as the probabilities of specific magnitudes of events are known. In the case of landsliding, the following relationship can be produced based on Carrara (1983), Varnes (1984) and Jones (1995b):

$$R_s = P(H_i) \times (E \times V \times E_x)$$

where R_s = *Specific Risk* or the expected degree of loss due to a particular magnitude of landslide (H_i) occurring within a specified area over a given period of time, $P(H_i)$ = *Hazard*, or the probability of a particular magnitude of landslide (H_i) occurring within the specified area and time frame, E = the total *value* of the *elements at risk* threatened by the landslide hazard, V = *Vulnerability* or the proportion of E likely to be affected detrimentally

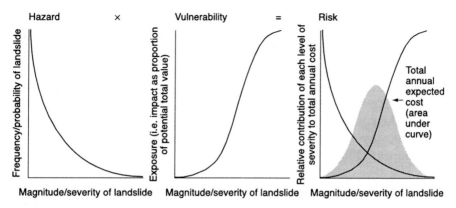

Fig. 1.2. *Diagrammatic representation to show how risk is the product of hazard and vulnerability (based on Coburn and Spence, 1992)*

by the given magnitude of landslide (H_i) expressed as either a percentage of E or on a scale of 0–1, E_x = *Exposure* or the proportion of total value likely to be present and thereby susceptible to being adversely impacted by the landslide, expressed on a scale of 0–1.

Clearly, in reality, the varied components of E have to be disaggregated and each considered separately, which is one reason why risk assessments are so complex. It also has to be recognised that exposure is both complex and variable because it consists of two distinct components (see Section 'Hazard and vulnerability' of this chapter):

- *fixed or static assets*, such as buildings, which change over time;
- *mobile assets*, such as humans or modes of transport, whose presence and concentration display marked short-term fluctuations, as well as long-term trends (see Sections 'Exposure' to 'Multiple outcome consequence models' of Chapter 5).

As a consequence, the above relationship should be rewritten

$$R_s = P(H_i) \times \sum (E \times V \times E_x)$$

Note that E, V and E_x have to be computed for every relevant category of element at risk.

Finally, it should be noted that the *total risk* (R) posed by landsliding is the sum of the calculations of specific risk for the full range of potential magnitudes of landslides:

$$R = \sum R_s \text{ (Landslide events } 1, \ldots, n)$$

Just as there are many types of hazard so too there are numerous categories of risk. An often quoted fundamental division is between natural and human-made risks, otherwise known as external and manufactured risks (Giddens, 1999). This clear-cut distinction does not survive close scrutiny. Perhaps a better division is into societal, technological and environmental risks, although there is much disagreement as to what should be included within each of these groups. The term environmental risk is especially problematic, with some arguing that it is concerned with adverse consequences to the environment (Royal Society, 1992), whereas others take a broader view by extending the definition of environmental hazard proposed by Kates (1978) (see Section 'Hazard and vulnerability' of this chapter) into the following definition of environmental risk: 'An amalgam of the probability and scale of exposure to loss arising from hazards originating in, or trans-mitted by, the physical and built environments'. Such a broad definition clearly includes a wide range of sources of risk, including those emanating from geohazards. It is normal to attribute risk to each particular hazard agent or cause of loss, so that landslide risk (i.e. the risk emanating from land-sliding) is merely one component of environmental risk, irrespective of whether it is *natural* or *manufactured*. But landslide risk can also fall within other categories of risk, such as earthquake risk and volcanic risk.

Viewed from a different perspective, each broad area of human activity is also affected by particular types of risk which may be termed risk domains. Thus there is economic risk, political risk, engineering risk, financial risk and so on. But to many the most important risks are those faced by humans themselves, in terms of death, injury, disability, incapacitation etc., which can be expressed in a number of different ways:

- *Societal risk* is the likelihood of death or injury within a society (usually a nation state or large administrative unit) due to a specified event (e.g. a landslide of particular magnitude) or a particular category of events (e.g. landsliding). It is usually defined as the product of the frequency of occur-rence of a specified hazard and the number of people in a given population suffering (or likely to suffer) from a specified level of harm, and is normally restricted to events potentially capable of causing large-scale loss of life, injury etc. (see Section 'Societal risk' of Chapter 6).
- *Individual risk* is the likelihood of death or injury to an individual and can be calculated by dividing societal risk by the number of individuals exposed to the hazard (although there are other measures of individual risk, see Section 'Individual risk' of Chapter 6).
- *Group risk* is the risk faced by particular groups within society, based on activity (e.g. climbers), occupation (e.g. farmers) or other relevant divi-sions (e.g. males).

Risk and uncertainty

Establishing risk in the case of geohazards such as landsliding requires both an estimation of the likely frequency and magnitude of future events (landsliding), and the likely adverse consequences that will flow for such events. These are difficult tasks (see Section 'Risk assessment' of this chapter) and raise the issue of the relationship between risk and uncertainty.

The classic distinction was made by Frank Knight in 1921 who observed that if one did not know for sure what will happen, but did know the odds (i.e. probabilities), then that is risk, but if one did not even know the odds, then that is uncertainty (Knight, 1964).

Clearly, if all possible outcomes of an event, action or circumstance are known along with the likelihood of each, then the computation of risk would appear to be relatively straightforward. However, it is more normal for there to be a lack of certainty regarding the full range of possible outcomes or the likelihood of each. This is uncertainty in the true sense and probability theory represents the scientific analysis of uncertainty with reference to *likelihood*. Calculations of risk can still be undertaken and will be of value, but the results will become more and more vague as the levels of uncertainty increase. Indeed, there will be a gradual gradation from quantitative estimations of risk to qualitative estimations of risk, with the latter based more and more on expert judgement and informed guesswork. But in some cases there will be so little information regarding possible outcomes and likelihoods that no meaningful guesses can be made. This is ignorance. Landsliding problems tend to be problematic with respect to the assessment of risk because many relevant aspects are still characterised by uncertainty and ignorance. The challenge, therefore, is to develop approaches and techniques so that at least rudimentary assessments of risk can be produced. This important subject will be returned to at the end of the chapter (see Section 'Uncertainty and risk assessment').

Risk assessment

Risk assessment has been defined as 'the structured gathering of the information available about risks and the forming of judgements about them' (DoE, 1995). It is widely claimed that its main function is to produce objective advisory information that effectively links science with decision-making, thereby providing the basis for better informed decisions. The extent to which any risk assessment process can actually be considered to achieve objective results is debatable, and will be considered later.

At the simplest level, risk assessment can be envisaged as addressing four main questions:

1 What can go wrong to cause adverse consequences?
2 What is the probability or frequency of occurrence of adverse consequences?
3 What is the range and distribution of the severity of adverse consequences?

The answers to these three questions then allow the fourth question to be addressed:

4 What can be done, at what cost, to manage and reduce unacceptable risks and detriment?

Several models of risk assessment exist and there is a varied, and sometimes confusing, set of terminology to accompany the various approaches. The range of models reflects the wide variety of contexts within which forms of risk assessment are applied. At a generic level it is possible to recognise the following stages based on DETR (2001):

1 *Hazard identification*: What are the possible problems?
2 *Hazard assessment*: How big might these problems be?
3 *Risk estimation*: What will be their effects?
4 *Risk evaluation*: Do they matter?
5 *Risk assessment*: What should be done about them?

A fuller scheme is presented below and is recommended for general use:

1 *Description of intention.*
2 *Hazard identification.*
3 *Estimation of magnitude and frequency/probability of hazards.*
4 *Identification of consequences.*
5 *Estimation of magnitude of consequences.*
6 *Estimation of frequency/probability of consequences.*

Combining the results of Stages 5 and 6 results in

7 *Risk estimation*, which may be described as a combination of the adverse outcomes or adverse consequences of an intention or event, and the probability of occurrence.

This has, in many cases, been considered to be the conclusion of the risk assessment process. However, the question *so what?* has resulted in the addition of

8 *Risk evaluation*, which is concerned with determining the significance of estimated risks for those affected.

This last stage in the process requires the gathering of information about how affected people feel about and value objects, actions, processes, amenities etc., an area of study generally referred to as *risk perception*. This element of risk assessment is becoming of increasing importance, for in many contexts decisions about risk issues are no longer seen to be the preserve of the scientific elite (experts) but a matter of legitimate concern for all those people that are, or could be, affected (stakeholders).

Risk estimation and risk evaluation together inform risk assessment, where the results should be combined with some form of cost–benefit analysis (or benefit–cost analysis) to result in decisions whether to accept, reduce or minimise the identified risks by the implementation of strategies. Risk management involves the actual implementation of the identified risk reduction strategies.

Building on the above, the framework advocated here (Fig. 1.3) is based loosely upon DoE (1995) and involves the stages described below. It is generally applicable in the case of landsliding. However, it has to be recognised that the form of risk assessments actually undertaken with reference to landsliding will differ greatly in terms of emphasis and details due to variations in scale, purpose and availability of data.

Description of Intention

This crucial initial stage of the risk assessment process involves two activities known as scoping and screening:

1 *Screening* is the process by which it is decided whether or not a risk assessment is required or whether it is required for a particular element, or elements, within a project.
2 *Scoping* defines the focus of enquiry and the spatial and temporal limits of the resulting risk assessment. These may be determined on purely practical grounds such as budget constraints, time constraints, staff availability or data availability. On the other hand, there are strong arguments for limiting the spatial and temporal dimensions of a risk assessment, especially in view of the uncertainties of predicting and valuing longer-term consequences. Very rare, high magnitude events may be excluded for similar reasons.

Scoping and screening should culminate in the production of a clear and detailed Statement of Intent/Description of Intention; an important document in an increasingly litigious society.

Hazard assessment

Hazard assessment is the first *active* stage of a risk assessment and is sometimes referred to as *Hazard Auditing* or *Hazard Accounting*. This stage focuses upon the development of hazard models, together with estimating the nature, size

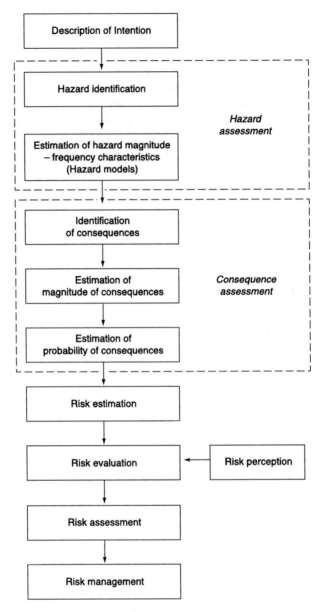

Fig. 1.3. Flow chart of the stages in the risk assessment process

(magnitude) and frequency characteristics of hazardous events (i.e. land-sliding) within the parameters established in the scoping process (see Sections 'Landslide susceptibility and hazard assessment' and 'Hazard models' of Chapter 2). Should this involve or even focus upon the possibility of a major, potentially highly destructive, event (i.e. catastrophic failure) then the hazard assessment must concentrate on establishing the likely magnitude, character, *time to on-set* and *speed of on-set* of the event, or sequence of events. Speed of on-set (i.e. suddenness) and violence of activity are both important in determining the extent to which risk can be reduced through the application of monitoring systems, the issuing of warnings, the establishment of emergency action plans and the adoption of evacuation procedures. Any activities that seek to reduce the element of surprise associated with hazardous events will reduce risk.

Hazard assessment can be a surprisingly difficult task, especially where complex or compound hazards are involved, and it is necessary carefully to analyse the phenomena that produce *hazard*. Geophysical events (geohazards), including landslides, simply do not just occur – there has to be preparatory activity which leads to the build-up of energy, following by an imbalance between forces and constraints which leads to the release of energy which, in turn, is followed by dynamic activity (i.e. the hazard event). This can be described in terms of a 3-phase model of hazard:

Incubation → Trigger → Event

However, this model can be an over-simplification in many situations, especially with reference to the post-trigger sequence of events. First, it is necessary to note that the trigger leads to the *initial event* which may not necessarily be the most important event in terms of size or impact and indeed may not have any great hazard potential. This initial event may, in turn, directly cause further events (i.e. landslides which may be far larger, more violent and have much greater impact potential). Thus the sequence has to be extended to include these *primary events* which flow directly from, and are intimately associated with, the initial event. Thus, a more realistic model of hazard is

Incubation → Trigger → Initial Event → Primary Event

This 4-phase sequence can be envisaged to result in outcomes (Fig. 1.4), which are divisible into three distinct groups: *further hazards, adverse consequences* and *benefits*. Focusing for the moment on *further hazards*, three main groups can be distinguished:

- *Post-event hazards*, which occur after the initial sequence and are a product of the specific geosystem returning towards stability (i.e. the system is relaxing), as is the case with the aftershocks following major earthquakes and the landslides that occur after major slope failure events.
- *Secondary hazards*, which are different geohazards generated by the main hazard event sequence. Examples of these are the destructive tsunami generated by earthquakes and major landslides and the often unexpected

15

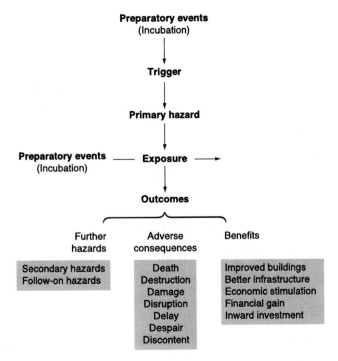

Fig. 1.4. The hazard event cascade. The diagram shows how incubation in both the 'physical' and 'human' systems can interact as a hazard event, resulting in a range of outcomes including further hazards, adverse consequences and benefits

floods caused by the failure of landslide-generated dams which can, under certain circumstances, generate further instability elsewhere due to slope undercutting. The crucial point about secondary hazards is that they occur in areas that are sometimes far removed from the primary hazard location and with a sometimes considerable time lag.

- *Follow-on hazards* are rather different forms of hazard, generated by the primary events but occurring after variable lengths of time. Fire caused by overturned stoves, electrical short-circuits and broken gas pipes, localised flooding caused by broken supply mains or sewers and disease, can all be grouped under this heading.

The hazard assessment stage of a risk assessment should not be restricted to ascertaining the magnitude and frequency characteristics of the principle damaging event (e.g. the major landslide), but should ideally seek to identify possible sequences, or chains, of hazards that could develop from an initial event. Establishing models for such *hazard sequences* can be difficult for medium- to large-scale events which lie intermediate between the simple, small scale, clearly defined and spatially confined event on the one hand

16

and the high-magnitude low-frequency catastrophic event which results in a zone of total destruction. Scenarios have to be developed regarding *what could happen* using a combination of hindsight reviews of relevant past impacts (analogues) and expert judgement.

There are various approaches for expressing the likelihood of hazard, with the following four most commonly encountered:

1 Identified but unspecified likelihood for areas/locations where hazard and vulnerability obviously co-exist but where no data exist, i.e. in the case of a new house built at the base of an un-investigated cliff or on an ancient landslide. Such subjective and vague statements are only of value at the reconnaissance level.

2 Qualitative expressions of likelihood using categories with word labels (maximum of 5). An example of such a scheme is

Category	Indicative probability
Frequent	High probability
Infrequent	Medium probability
Uncommon	Low probability
Rare	Negligible probability

MAFF (2000) advocates the following five category schemes:

Frequent = likely to occur many times during the period of concern.

Probable = likely to occur several times during the period of concern.

Occasional = likely to occur some time during the period of concern.

Remote = unlikely, but possible.

Improbable = can be assumed, for most purposes, that it will not occur.

3 Qualitative expressions of likelihood using numbered categories, ranging from 1 to 5 up to 1 to 10, where 1 represents very rare (exceedingly low probability) to either 5 or 10 which is very high likelihood/certainty.

4 Quantitative estimations of frequency and probability. The former are usually expressed in terms of average number of occurrences per unit time, or return periods/recurrence intervals. Probabilities are also determined with reference to time periods and expressed either as percentage probability or as a decimal value between 0 and 1.

Consequence assessment
The next major stage in the risk assessment process and one that is crucial to the determination of risk is *consequence assessment* which, as its name

suggests, attempts to identify and quantify the full range of adverse consequences arising from the identified patterns and sequences of hazard.

Reference to Section 'Hazard and vulnerability' of this chapter and to Fig. 1.4 indicates that in most situations it is possible to envisage a cascade of hazards and consequences flowing from an initiating event. In the case of major geophysical events, losses are usually attributed to *primary*, *secondary* and *tertiary* processes. For example, earthquake impacts can be grouped into those resulting directly from ground shaking, those produced by the subsequent fires and those associated with delayed economic effects. But even in the case of simple situations, such as a cliff fall, it is possible for the ramifications to spread far and wide due to chains of events. Such chains are known as *event sequences* (Fig. 1.5) and illustrate how chance plays a major role in determining consequences. However, where failures are large or extensive and affect concentrations of human activity and wealth, such as urban areas or industrial complexes, the numbers of identifiable event sequences are huge and produce complex webs of consequences that are exceptionally difficult to unravel and virtually impossible to predict with any degree of accuracy, even with scenarios produced by experts.

Consequence assessment will, except in the cases of small and clearly defined problems and projects, inevitably have to be broad-brush and based on a combination of:

- analogies with known patterns of consequences from similar situations elsewhere (analogues);
- a range of scenarios produced by panels of experts;
- computer modelling.

It is important to recognise that exposure and vulnerability are dynamic (see Section 'Hazard and vulnerability' of this chapter). As risk resides in the future, it is necessary to recognise that hazard events (i.e. landslides) will probably be impacting patterns of land use, socio-economic activity and wealth that have changed from those that exist at the time of analysis. So it is important to envisage that the incubation of hazard and the incubation of exposure evolve on separate paths until they intersect when a hazard event (i.e. slope failure) materialises (see Fig. 1.4).

The problems associated with consequence assessment can be constrained to some extent by the careful use of scoping and clearly specified descriptions of intention. However, care needs to be taken not to make both the scoping and scenario development processes too restrictive. It has to be recognised that even the apparently most unlikely events or combinations of circumstances can, and indeed do, occur. The media has a tendency to refer to such events as *freak* occurrences, somehow implying that they are outside rational explanation and are, therefore, unpredictable. Good scenario

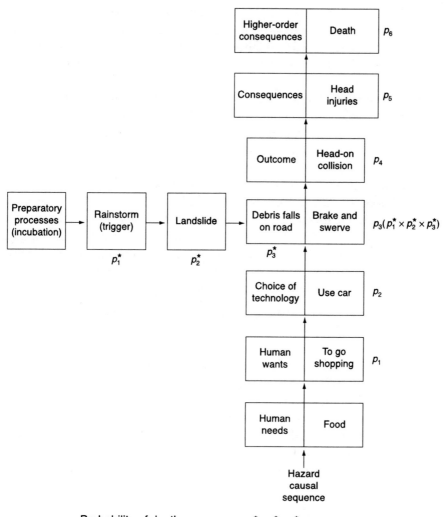

$$\text{Probability of death} = p_1 \times p_2 \times p_3(p_1^* \times p_2^* \times p_3^*) \times p_4 \times p_5 \times p_6$$

Fig. 1.5. Diagram showing how a 'hazard event sequence' (horizontal series of boxes) can interact with human activity to result in an 'accident sequence' (vertical sequence of boxes)

development should, nevertheless, be able to envisage such occurrences within the category of *worst case scenarios*, although computing likelihood and actual levels of consequence may prove to be very difficult.

In this context it has to be remembered that there is usually a very poor correlation between magnitude of primary event and magnitude of adverse consequences. Very minor events can, through chance sequences of inter-action which serve to amplify consequences, result in huge impacts. For

19

example, a minor slump derails a freight train carrying toxic and inflammable chemicals so that it plunges into a river upstream of a town etc. Modelling the likelihood of these chance sequences of interactions can involve the use of similar approaches to those outlined in Section 'Hazard assessment' in this chapter.

The possibility of huge impacts is known as *catastrophe potential*. As societies are increasingly dominated by blame cultures that seek redress and retribution when *things go wrong*, it is important that risk assessments do not ignore the possibilities of catastrophic outcomes, even though the probabilities of occurrence may appear incredibly small.

The end product of consequence assessment should be an expression of the likelihood of differing magnitudes of adverse outcomes resulting from the previously identified hazards. Likelihood of adverse consequences can be expressed in the same ways as likelihood of hazard (see Section 'Hazard assessment' of this chapter). Expressions of the magnitude of adverse consequences, on the other hand, tend to fall within the four groups described below, commencing with the most vague:

1 Where time and/or data are extremely limited, magnitude can be expressed as exceeding a certain threshold, but with no detail. Thus the resulting estimations will be concerned with the likelihood of occurrence of a *damaging landslide, adverse consequences* or *serious adverse consequences*. Such simple *will/will not* alternatives are highly subjective and only of value at a reconnaissance level.

2 Magnitude can be expressed qualitatively using categories described by words to represent differing levels of adverse consequences. It is recommended that no more than five categories are used, as larger numbers tend to result in confusion. The following four categories are advocated in DoE (1995):

Severe

Moderate

Mild

Negligible

Other terms that might be employed include *extremely severe, very severe* and *minor*. Such terms have a wide spectrum of applicability including consequences for human populations, physical structures, infrastructure, environmental quality, ecological status and socio-economic activity, but need to be defined clearly and carefully.

3 It is also possible to express qualitative estimations in terms of numerical gradings or numerical scoring scales, usually from 0–10, with zero meaning no observable effects and 10 signifying total destruction. Alternatively, the

principal adverse consequences can be weighted according to nature/level of harm posed by differing magnitudes of events; for example, ranging from 1 (no significant damage) via 50 (minor structural damage) and 500 (major structural damage or injury) to 2000 (loss of life) (DETR, 2000).

4 The desired objective of risk estimation is for magnitudes of adverse consequences to be expressed quantitatively in terms of numbers of deaths, severe injuries, minor injuries and people displaced or otherwise affected, together with the costs in terms of buildings, infrastructure, economic activity, environment, ecology etc. The ultimate aim here is to get all losses expressed in comparable units and preferably the same units, i.e. monetary terms, but this remains extremely problematic because many adverse consequences are difficult to value (see Section 'Risk evaluation' of this chapter). As a result, it is sometimes the case that one outcome is highlighted, usually death.

Risk estimation

Risk estimation represents an amalgam of magnitude and probability of the hazard(s) and the magnitude and probability of the identified potential adverse consequences. The estimations have to be produced in such a way that the results are understandable and comprehensible by others (i.e. non-specialists) and be capable of comparison with other risks. This implies the need for some standardisation of units and nomenclature.

Achieving this desirable goal is problematic because risk assessments can be undertaken for a great variety of reasons and at very different scales. For example, landslide risk assessments may be undertaken in the contexts of:

1 Specific small-scale engineering/construction projects such as road cuttings, individual buildings or other structures.
2 Clearly defined small sites of known instability.
3 Large-scale engineering projects such as dam sites.
4 Linear construction projects through landslide prone areas, such as roads, railways and pipelines.
5 Large areas of known or potential instability, such as stretches of coastal cliffs or major landslide complexes.
6 Investigations of the possible areally-extensive effects of major engineering projects, such as the widespread response of slopes to new road-lines or the creation of a reservoir lake.
7 A really very-extensive assessment concerned with evaluating landslide risk due to potentially high-impact geohazard events such as earthquakes, widespread changes in land use (e.g. urbanisation) or as a consequence of climate change.

21

	Consequences			
	Severe	Moderate	Mild	Negligible
Probability				
High	high	high	medium/low	Near zero
Medium	high	medium	low	Near zero
Low	high/medium	medium/low	low	Near zero
Negligible	high/medium/low	medium/low	low	Near zero

Fig. 1.6. Example of a risk matrix

Combining the different approaches to estimating the likelihood of the hazard and the magnitude and likelihood of adverse consequences produces a spectrum of risk estimations, extending from the exceptionally vague at the one extreme, to the relatively precise at the other. Vague risk estimations are often heavily based on hazard assessments which are extended to indicate risk. Thus maps of landslide hazard can be interpreted in the context of risk. Similarly, predictions of major hazard events can also form basic risk estimations if specific levels of adverse consequences are assumed to follow. However, it is this use of hazard assessments as a proxy for risk assessment that has led to a blurring of the distinction between hazard and risk.

Disregarding the primitive forms of risk identification mentioned above results in the recognition of three broad groups of risk estimation:

1 *Qualitative risk estimations* are those where both likelihood and adverse consequences are expressed in qualitative terms. They are, therefore, highly subjective estimations. One of the most widely used approaches is to combine the qualitative scoring of adverse consequences and the qualitative scoring of likelihood/probability in order to produce a matrix (Fig. 1.6). Use of four categories for each results in 16 cells; five produces a rather unwieldy 25. The cells of the matrix can be labelled *high, medium, low* or *near zero/negligible* to indicate the level of risk (Fig. 1.6) or, if a numerical scheme is adopted, the numbered co-ordinates of each cell can be summed to yield the same end product, even though the number of cells may be greatly increased. Such processes facilitate the comparison of different risks and provide an indication of relative levels of risk, but it must be recognised that the results represent a gross oversimplification of reality. Indeed, this whole approach is capable of producing serious errors unless used critically and subjected to expert review.

2 *Semi-quantitative risk estimations* are, as the name implies, combinations of qualitative and quantitative measures of likelihood and consequence. More usually it is probabilities of frequency that are known, or assumed, while levels of consequence remain elusive.

3 *Quantitative risk estimations*, where values of detriment are combined with probabilities of occurrence. It must be noted that such an approach does not always produce a single answer, for there can be probability distributions for both magnitude and frequency of the hazard event, as well as a range of possible losses.

The relative merits of *quantification* versus *qualitativism* is one of the seven major debates within risk management discussed in Royal Society (1992) and expanded upon in Hood and Jones (1996). It is quite natural for scientists and engineers to seek to employ numerical values and mathematical relationships in their work on risk and, therefore, to extol the virtues of quantification. Numbers also provide a clear impression of detailed consideration, accuracy and precision; attributes much appreciated by decision-makers. However, it has to be recognised that quantitative risk estimation is much more demanding than qualitative estimation, in that it can require enormous amounts of data and considerable computational effort. The use of numbers also conceals the fact that the potential for error is great because of the assumptions made and the computations involved. Thus, while the value of the quantitative approach is widely accepted, it should not be seen to be the only valid or acceptable form of risk estimation. In many situations, constraints of time, resources and lack of data will make it impossible to produce anything other than a qualitative estimation. Such estimations are useful in their own right and it also must be recognised that qualitative estimations can often be improved by inputs of expert judgement.

The extent to which risk estimation can be viewed as an objective process has also been a matter of much debate. Many scientists still hold to the view that there are two categories of risk assessment based on *objective risk* and *subjective risk*, with the former produced by scientific investigation and the latter resulting from the risk evaluation process (see Section 'Risk evaluation' of this chapter) or what is sometimes called *the human factor*. As a result, quantitative risk estimation was long seen to be the result of the objective process. However, the very fact that the scoping process excludes certain risks and emphasises others, the fact that assumptions are made which reflect investigators' backgrounds, experience or emphasis of the project, together with the recognition that the value of *elements at risk* is often disputed (e.g. what is the value of an ancient monument?), indicates that all risk estimations are, to varying degrees, subjective (e.g. Stern and Fineberg, 1996).

Thus a better distinction is between *statistical* and *non-statistical* evaluations of risk.

Risk evaluation

Risk evaluation is, quite simply, a judgemental process designed to ascertain just how significant the estimated risks are and to establish the best course of future action, including the nature of risk management required. However, the distinction between *risk estimation* and *risk evaluation* has become blurred and subject to different interpretations. Four distinct, but interrelated, elements of risk evaluation can be distinguished:

- Placing values on all aspects of detriment so that they can be compared and combined.
- Comparing the risks and benefits of proposed activities and developments.
- Ascertaining how people who either are, or could be, affected by risk view the risks they face.
- Deciding on what constitutes tolerable levels of risk and acceptable courses of future action.

Many of these aspects will fall beyond the competence of the scientists or engineers responsible for developing the risk estimation but it is, nevertheless, important to recognise their existence and to understand that they form crucial elements of the risk assessment process.

The ultimate aim of risk assessment is to get all losses expressed in the same units, e.g. monetary units. This is extremely problematic because many aspects of adverse consequences or detriment cannot be expressed in monetary terms because there is no market. For example, it is difficult to place a value on *a view* or *the environment* or *amenity*. In addition to these so-called *intangibles*, there are artefacts, monuments, buildings and elements of cultural history that have values to individuals and groups that differ greatly from their replacement costs (e.g. a Norman church, Stonehenge etc.). Economists use a number of techniques, including contingent valuation, in an attempt to provide values for these items (see DoE, 1991; DTLR, 2002), but problems remain. This is especially true in the case of human life where techniques employed to establish what is known as *the value of a statistical life* (see Marin, 1992 and Section 'Loss of life and injury' of Chapter 5) have provoked outrage on the grounds of being morally offensive and leading to discriminatory results. Nevertheless, such approaches are necessary if risk estimations are to be realistic, despite the fact that there is huge scope for disagreement. It is this disagreement that has led to some authors placing the process within risk evaluation, while others see it as part of risk estimation.

Comparing risks and benefits is another important aspect of risk evaluation and suffers from the problem that benefits, as is the case with risk, include both tangible and intangible elements. Therefore, while it might be relatively easy to compare the risks and benefits associated with the construction of a simple structure, for example a retaining wall, such comparisons are normally more difficult and involve the risk perceptions of those affected and the resulting trade-offs between perceived risks and perceived benefits. As stated in DoE (1995):

> ... perceptions of risk and benefit, and of the values of intangibles such as quality of life will lead to different views on where to strike the balance between risks, costs and benefits which will vary from group to group.

Although this statement was made in the context of more general environmental considerations, it has some relevance in the broader aspects of landslide risk assessment.

How people view potential risks and benefits is an important aspect of risk evaluation. This is an aspect of the risk assessment process that many scientists consider problematic because it involves the risk perception of non-specialists. At the simplest level, perception is how people's knowledge, beliefs and attitudes (i.e. their socio-cultural make-up) lead them to interpret the stimuli and information that they receive.

Risk perception, according to Royal Society (1992), 'involves people's beliefs, attitudes, judgements and feelings, as well as the wider social or cultural values and dispositions that people adopt, towards hazards and their benefits'. The result is perceived risk, which is 'the combined evaluation that is made by an individual of the likelihood of an adverse event occurring in the future and its likely consequences' (Royal Society, 1983).

Perceived risk can, therefore, be likened to a risk estimation made by an individual within a bounded rationality framework; in other words, a subjective assessment based on an imperfect view of probable outcomes, biased by belief, experience and personal disposition towards risk. That said, it has to be noted that because risk is a cultural construct, then it follows that if risk is perceived then it is real. As a consequence, perceptions of risk cannot simply be considered as irrelevant where losses may be incurred by individuals. Any reader interested in further developing an understanding of risk perception is directed to Royal Society (1983, 1992), Lofstedt and Frewer (1998) and Slovic (2000).

There has been much work on establishing what are the main influences on human perceptions of risk and attitudes towards risk. Starr (1969) developed three 'laws of acceptability' of risk which have survived in modified form up to the present:

1 risks from an activity are acceptable if they are roughly proportioned to the third power of benefits for that activity;
2 the public will accept risks from voluntary activities, or if chosen voluntarily, that are roughly 1000 times as great as they would tolerate from involuntary hazards, or from hazards imposed upon them, even though the same level of benefits is provided;
3 risk acceptability is inversely proportional to the number at risk.

In the context of point 2 above it has to be noted that natural hazards (geohazards) have traditionally been viewed as events outside *normal life* (hence the term *Acts of God*) and, therefore, as not voluntarily chosen. Despite scientific studies that have produced explanations for geohazards and thereby led to their internalisation within *normal life* (Jones in Hood and Jones, 1996), this perception persists for most geohazards including landslides.

Subsequently, Lichtenstein *et al.* (1978) showed that people tend to overestimate the risk of rare adversities and to underestimate the risk of more common ones; in other words, people underestimate the risks associated with familiar events. By the late 1980s and based largely on the work of Slovic *et al.* (1980), it was generally considered that there were three main influences on risk perception which were, in rank order:

- The horror of the hazard and its outcomes, the feeling of lack of control, fatal consequences, catastrophe potential (*dread*);
- The unknown nature of the hazard, unobservable, unknown, new, delayed in terms of its manifestation (*unknown*);
- The number of people exposed to the risk (*exposure*).

The opposite of *dread* is *controllability* or *preventability*, the opposite of *unknown* is *familiar*, and exposure is simply measured from high to low.

Landsliding normally scores low on all three measures, except in those areas where there is a history of active landsliding, widespread landslide reactivation or progressive coastal cliff retreat, or in the vicinity of a recent landslide disaster. It appears that the statement 'out of sight, out of mind' is genuinely applicable in the case of landsliding.

As with most lines of research, later studies have introduced greater complexity and the risk perception of individuals is currently considered to be influenced by the following factors:

- experience, especially of the activities and hazards involved;
- environmental philosophy;
- world view;
- race, gender and socio-economic status;
- catastrophe potential;

- voluntariness;
- equity and nature of threat to human generations.

The significance of *world view* arises from the work of anthropologists in exploring how risk is culturally constructed. It essentially began with the work of Douglas and Wildavsky (1983), who addressed the question as to why some cultures select certain dangers to worry about, while other cultures see no cause for concern, and has been refined in later studies by Schwarz and Thompson (1990) and Thompson *et al.* (1990) (see Adams, 1995 for general discussion). Of crucial importance were the studies of the ecologist Holling (1979, 1986) who noted that managers of managed ecosystems in the Developing World, when confronted with similar problems, often adopted different management strategies. Such variations appeared explicable only in terms of the respective managers' *beliefs* about nature. Thus, when faced with the need to make decisions based on insufficient information, they appeared to assume that nature behaves in certain ways. Holling reduced these sets of assumptions to three *myths of nature* – *nature benign, nature ephemeral* and *nature perverse/tolerant* – and a fourth, *nature capricious*, was subsequently added by Schwarz and Thompson (1990).

Schwarz and Thompson also produced the typology of two intersecting axes and the illustrative devices of a ball in a landscape to illustrate each of the four myths, as shown in Fig. 1.7. They also argued that a four-fold typology of human nature could be mapped onto the typology of physical nature – *individualists, hierarchists, egalitarians* and *fatalists*, which are portrayed on Fig. 1.7 and adapted for use with reference to attitudes towards geohazards. The four myths of nature represent beliefs about nature and about humankind's place in nature. They are, it is argued, *world views* and the basis for four different *rationalities*. It has to be noted that opinion is divided as to the validity of the approach and its practical use in risk management.

It is clear from the above that lay people's perceptions of risk are both complex and varied. As a consequence, their reactions to risk often differ considerably, both within a population but, more importantly, with the judgements of environmental managers and decision-makers based on estimations of statistical risk involving probabilities produced by scientific investigations. This has, understandably, led to remarks that lay people's subjective assessments have no role in the risk assessment process. However, as virtually all risk assessments are now seen to be subjective to a greater or lesser degree, and the debate about *narrow participation* versus *broad participation* (Royal Society, 1992; Hood and Jones, 1996) has led to an increased tendency for inclusivity in discussions on risk issues, so public participation has come to be seen to be both necessary and useful. As DoE (1995) clearly stated, the

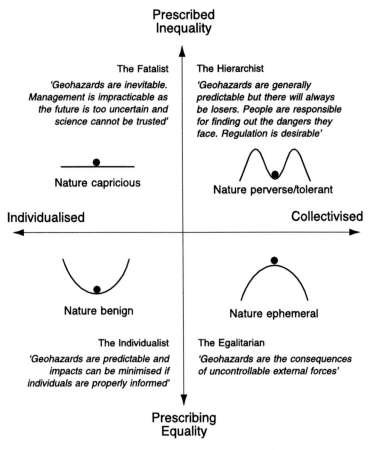

Fig. 1.7. A so-called grid/group matrix showing the 'four myths of nature' (as illustrated by the behaviour of a ball in the landscape) and the associated four 'rationalities', the latter supported by typical statements regarding attitudes to geohazards

view that risk perception has nothing to do with risk evaluation because it is subjective while the scientific approach is objective '... is an idealised view which does not correspond to the world as it is and how decisions are taken'.

This view is reinforced in DETR (2000) 'while risk management decisions should be based on the best scientific information available ... an important step is the creation of a constructive dialogue between stakeholders affected by or interested in risk problems', to which can be added the caveat that all the parties involved must study the relevant data so that outcomes will not be swayed on the basis of beliefs, dogmas and misinformation.

The need for public involvement and the establishment of good dialogues is brought sharply into focus when it is appreciated that risk and *trust* are intimately associated (e.g. Cvetkovich and Lofstedt, 1999). If there is no trust between the public and those undertaking the risk assessment, then the results may not be believed and the proposed outcomes could well be opposed.

Consideration of risk perceptions and dialogues leads on, quite naturally, to a brief examination of *risk communication*. Risk communication evolved from risk perception to gain pre-eminence and has been defined as:

'... Any purposeful exchange of information about health or environmental risks between interested parties. More specifically, risk communication is the act of conveying or transmitting information between parties about:
(a) levels of health or environmental risks;
(b) the significance or meaning of health or environmental risks;
(c) decisions, actions or policies aimed at managing or controlling health or environmental risks.
Interested parties include government agencies, corporations and industry groups, unions, the media, scientists, professional organisations, public interest groups, and individual citizens (Renn in Kasperson and Stallen, 1991).

It is important to recognise that risk communication involves the multiple flows of information between scientists, decision-makers, the media and the public, and occurs at all stages from preliminary hazard assessment through to the development of zoning policies, the establishment of building codes/ordinances and the issuing of forecasts and warnings. Improvements in risk communication will, inevitably, result in improved perceptions of risk and reductions in the levels of risk.

The complex and dynamic nature of risk assessment

It has been shown that risk assessment is a complex, multi-stage process that combines scientific investigation, expert judgement and human values. As both hazard, exposure and vulnerability are dynamic in space and time, risk assessments may have to be periodically up-dated or even re-done in order to accommodate change. Therefore the linear model portrayed in Fig. 1.3 should, in many circumstances, be replaced with a circular model, where risk management is followed by risk monitoring which reveals changes in risk levels or even the recognition of new risks, thereby resulting in the need for new risk assessments.

Landslide risk assessment

Detailed slope investigation and analysis has long been the domain of the geotechnical engineer, and no doubt the question that such specialists will ask is 'why do risk assessment?' when standard approaches to stability analysis have served the engineering profession well over the years. There are, in fact, a number of good reasons, many of which have been detailed by Ho *et al.* (2000), and essentially concern the need to consider broader issues and to build upon and extend what they call the current *pragmatic* or deterministic approach. Indeed, they argue that the traditional approach of obtaining indices from standard tests, the pragmatic discounting of minor factors as unimportant and extreme factors as unlikely, and the use of the Factor of Safety (see Section 'Introduction' of Chapter 4) which they describe as 'an experienced-based index, intended to aid judgement and decision-making', have together only been adequate for routine situations and problems. To further their argument, they go on to point out that:

- geotechnical failures are not a rarity and are sometimes disastrous;
- conventional stability analysis with traditional Factors of Safety are not always capable of averting undesirable performance (Morgenstern, 1991);
- over-designing costs money and often mitigates against achieving an elegant solution;
- excessive use of Codes of Practice can result in unsatisfactory performance if used by the inexperienced. For example, over conservatism in the assessment of natural hillsides could prohibit new development, result in reductions in land values and raise fears regarding the integrity of existing developments, while lack of recognition of potential failures could result in dire consequences.

In the last 5–10 years, risk assessment methods have begun to be applied to quantify the risk of slope failures as a result of growing pressure on the geotechnical community from the following sources:

- clients who want to know their exposure to risk and assign priorities;
- regulatory requirements by governments (e.g. DRM, 1990; Cave, 1992; Graszk and Toulemont, 1996; Garry and Graszk, 1997; Besson *et al.*, 1999);
- public bodies who are increasingly concerned about the adequacy of safety systems or measures, especially after disasters.

Notable advances have occurred in Australia (e.g. Fell, 1994; Moon, 1997; Michael Leiba *et al.*, 2000; Australian Geomechanics Society, 2000), France (e.g. Mompelat, 1994; Rezig, 1998) and Italy (e.g. Eusebio *et al.*, 1996; Aleotti *et al.*, 2000; Cardinali *et al.*, 2002), together with the application of risk-based methods for economic evaluation of coastal landslide and cliff recession problems in the UK (e.g. Hall *et al.*, 2000; Lee *et al.*, 2000, 2001b). Although some of the advances have been the product of research and development studies, risk assessment methods have also been applied to specific sites to tackle real problems (e.g. the risk assessment of ground movement problems in the Ventnor Undercliff, UK; Halcrow, 2003).

Perhaps the greatest advances have been in Hong Kong, where there has been a significant move towards the development of a risk-based approach to supplement the conventional *geotechnical* approach for particular types of slope problems. Ho *et al.* (2000) suggest that the increasing use of risk assessment in Hong Kong was due to a number of factors, including:

(a) There is a growing realisation that there are considerable uncertainties associated with the ground and groundwater conditions, especially given the inherent variability of weathered profiles and tropical rainstorm characteristics; even slopes or other geotechnical structures which have previously been assessed as being up to the required standards can have a fairly high failure rate (e.g. Whitman, 1984, 1997; Wong and Ho, 2000; Morgenstern, 2000).

(b) A risk-based approach assists in the prioritisation of the retrofitting of smaller-sized slopes with less serious failure consequences and the development of a rational strategy to deal with such a category of slopes.

(c) A risk-based approach facilitates the communication of the realities of landslide risk to the public.

Risk assessment is not, therefore, merely a new *fad* or *fashion*, but a broader framework for considering the threat and costs produced by landsliding and for examining how best to manage both landslides and the risk posed by landslides.

As risk assessment is a relatively new procedure, the model presented earlier (Fig. 1.3) is essentially an idealised sequence of stages. It is suggested that landslide risk assessment (LRA) should conform to this sequence, although the actual number of stages undertaken will depend upon the circumstances and contexts within which the risk assessment is undertaken (see Section 'Risk assessment' of this chapter). For example, the LRA for a coastal town built on pre-existing landslides prone to reactivation could involve greater risk evaluation and public participation than one undertaken for a dam site in an arid region or for an isolated construction site. Similarly, the LRA for the coastal town may involve much vaguer and more subjective risk estimation procedures than an investigation of cliff instability threatening a major hotel. However, the sequence of steps in the procedure should be the same, even though the level of detail, the investigatory effort and the techniques employed may vary from case to case. It is these varied approaches and techniques that are examined in detail in the following chapters.

But before embarking on the detail, it needs to be emphasised that the production of an LRA that will withstand close scrutiny requires that the

31

following points be properly addressed (based on DoE, 1995):

1 scoping should be carried out carefully and reasons for limiting the areal extent of the risk assessment, restricting the time-frame or excluding certain risks, should be clearly stated. The Australian Geomechanics Society (2000) have suggested that before embarking on LRA it is important to ensure that the process is focused on relevant issues and that the limits or limitations of the analysis are recognised. This should involve setting clear objectives that define:
   • the site, being the primary area of interest,
   • geographic limits involved in the study of processes that may affect the site,
   • whether the analysis will be limited to addressing only property loss or damage, or will also include injury to persons, loss of life and other adverse consequences,
   • the extent and nature of investigations that will be completed,
   • the type of analysis that will be carried out,
   • the basis for assessment of acceptable and tolerable risks;
2 the description of intention should be clear and precise;
3 assumptions should be made explicit and recorded;
4 the nature of uncertainties should be made explicit and recorded;
5 the process of identifying hazards must be appropriate to the description of intention but take account of the possibility of unintended or extremely rare events;
6 use should be made of historical studies and relevant analogues in establishing the nature of hazardous events;
7 if the magnitude of consequences cannot be estimated directly, then careful use should be made of comparable precedents;
8 risk estimation will nearly always be judgemental and the careful use of expert judgement is to be recommended;
9 the main purpose of risk evaluation should be to identify those risks that are considered to be unacceptable.

Uncertainty and risk assessment

Risk assessment should present a view of the world that recognises uncertainty in the future rather than presenting an over-confident *this will happen* view of what is known. Uncertainty will inevitably be present because of incomplete knowledge of either or both the probability of events and their consequences. As illustrated in Fig. 1.8, where there is good knowledge of both, it is possible to characterise the risk, both quantitatively and reliably. In most instances, however, there will be gaps in

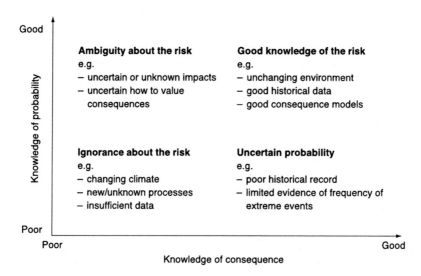

Fig. 1.8. Impact of uncertainty on risk assessment (adapted from Willows et al., 2000)

the available knowledge and the characterisation of the risk will be less reliable.

Uncertainty is often described as comprising three key components; fuzziness, incompleteness and randomness (e.g. Blockley, 1995).

Fuzziness relates to imprecision in parameter definition. For example, a potential landslide event might be described as 'very large', around '1 Mm^3' or 1.76 Mm^3. Each description has a degree of precision attached to it that should be explicitly recognised and defined. It is, of course, important to choose the level of precision (or fuzziness) that is appropriate to the particular problem, so that extremes or spurious precision (e.g. Annual Probability of a Landslide = 0.000 001 35) or unhelpful vagueness (*it might happen*) are avoided.

Incompleteness exists because models are, by definition, simplifications of reality. Models may be incomplete for a number of reasons (Davis and Blockley, 1996):

- factors that are ignored in error, that is the effects should be taken into account but have been missed by mistake;
- factors that have been ignored by choice as their effects are considered to be unimportant;
- factors ignored because of lack of resources, that is their effects are too complex to model with the time available, or the data required are too expensive/difficult to collect;
- factors that cannot be foreseen, including phenomena that were previously unknown.

33

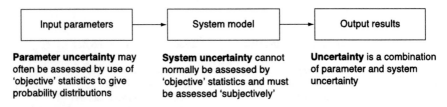

Fig. 1.9. *System and parameter uncertainty (after Hall et al., 1997)*

Even if completely reliable parameter values were input into a model it would still yield imperfect results if the model was incomplete or biased towards a particular process (Fig. 1.9; Blockley, 1980, 1985). Examples of model uncertainty in landslide studies include progressive failure, the influence of lessening of the strength of blocky materials, and the time-dependent softening process (Morgenstern, 1997). One approach to deal with incompleteness is through the explicit acknowledgement of the limitations of current understanding of the hazards and risks in the area of concern. These limitations should be documented and an attempt made to evaluate their significance in terms of the reliability of the risk assessment. An alternative strategy involves the use of more structured and sophisticated methods such as *interval probability theory* (Cui and Blockley, 1990). An interval number, on the range [0, 1], is used to represent the belief in the dependability of a concept (e.g. the probability of an event):

$$P(E) = [S_n(E), S_p(E)]$$

where $P(E)$ is the measure of belief in the dependability of the concept; $S_n(E)$ represents the extent to which it is believed with certainty that concept E is dependable; $1 - S_p(E) = S_n(\bar{E})$ represents the extent to which it is believed with certainty that concept E is not dependable; $S_n(E) - S_p(E)$ represents the extent of uncertainty of belief in the dependability of the concept E.

Three extreme cases illustrate the meaning of this interval measure of belief:

$P(E) = [0, 0]$ represents a belief that E is certainly not dependable;

$P(E) = [1, 1]$ represents a belief that E is certainly dependable;

$P(E) = [0, 1]$ represents a belief that E is unknown.

The interval $S_n(E) = S_p(E)$ implies that there is no uncertainty in the evidence. The value of the approach is that it can be used to express situations where incompleteness is an important issue (Hall *et al.*, 1997).

Randomness relates to the absence of a specific pattern in observations or a series of measurements. When variations in a factor, such as wave height, cannot be explained in terms of a deterministic cause and effect, it is usually referred to as random. Wave height is, however, influenced by weather patterns which, in turn, are a function of many other complex factors. Randomness is generally modelled using *probability theory*.

The types of uncertainty that will impact on most landslide risk assessments include:

1 *data uncertainty*; there will inevitably be limitations to the accuracy and precision of models of complex physical systems such as slopes. Data uncertainty can arise because of measurement errors (random and systematic) or incomplete data. In many situations the available measurements will not correspond to the processes or event types that need to be addressed. As a result, the conditions will need to be inferred (e.g. interpolated, extrapolated or analytically derived) from other information. However, there may be imperfect understanding regarding the processes involved or the applicability of transferring knowledge from one site to another;
2 *environmental (real world) uncertainty*; some aspects of the risk context may defy precise predictions of future conditions. For example:
   - it may not be possible to predict the full range of landslide events that might occur because of the limited understanding of the nature and behaviour of complex physical systems,
   - future choices by governments, businesses or individuals will affect the socio-economic and physical environments in which landslide hazards operate. There is little prospect of reliably predicting what these choices will be.

Uncertainty tends to increase as time frames of analysis extend further and further into the future, or the past. It may be possible to assess the main features of landslide hazard over the next few decades and use this knowledge to develop quantitative assessments of risk. However, as uncertainties accumulate in the future, especially so-called *non-probabilistic* uncertainties such as socio-economic change, so it becomes less and less appropriate to attempt to construct the probability distributions or consequence models that are needed to support quantitative risk assessment.

Stern and Fineberg (1996) suggest that risk assessments often present misleading information about uncertainty. For example, they might give the impression of more scientific certainty or agreement than is the case, or suggest that the uncertainty is a reflection of data availability, when in fact there are differences in interpretation of the ground conditions and in judgement about the significance of the features that have been identified. It is also possible to give the impression that particular risks do not exist, when in fact they simply have not been analysed.

Risk assessment needs to be supported by a clear statement of the uncertainties in order to 'inform all the parties of what is known, what is not known, and the weight of evidence for what is only partially understood' (Stern and Fineberg, 1996). However, this is not a straightforward process, as it is difficult to characterise uncertainty without making the risk appear larger or smaller than the experts believe it to be (Johnson and Slovic, 1995). Careful and elaborate characterisation of the uncertainties might be incomprehensible to non-specialists and unusable by decision-makers. Clearly a balance needs to be found between providing sufficient information on uncertainty to enable decision makers and other participants in the risk management process to be aware of the issues, and diverting attention away from the reality of the situation by dwelling on the unknown.

Risk assessment as a decision-making tool

Risk assessment techniques can be applied at different stages in the decision-making process, from strategic planning to site evaluation. The nature of the risk assessment will generally vary according to the stage in the decision-making process, from a general indication of the threat across an area or region, to statements about the level of risk at a particular site. Wong *et al.* (1997) have termed the different applications as:

1 *global risk assessment* to determine the overall risk to a community posed by landslides. This level of approach can help define the significance of landslide hazards in relation to the overall risks faced by society and allow decision-makers to make a rational allocation of resources for landslide management;
2 *site-specific risk assessment* to determine the nature and significance of the landslide hazards and risk levels at a particular site. This approach can help decision-makers decide whether the risk levels are acceptable and if risk reduction measures are required.

It is important to appreciate that risk assessment is a decision-making tool, not a research tool. It can provide information of value at many levels, ranging from crude qualitative approximations to sophisticated quantitative analyses involving complex computations. The level of precision and sophistication required merely needs to be sufficient for a particular problem or context, so as to enable an adequately informed decision to be made. Thus both qualitative and quantitative approaches have a role to play. Decisions will have to be made despite uncertainty, provided due acknowledgement is made of the limitations. In many instances this may mean ensuring that the estimated level of risk is in the right *ballpark*. With reference to Fig. 1.10, it may be more important to know which of the range of risk levels a site or area falls within than a precise risk value. To illustrate this point, the assessed risk levels at a number of sites have been plotted on

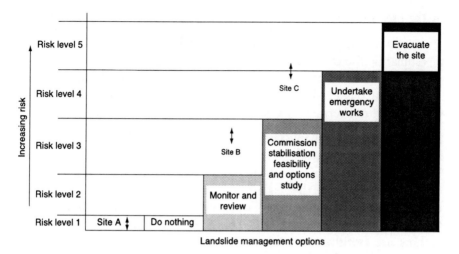

Fig. 1.10. A schematic illustration of the link between risk levels and decision making

Fig. 1.10. Site A clearly falls within risk range 1, suggesting that the most appropriate management strategy might be to do nothing. Site B has a higher estimated risk, indicating that an appropriate strategy would be to commission a feasibility and options study to determine the best ways of reducing the risk. The management decision at Site C may be more problematic as the estimated risk appears to be close to the boundary of two risk levels. The decision maker may choose to be cautious and evacuate the site or seek more detailed evaluation of the risks before having to make a very difficult and politically-sensitive decision.

Structure of the book

It is the intention of this book to examine a variety of approaches to landslide risk assessment and management using relevant examples from across the world. It is not intended to be a manual, but rather to appeal to all people interested in landslides and the problems they cause. The chapters have been arranged to cover the main stages in the risk assessment process. Chapter 2 presents an introduction to the nature of landslide hazard, highlighting the importance of understanding the problem before trying to determine the risk. Qualitative and semi-quantitative risk assessment methods are covered in Chapter 3, including the use of risk registers in the scoping or screening process. However, as the main focus of the book is on quantitative risk assessment, the subsequent chapters are intended to provide an introduction to the key challenges that will need to be overcome by the practitioner:

- estimating the probability of landsliding (Chapter 4);
- estimating the consequences (Chapter 5);
- combining the hazard and consequence estimates to develop a measure of the risk (Chapter 6);
- the transition between risk assessment and risk management (Chapter 7).

Each chapter is accompanied by a series of examples to illustrate the different approaches that are available at each stage of the risk assessment process. However, as each site or problem will tend to be unique, these examples can only hope to give the reader an appreciation of how a particular issue *might* be addressed, rather than how it *should* be addressed. There is a bias towards examples from the UK. This is perhaps inevitable, as both the authors are British. However, every attempt has been made to include examples from other landslide environments, drawing both on personal experience and that of colleagues, as well as from the published literature.

The book is certainly not exhaustive in its coverage as this would be a monumental task. Thus while all kinds of slope failure will be considered, ranging from the small to the catastrophic and including volcanic landslides and earthquake-generated landslides, as well as more standard coastal and inland examples, the various individual types of landslide will not be examined in any great detail. There are many books on landslides that provide such information. This book is about landslide risk and how it can be assessed, and aims to bring together ideas and examples that illustrate the objectives, approaches, difficulties and short-comings of landslide risk assessment.

By way of conclusion it is important to re-emphasise comments made earlier regarding landslides, the risk generating agent that forms the subject of this book. As with all geohazards it is easy to get engrossed in the complexities of landslide form and generation and in the details of investigated failures. Clearly an understanding of landsliding is crucial to the formation of landslide hazard assessment, but it has to be remembered that this is only the first step in the development of a risk assessment. It is always important, therefore, to focus on the central issue which is that landslide risk assessment is concerned with establishing the likelihood and extent to which slope failures could adversely impact humans and the things that humans value. Therefore it is size, depth, suddenness, run-out or travel distance, magnitude–frequency characteristics and secondary hazards that figure among the relevant parameters, together with the distribution of humans and the disposition, value and vulnerability of objects valued by humans. The fact that two potential landslides can be identical in all physical respects yet pose very different levels of risk due to location, shows that the agent is less crucial than the consequences; in other words, emphasis should not focus on the *what* but on the *so what*!

2

Landslide hazard

Introduction

To many people the term 'landslide hazard' evokes images of sudden, dramatic and violently destructive events that cause loss of life and widespread devastation. In the UK, the 1966 Aberfan disaster fits this stereotype. A flowslide from a colliery spoil tip travelled down the valley side at 4.5 m per second, engulfing the Pantglas Infants' School; 144 people died, including 107 out of 250 pupils who had gathered for the morning assembly (Bishop *et al.*, 1969; Miller, 1974).

One of the greatest disasters of the last century occurred in Tadzhikistan in 1949, when a large earthquake (magnitude $M = 7.5$) triggered a series of debris avalanches and flows that buried 33 villages. Estimates of the death toll range from 12 000 to 20 000 (Wesson and Wesson, 1975). Around 25 000 people were killed in the town of Armero, Colombia, by debris flows generated by the 1985 eruption of the Nevado del Ruiz volcano. The flows are believed to have been initiated by pyroclastic flows and travelled over 40 km from the volcano (Herd *et al.*, 1986; Voight, 1990). The 1970 Huascaràn disaster destroyed the town of Yungay, Peru, killing between 15 000 and 20 000 people (Plafker and Ericksen, 1978). An offshore earthquake ($M = 7.7$) triggered a massive rock and ice avalanche from the overhanging face of a mountain peak in the Andes. The resulting turbulent flow of mud and boulders (estimated at 50–100 Mm^3) descended 2700 m, passing down the Rio Shacsha and Santa valleys as a 30 m high wave travelling at an average speed of 270–360 km/hour in the upper 9 km of its path. Perhaps the fastest moving known event was the Elm rock slide which occurred in Switzerland in 1881. It is believed that quarrying operations undermined the mountain slope and caused a 10 Mm^3 landslide that moved across the valley towards the town of Elm at over 80 m per second; 115 people were killed (Buss and Heim, 1881; Heim, 1882, 1932).

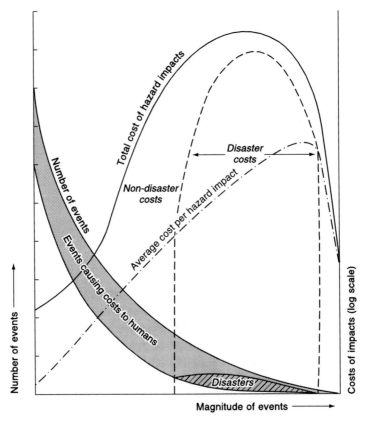

Fig. 2.1. Hypothetical diagram to show the contemporary magnitude–frequency distribution of landslides divided into non-cost-inducing events, cost-inducing hazards and disasters, and the postulated relationships with average and total costs of impact

The landslides described above are examples of high-magnitude low-frequency events (Fig. 2.1) that occur with an estimated frequency of 40–45 per century. Not all have disastrous impacts on humans. For example, the 1965 Hope slide in British Columbia, Canada, had a limited impact despite being one of the largest failures in recorded history (47 Mm^3), while the conspicuous rock avalanche caused by the collapse of part of the summit of Mount Cook, New Zealand, on 14 December 1991 (McSaveney et al., 1991, 1992) caused minimal adverse consequences. However, growing global population and the spread of human activity both increase the potential for such events to produce disastrous outcomes. As a consequence, the published lists of major landslide disasters which normally show fewer than 30 events since 1900 (see D. K. C. Jones, 1992) and the relative unimportance of landsliding in the statistics produced on geohazard disasters, as illustrated by the figure of 3006 deaths in landslide disasters over the period 1900–1976 (39 per year) out of a total of 4.8 million killed in all natural disasters (Red Cross, quoted in Crozier, 1986), have done little to emphasise the significance of landsliding as a hazard. More recently, the Red Cross/Red Crescent (IFRCRCS, 1999), using a different threshold for disaster, quote an average of 14 landslide disasters per year for the decade 1988–97 out of a total of 269 natural disasters per year, with average annual death tolls of 790 (out of 85 000), injured (267/65 000), affected (138 000/144 million) and made homeless (107 000/48 million), but even these figures must be seen as gross underestimations of both present and future significance, especially as landslide impacts are often subsumed under more conspicuous hazards such as earthquakes, volcanoes and tropical revolving storms.

It is crucial to recognise that these conspicuous disasters are only a small part of the landslide story. Although some landslides may be fast moving and highly destructive, the overwhelming majority involve much smaller masses, slower displacements and individually have considerably less impact. However, even relatively minor failures or ground displacements can cause substantial economic losses through property damage, disruption of activity and services, delay and land degradation. The full spectrum of landslide events, from catastrophic failures to periodic slow ground movement, form part of the risk profile and have to be considered as part of the preliminary stage of a risk assessment. Indeed, the greatest total impact of landsliding is probably caused by medium-sized events because of their increased frequency (Fig. 2.1). Reducing the problems associated with small- and medium-scale events will often result in the most cost-effective reduction in risk in an area (Jones, 1995a).

Unfortunately, public perception of landslide hazard does not correspond to reality. While the catastrophe potential of dramatic major failures focuses attention on low frequency or rare events, the cumulative effects of ubiquitous minor failures tends to go little noticed. Thus McGuire *et al.* (2002) were correct in observing that 'landslides are the most widespread and undervalued natural hazard on Earth', although it has to be pointed out that an increasing proportion are quasi-natural or hybrid in origin due to human activity. It is important, therefore, that proper attention be directed to evaluating landslide hazard for the purposes of risk assessment.

Landslide mechanisms and type

The full range of very rapid to extremely slow movements of ground or earth materials can be considered to be landslides because that they all involve the 'movement of rock, debris or earth down a slope' (Cruden, 1991). The reference to 'down a slope' is important as it distinguishes landsliding from

41

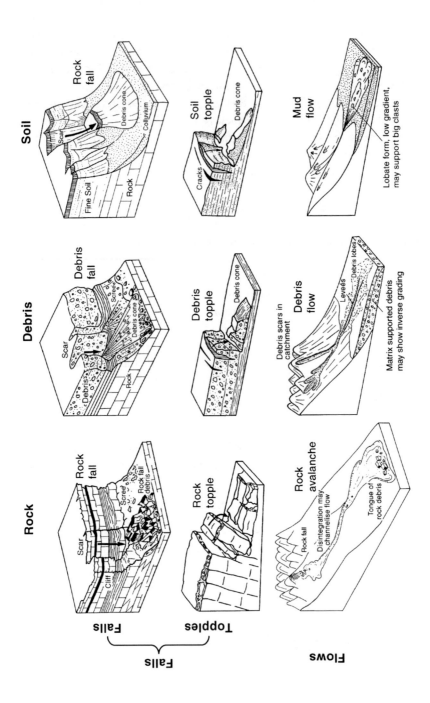

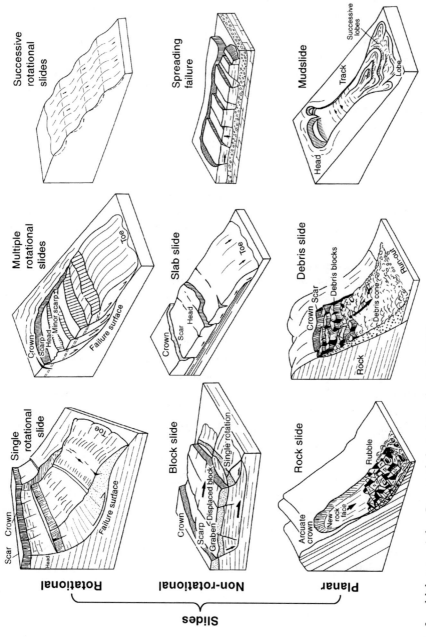

Fig. 2.2. *Landslide types (after Geomorphological Services Ltd, 1986)*

subsidence, although the two may often be intimately associated. Perversely, phenomena described as landslides are not restricted to the land and do not necessarily involve sliding. Phenomena that could be described as landslides can involve five distinct mechanisms (Cruden and Varnes, 1996):

1 *Falling*; involving the detachment of soil or rock from a steep face or cliff, along a surface on which little or no shear displacement occurs. The material then descends through the air by falling, rolling and bouncing.
2 *Toppling*; involving the forward rotation out of a slope of a mass of soil or rock about a point or axis below the centre of gravity of the displaced mass.
3 *Spreading*; the extension of a cohesive soil or rock mass combined with a general subsidence of the fractured mass into softer underlying material. The surface of rupture is not a surface of intense shear. Spreads may result from liquefaction or the flow and extrusion of the softer material.
4 *Flowing*; the turbulent movement of a fluidised mass over a rigid bed, with either water or air as the pore fluid (e.g. like wet concrete or running dry sand; see Hungr *et al.*, 2001). There is a gradation from flows to slides depending on water content and mobility.
5 *Sliding*; the downslope movement of a soil or rock mass as a coherent body on surfaces of rupture or on zones of intense shear strain. Slides are characterised by the presence of a clearly defined shear surface at the contact between the moving mass and the underlying soil or rock.

These basic mechanisms combine with site factors such as topography, lithology, geological structure, hydrogeology, climate and vegetation to produce a remarkable diversity of landslide types (Fig. 2.2); a diversity which Terzaghi (1950) observed 'opens unlimited vistas for the classification enthusiast'. Numerous classifications of landslide types have been developed, most based on shape (morphology), the materials involved and some aspect of the principal mechanisms. The most widely accepted and comprehensive scheme in the English-speaking world is that of Hutchinson (1988). Simpler alternatives are those of Varnes (1978, with modifications proposed in Cruden and Varnes, 1996) and the system developed by the EPOCH project for use in Europe (Dikau *et al.*, 1996; Table 2.1). The overwhelming importance of local site conditions often means that the landslide features observed on the ground are difficult to classify. As a consequence, great care is needed to ensure that recognised landslides are correctly classified.

Landslide behaviour

The focus on what landslides look like must not be allowed to divert attention away from those aspects that influence landslide risk. As will be discussed in Chapter 5, the destructive intensity of a landslide is mainly

Table 2.1. Classification of landslide types proposed by the EPOCH project (EPOCH, 1993; Dikau et al., 1996)

Type	Material		
	Rock	Debris	Soil
Fall	Rockfall	Debris fall	Soil fall
Topple	Rock topple	Debris topple	Soil topple
Slide (rotational)	Single (slump) Multiple Successive	Single Multiple Successive	Single Multiple Successive
Slide (translational). Non-rotational	Block slide	Block slide	Slab slide
Planar	Rock slide	Debris slide	Mudslide
Lateral spreading	Rock spreading	Debris spread	Soil (debris) spreading
Flow	Rock flow (sackung)	Debris flow	Soil flow
Complex (with run-out or change of behaviour downslope; note that nearly all forms develop complex behaviour)	E.g. Rock avalanche	E.g. Flow slide	E.g. Slump-earthflow

Note. A compound landslide is one that consists of more than one type, for example a rotational–translational slide. This should be distinguished from a complex slide where more than one form of failure develops into a second form of movement, that is a change in behaviour downslope by the same material.

related to kinetic parameters, such as velocity and acceleration, along with its dimesions and the material characteristics (Leone *et al.*, 1996). It follows that an understanding of the mechanical behaviour of a landslide at the different stages of its development is needed in order to properly support the development of a risk assessment.

Four stages of landslide movement can be defined (Fig. 2.3; Leroueil *et al.*, 1996):

1 *Pre-failure movements* can involve small displacements that reflect the progressive development of shear surfaces, from isolated shear zones to continuous displacement surfaces. Soils generally behave as viscous materials and can creep under constant stresses. This progressive creep is of considerable importance as it provides forewarning of impending failure (i.e. precursors), involving the development of tension cracks or minor settlement behind a cliff face, and bulging on the slope or at the slope foot. These movements equate with the late incubation stage of hazard outlined in Chapter 1 and it should be noted that the correct

45

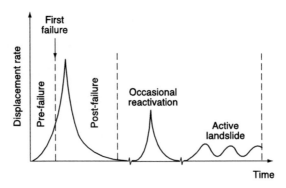

Fig. 2.3. Different stages of slope movement (after Leroueil et al., 1996)

interpretation of precursors can reduce risk by stimulating monitoring and emergency action. Indeed, many major landslides are preceded by months to decades of accelerating creep (Voight, 1978). For example, a hillslope above the Swiss village of Goldau failed in 1806, creating a 10–20 Mm^3 landslide that destroyed the village and killed 475 people. The last words of one of the victims were (Hsu, 1978): 'For thirty years we have waited for the mountain to come, well, now it can wait until I finish stuffing my pipe'.

2 *Failure* occurs when the disturbing forces acting on the slope exceed the forces resisting failure (i.e. it corresponds to the 'trigger' in the model presented in Chapter 1). As disturbing forces increase due, for example, to the oversteepening of a cliff by marine erosion, or the build up of water pressure in the soil, deformations occur as shear strength is mobilised. For a non-brittle material the shear strength will increase to an ultimate value and will then remain constant. However, for a brittle material, such as an overconsolidated clay, the shear strength will rise to a peak value and then reduce, as deformation continues, to a residual value. The residual strength of clay soils can be significantly less than the peak strength. When there is insufficient strength available to counter the disturbing forces the slope or cliff will fail. Movement will occur until equilibrium is restored.

Failure may involve:

• *Peak strength failures*, in which the peak strength (the maximum stress that the material can withstand) is mobilised during failure. After failure the shear resistance mobilised along the shear surface decreases to a lower, residual value. Such slides are often characterised by large, rapid displacements; the velocity is proportional to the difference between the peak and residual strength values.

- *Progressive failure,* in which the fully softened strength or the residual strength is mobilised during first-time failure. The loss of strength is due to strain softening as a result of high lateral stresses developing within the slope. The in-situ materials will expand to relieve this stress. If this expansion is large and the strains concentrated at a particular horizon, for example on bedding planes, then peak strength may be exceeded and a shear surface formed along which strength is lost. This weakened surface can then lead to the development of a deep-seated slide as a shear surface propagates upwards towards the slope crest. Plastic clays (PI > 25%) are prone to strain softening and progressive failure. This style of landsliding generally involves slower and less dramatic movements than peak strength failures.

3 *Post failure movements* include movement of the displaced mass from just after failure until it stops and are, therefore, equivalent to the 'initial' and 'primary' events of the hazard model outlined in Chapter 1. Some of the potential energy of the landslide (a function of slope height and geometry) is lost through friction as the material moves along the shear surface. The remainder is dissipated in the break-up and remoulding of the moving material and in accelerating it to a particular velocity (*kinetic energy*). In brittle materials, where there is a large difference between the peak and residual strengths, the kinetic energy can be very large, giving rise to long run-out landslides.

Important mechanisms involved in the post failure stage include:

- *Mass liquefaction* occurs when the soil structure suddenly fails without exerting its frictional shear resistance, for example as a result of rapid seismic loading. Channelised debris flows can develop as a result of the liquefaction of saturated stream channel or valley floor infills in response to rapid loading by hillside failures. Loess (wind blown silt deposits) are particularly prone to liquefaction. For example, in China around 200 000 people were killed when a magnitude 8.5 earthquake triggered numerous landslides in December 1920 (Close and McCormick, 1922), and at least twice that number are thought to have been buried by the Shansi earthquake of 1556 which killed an estimated 830 000 people (Eiby, 1980). The impact of one of the 1920 loes flows was described by Close and McCormick, 1922):

  The most appalling sight of all was the Valley of the Dead, where seven great slides crashed into the gap in the hills three miles long, killing every living thing in the area except three men and two dogs. The survivors were carried across the valley on the crest of the avalanche, caught in the cross-current of two other slides, whirled in a gigantic vortex, and catapulted to the slope of another hill. With them went house, orchard, and threshing floor, and the farmer has since placidly begun to till the new location to which he was so unceremoniously transported.

In 1988 an earthquake (magnitude 5.5) southwest of Dushanbe, Tajikistan, triggered a series of landslides in a loess area. The slides turned into a massive mudflow ($20 \, \text{Mm}^3$) which travelled about 2 km across an almost flat plain (Ishihara, 1999). It buried more than 100 houses in Gissal Village under 5 m of debris; 270 villagers were killed or reported missing.

- *Sliding surface liquefaction*, which occurs when a shear surface develops in sandy soils and the grains are crushed or comminuted in the shear zone (Sassa, 1996). The resulting volume reduction causes excess pore pressure generation which continues until the effective stress becomes small enough so that no further grain crushing occurs. This mechanism is common in earthquake-triggered debris slides and flows, and can result in devastating large run-out landslides.

- *Remoulded 'quick-clay' behaviour* of sensitive Late Glacial and Post-Glacial marine clays of Scandinavia and eastern Canada (Bentley and Smalley, 1984; Torrance, 1987). Failure of a sensitive clay slope can result in the material remoulding to form a heavy liquid that flows out of the original slide area, often supporting rafts of intact clay. The affected area can quickly spread retrogressively, as the unsupported landslide backscar fails. The quick clay slide that occurred in 1971 at St Jean Vianney, Quebec, involved $6.9 \, \text{Mm}^3$ of material, destroyed 40 homes and killed 31 people (Tavenas *et al.*, 1971).

- *Impact collapse flow slides* can occur as a result of large cliff falls involving weak, high porosity rock. The impact of the fall generates the excess pore fluid pressures required for flow sliding. Hutchinson (1988) has demonstrated that, on chalk coastal cliffs, the susceptibility to flow sliding increases with cliff height. Falls from cliffs below 50 m high tend to remain at the cliff foot as talus. However, large falls from higher cliffs can generate flow slides that travel up to 500 m.

- *Sturzstroms*, or high-speed flows of dry rockfall or rock slide debris. These flows can attain velocities of 30–50 m per second with enormous volumes of material ($10 \, \text{Mm}^3$ to $160 \, \text{Mm}^3$; Hutchinson, 1988), as was displayed in the Huascaran failure of 1970, described earlier. Sturzstroms are believed to involve turbulent grain flow, with the upward dispersive stresses arising from momentum transfer between the colliding debris (Hsu, 1975).

4 *Reactivation* is when part or all of a stationary, but previously failed, mass is involved in new movements, along pre-existing shear surfaces where the materials are at residual strength and are non-brittle. Reactivation can occur when the initial failed mass remains confined along part of the original shear surface. Such failures are generally slow moving with relatively

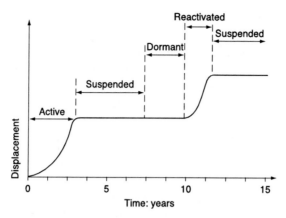

Fig. 2.4. Landslide displacement in different states of activity (after Cruden and Varnes, 1996) (see Table 2.2 for definition of terms)

limited displacements associated with each new phase of movement, although there are circumstances when larger displacements can occur (Hutchinson, 1987). Reactivation is an episodic process, with phases of movement, often associated with periods of heavy rainfall or high ground-water tables, separated by inactivity (Fig. 2.4). Reactivation also results in larger landslide masses becoming broken down into smaller units, a process sometimes called 'block disruption' (Brunsden and Jones, 1972).

An area can contain a mosaic of landslide features of different ages and origins, many of which may be subject to frequent reactivation. Statements about landslide *age* can, therefore, be both confusing and difficult to define; for example, does the age refer to the date of the first-time failure or a significant reactivation? Eyewitness accounts are likely to provide the most reliable dates, although it may not be clear as to whether the event described was a first-time failure or movement within an area of pre-existing landsliding. The so-called Wonder Landslide at Marcle (Hall and Griffiths, 1999) is probably the earliest reliable recorded landslide event in Britain:

on the 17th of February, at six o-clock in the Evening, the Earth began to open, and a Hill with a Rock under it . . . lifted itself up a great height, and began to travel, bearing along with it the Trees that grew upon it, the Sheep folds and Flocks of Sheep abiding there at the same time. In the place from where it was first mov'd it left a gaping distance forty foot broad, and fourscore ells long; the whole Field was about 20 acres. Passing along, it overthrew a Chapel standing in the way. . . (Baker, 1674).

For areas where first-time failures are a regular occurrence, landslides can be dated with a reasonable degree of precision through their presence or absence on aerial photographs taken at different times. For example, the Natural Terrain Landslide Inventory in Hong Kong (see Example 2.1, below) used vegetation cover and re-vegetation rates to provide an indirect measure of the date of shallow hillside failures. Evans and King (1998) report that in Hong Kong, landslide source areas start to re-vegetate within 2–3 years and reach 70% vegetation after 15 to 30 years. This allowed a 4-fold classification to be developed (Evans *et al.*, 1997):

- Class A; totally bare of vegetation, assumed to be less than 2–3 years old;
- Class B; partially bare of vegetation, assumed to be between 2–3 and 30 years old;
- Class C; completely covered in grasses, assumed to be more than 15 years old;
- Class D; covered in shrubs and/or trees, assumed to be more than 25 years old.

The dating of older landslides can, however, prove extremely problematic. In some instances, it might be possible to use methods such as radiocarbon dating to define the inception of landsliding, as has been achieved at Mam Tor, UK, where the first-time failure is believed to have been of Sub-Boreal age (around 3600 BP), as the landslide deposits have overridden tree roots dated at about 3200 BP (Skempton *et al.*, 1989). However, as Voight and Pariseau (1978) commented 'the larger the mass movement (usually) the further back in time the event occurred, and in consequence, the more descriptive and less quantitative is our knowledge of the specifics of the event'. As a result, it can be more useful to focus on the timing of movements (i.e. the *activity state* of the landslide). Cruden and Varnes (1996) have proposed six activity states (Table 2.2) which serve as a useful starting point. However, the definition of 'active' can be too restrictive and

Table 2.2. Landslide activity states (from Cruden and Varnes, 1996)

Activity state	Description
Active	Currently moving
Suspended	Moved within the last 12 months but not currently active
Dormant	An inactive landslide that can be reactivated
Abandoned	A landslide which is no longer affected by its original cause and is no longer likely to be reactivated. For example, the toe of the slide has been protected by a build up of material, such as a floodplain or beach
Stabilised	A landslide which has been protected from its original causes by remedial measures
Relict	An inactive landslide developed under climatic or geomorphological conditions different from those at present

Table 2.3. Landslide activity states (based on Jones and Lee, 1994)

Activity	Description
Active	Currently moving, or a currently unstable site such as an eroding sea-cliff or a site which displays a cyclical pattern of movement with a periodicity of up to 5 years
Suspended	Landslides and sites displaying the potential for movement, but not conforming to the criteria for 'Active' status
Dormant	A landslide or site that remains stable under most conditions, but may be reactivated in part or as a whole by extreme conditions
Inactive	A landslide or site of instability which is stable under prevailing conditions. Four sub-divisions can be identified:
	Abandoned A landslide which is no longer affected by its original cause and is no longer likely to be reactivated. For example, the toe of the slide has been protected by a build up of material, such as a floodplain or beach
	Stabilised A landslide which has been protected from its original causes by remedial measures
	Anchored A landslide that has been stabilised by vegetation growth
	Ancient An inactive landslide developed under climatic, environmental or geomorphological conditions different from those prevailing at present

may become confused with 'suspended'. The failure to define 'inactive' is also problematic, and whether a landslide has to be capable of reactivation in whole or only in part to qualify for being classified as 'dormant' is unclear. As regards the term 'relict', it would appear better to add environmental to the list of conditions so as to include changes in vegetation and human influences. As a consequence, an alternative terminology is presented in Table 2.3, based partly on Jones and Lee (1994).

Potential for landsliding

The risk associated with landslide processes in an area is determined by the type of movements which can be expected to occur and their potential to produce adverse consequences. A wide variety of factors (e.g. material characteristics, geological structure, pore water pressures, topography, slope angle etc.) and causes (e.g. river and coastal erosion, weathering, seepage erosion, high groundwater levels etc.) are important in determining the occurrence of landsliding (see, for example, Crozier, 1986). However, it is the presence or absence of pre-existing shear surfaces or zones within a slope that control the character of landslide activity.

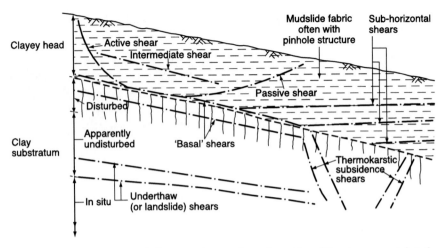

Fig. 2.5. *Principal types of shear surface within clay-rich solifluction material (after Hutchinson, 1991b)*

Three main categories of landslide development have been recognised (Hutchinson, 1988, 1992):

- *First-time failures* of previously unsheared ground, often involving the mobilisation of the peak strength of the material. Such slides are often characterised by large, rapid displacements, particularly if there are significant differences between the peak and residual strength values.
- *Failures on pre-existing shear surfaces of non-landslide origin.* Probably the most important processes which have created such shears are flexural shearing during the folding of sequences of hard rocks and clay-rich strata, and periglacial solifluction (Fig. 2.5).
- *Reactivation of pre-existing landslides,* where part or all of a previous landslide mass is involved in new movements, along pre-existing shear surfaces.

This sub-division leads on to perhaps the simplest yet most profound concept in landslide hazard and risk studies, namely the sub-division of the land surface into (Hutchinson, 1992, 1995):

- areas where pre-existing landslides are present (the *slid areas*);
- areas which have not been affected by landsliding (the *unslid areas*).

In the former there could be potential for reactivation; in the latter new, first-time, landslides may occur in previously unsheared ground.

The importance of this distinction between slid and unslid areas is that once a landslide has occurred it can be made to move again under conditions that the slope, prior to failure, could have resisted. Reactivations can be triggered much more readily than first-time failures. For example, they may be associated with lower rainfall/groundwater level thresholds than first-time failures in the same materials. However, first-time failures may be more frequent than reactivations in those situations where the failed material is remobilised and transported beyond the landslide system or deposited at depths and angles which preclude further instability (e.g. Crozier and Preston, 1998).

Nature of landslide hazards

The hazards associated with landsliding are many and varied because of the great variety of phenomena classified as landslides. Nevertheless, building on the discussion in Chapter 1 and in the preceding sections of this chapter, four phases of hazard can be recognised:

1 Hazards associated with pre-failure movements;
2 Hazards associated with the main phase(s) of movement;
3 Secondary hazards generated as a consequence of movement;
4 Hazards associated with subsequent movements (i.e. reactivation).

Pre-failure hazards

During this phase, actual physical hazard is usually restricted to minor falls, the formation of tension cracks and minimal displacements which can affect infrastructure and buildings. However, people's perceptions of hazard may be raised because of the uncertain threat of impending events of unknown magnitude, violence and timing which may result in alarm. This is the time for monitoring, the development of hazard scenarios and the formulation and implementation of emergency actions.

Main phase hazards

In this phase the main attributes that contribute to hazard are:

- rate of onset (suddenness);
- volume involved in movement;
- depth of movement;
- areal extent of movement;
- speed of movement;
- displacement distance;
- scale of surface distortion;
- development of missiles;
- destructive force mobilised.

53

Table 2.4. Velocity classes for landslides (after WP/WLI, 1995; Cruden and Varnes, 1996)

Velocity class	Description	Velocity: mm/sec	Typical velocity	Nature of impact
7	Extremely rapid			Catastrophe of major violence; exposed buildings totally destroyed and population killed by impact of displaced material or by disaggregation of the displaced mass.
		5×10^{3}	5 m/s	
6	Very rapid			Some lives lost because the landslide velocity is too great to permit all persons to escape; major destruction.
		5×10^{1}	3 m/min	
5	Rapid			Escape and evacuation possible; structure, possessions and equipment destroyed by the displaced mass.
		5×10^{-1}	1.8 m/hr	
4	Moderate			Insensitive structures can be maintained if they are located a short distance in front of the toe of the displaced mass; structures located on the displaced mass are extensively damaged.
		5×10^{-3}	13 m/month	
3	Slow			Roads and insensitive structures can be maintained with frequent and heavy maintenance work, if the movement does not last too long and if differential movements at the margins of the landslide are distributed across a wide zone.
		5×10^{-5}	1.6 m/yr	
2	Very slow			Some permanent structures undamaged, or if they are cracked by the movement, they can be repaired.
		5×10^{-7}	16 mm/yr	
1	Extremely slow			No damage to structures built with precautions.

Any hazard assessment will need to establish the likely importance of these attributes in the context of any specific landslide scenario, especially as the phenomena that threaten human life are not necessarily the same as those that disrupt underground infrastructure. As many of these attributes are examined further in Chapter 5, the present discussion will be brief.

Volume and speed combine to determine kinetic energy which is some-times equated with a landslide's destructive potential. Recorded landslide volumes range from around $1 m^3$ up to the $12 km^3$ of the Flims landslide (Switzerland) and the $20 km^3$ of the Saidmarreh landslide (Iran), both of which have been exceeded by failures on young volcanoes (e.g. $26 km^3$ at Shasta, USA). Speeds range from the imperceptible up to 400 kph (Table 2.4), and have been used to define three broad categories of landslide inten-sity (Cruden and Varnes, 1996): *sluggish*, *intermediate* and *catastrophic*.

While slides tend to be associated with the sluggish and intermediate categories, flows characterise the intermediate to catastrophic end of the spectrum. This is not merely because flows have high kinetic energy but also because they move over the ground surface, thereby mobilising destructive force to overwhelm objects in their path, sometimes at very great distances from their source (travel distance), especially in the case of volcanic mudflows (*lahars*). For example, over 200 lahars occurred during the first rainy season after the June 1991 eruption of Mount Pinatubo, Philippines (Pierson, 1999). The eruption had deposited $5-7 km^3$ of pyroclastic flow deposits and $0.2 km^3$ of tephra on the volcano flanks, filling many steep catchments with up to 220 m depth of material. Typical monsoon lahars were 2–3 m deep, 20–50 m wide, with peak discharges of $750-1000 m^3$/sec. However, lahars triggered by intense typhoon rainfall were up to 11 m/sec, with peak discharges reaching $5000 m^3$/sec. Lowland areas as far away as 50–60 km from the volcano were seriously affected (Pierson *et al.*, 1992). River channels were infilled, leading to widespread flooding, and agricultural land was buried beneath several metres of sand and gravel.

Slides, on the other hand, although not as rapid as flows and with smaller displacement distances, can, nevertheless, prove exceedingly hazardous because of profound disturbance (vertical displacement and tilt) to extensive tracts of ground. Both slides and flows are generally considered more of a threat than falls, despite the fact that falls produce significant missiles, some-times of enormous size. Although falls are confined to specific and somewhat restricted locations (i.e. cliffs and crags) their occurrence can be the initiating event for the generation of destructive flows and slides downslope (e.g. Huascaràn in 1970), illustrating how one type of landslide can turn into other types of landsliding downslope during one event sequence.

Finally, mention must be made of 'speed of onset', which helps to determine 'surprise' and the extent to which hazard potential is realised because people and valued objects have not been removed, protected or otherwise safeguarded.

Four of the more important of these main phase hazards are worth further elaboration at this point, with specific reference to the landslide portrayed in Fig. 2.6 (see also Table 2.5):

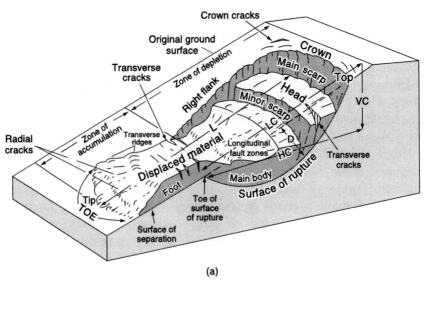

(a)

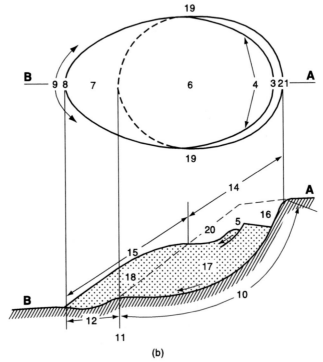

(b)

Fig. 2.6. Landslide morphology: (a) block diagram of an idealised rotational failure-earthflow; (b) landslide features – see Table 2.5 for definitions of numbers (after Cruden and Varnes, 1996; IAEG Commission on Landslide, 1990)

Table 2.5. Definitions of landslide features referred to in Fig. 2.6 (after Cruden and Varnes, 1996)

No.	Name	Definition
1	Crown	Practically undisplaced material adjacent to highest parts of main scarp
2	Main scarp	Steep surface on undisturbed ground at upper edge of landslide caused by movement of displaced material (13, stippled area) away from undisturbed ground, i.e. it is visible part of surface of rupture (10)
3	Top	Highest point of contact between displaced material (13) and main scarp (2)
4	Head	Upper parts of landslide along contact between displaced material and main scarp (2)
5	Minor scarp	Steep surface on the displaced material of landslide produced by differential movements within displaced material
6	Main body	Part of displaced material of landslide that overlies surface of rupture between main scarp (2) and toe of surface of rupture (11)
7	Foot	Portion of landslide that has moved beyond toe of surface of rupture (11) and overlies original ground surface
8	Tip	Point on toe (9) farthest from top (3) of landslide
9	Toe	Lower, usually curved margin of displaced material of a landslide, most distant from main scarp (2)
10	Surface of rupture	Surface that forms (or that has formed) lower boundary of displaced material (13) below original ground surface (20)
11	Toe of surface of rupture	Intersection (usually buried) between lower part of surface of rupture (10) of a landslide and original ground surface (20)
12	Surface of separation	Part of original ground surface (20) now overlain by foot (7) of landslide
13	Displaced material	Material displaced from its original position on slope by movements in landslide; forms both depleted mass (17) and accumulation (18): stippled on Fig. 2.6b
14	Zone of depletion	Area of landslide within which displaced material (13) lies below original ground surface (20)
15	Zone of accumulation	Area of landslide within which displaced material (13) lies above original ground surface (20)
16	Depletion	Volume bounded by main scarp (2), depleted mass (17) and original ground surface (20)
17	Depleted mass	Volume of displaced material (13) that overlies surface of rupture (10) but underlies original ground surface (20)
18	Accumulation	Volume of displaced material (13) that lies above original ground surface (20)
19	Flank	Undisplaced material adjacent to sides of surface of rupture (right and left is as viewed from crown)
20	Original ground surface	Surface of slope that existed before landslide took place

1 *Loss of cliff top land.* Active debris removal and undercutting at the landslide *toe* can generate repeated sequences of pre-failure movements, failure and reactivation events. Each first-time failure will involve the detachment of sections of cliff top, leading to the progressive retreat of the cliffline. Average cliff top recession rates of up to 2 m a year can occur on exposed soft rock coastal cliffs, although losses within an individual year can be much greater (Lee and Clark, 2002). Although individual failures generally tend to cause only small amounts of cliff retreat, the cumulative effects can be dramatic. For example, the Holderness coastline of the UK has retreated by around 2 km over the last 1000 years, resulting in the loss of at least 26 villages listed in the Domesday survey of 1086; 75 Mm^3 of land has been eroded in the last 100 years (Valentin, 1954; Pethick, 1996). A review of the variety of approaches available for predicting cliff recession rates is presented in Lee and Clark (2002).

2 *Differential ground movement and distortion within the main body.* There is a close association between landslide type and the style of ground disturbance that can be experienced. For example, in the single rotational slide shown in Fig. 2.6,
   - vertical settlement and contra-tilt occurs at the slide *head*;
   - the downslope movement of the *main body* can generate significant lateral loads;
   - deferential horizontal and vertical settlement occurs between the individual blocks within the *main body*;
   - compression and uplift (heave) occurs in the *toe area.*

In general, the amount of ground disturbance and the severity of the hazard will be greatest during first-time failure.

In slow moving slides, the mass can be assumed to move down the slope until it reaches a new stable position with a Factor of Safety of 1.0. The displacement can be calculated from (Khalili *et al.*, 1996)

$$\sin \theta_f = \text{FS}_{\text{residual}} \sin \theta_i$$

where θ_f and θ_i are the final and initial positions of the centre of gravity, $\text{FS}_{\text{residual}}$ is the factor of safety using residual shear strength parameters. The rates of displacement can be controlled by the geometry of the shear surface. Compound slides, for example, may suffer violent renewal of movement as a result of brittle failure within the slide mass, on the internal shears that are needed to make this type of movement possible (Hutchinson, 1995). Rotational slides tend to restabilise themselves by further movement, because of their geometry.

Other landslide features (generally hard rock slope failures or flows) can simply remain as stable debris fans or boulder ramparts at the base of the original failure site. For many large run-out landslides, the debris fan can have a very high long-term margin of stability against renewed movement. For example, Hutchinson (1995) records that after the Mayunmarca sturzstrom, the emplaced debris was dry and had a factor of safety of at least 2.9. Similarly, once the Aberfan debris fan had fully drained, its overall factor of safety was 3.7.

3 *Run-out beyond the source area.* The run-out of landslide debris can cause considerable damage. The distance the slide mass will travel and its velocity determine the extent to which the landslide will affect property and people downslope, and their ability to escape. The *travel distance* of landslide debris is generally estimated in terms of a travel angle (defined as the slope of a line joining the *tip* of the debris to the crest of the landslide *main scarp*; this is also termed the *shadow angle*). The key factors that influence the travel distance include (Wong and Ho, 1996):

- *Slope characteristics*, including slope height and gradient, the slope forming materials.
- *Failure mechanism*, including collapse of metastable or loose soil structures leading to the generation of excess pore water pressures during failure, the degree of disintegration of the failed mass, the fluidity of the debris, the nature of the debris movement (e.g. sliding, rolling, bouncing, viscous flow etc.) and the characteristics of the ground surface over which the debris travels (e.g. the response to loading, surface roughness etc.).
- *The condition of the downhill slope*, including the gradient of the deposition area, the existence of irregularities and obstructions, and the presence of well-defined channels etc.

A range of methods are available to assess landslide travel distance, including single body models, continuum models based on numerical methods and observational methods. Amongst the most widely used single body models is the 'sled' model proposed by Sassa (1988). In this model it is assumed that all the energy losses during landslide movement will be dissipated through friction. The apparent angle (or coefficient) of friction, as measured in high-speed ring-shear tests, is used as a measure of the amount of friction loss that can occur during movement and the fluidity of the debris. However, it should be noted that the deposition of the mass during motion can have a marked influence on run-out (Wong and Ho, 1996). As many landslides change mass during motion, the use of a sliding block model in which there is a constant mass can lead to underestimation of the run-out. A number of models have been developed that take account of deposition during motion (e.g. Hungr *et al.*, 1984; Hungr and McClung, 1987; Van Gassen and Cruden, 1989).

The use of advanced numerical models can help determine travel distance in complex failure and irregular topographic settings. Wong and Ho (1996), for example, describe the use of this approach to model

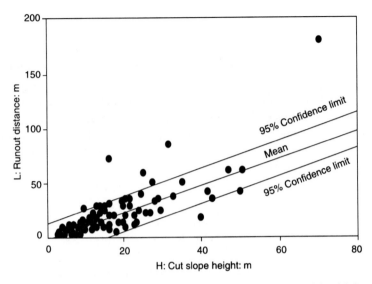

Fig. 2.7. Landslide run-out: relationship between cut slope height and landslide run-out, Hong Kong (after Wong et al., 1997)

the travel distance of major landslides in Hong Kong. The landslide debris was modelled as small discrete elements, among which sliding, impact and separation were allowed to take place. The energy losses associated with sliding, impact and viscous damping were also taken into account.

Empirical observations can also be used to develop a relationship between travel distance and particular slope and material properties. Typical data on debris run-out for different mechanisms and scale of landslides in Hong Kong are given in Fig. 2.7. For slides which break up, and in some cases become flows, the travel distance is usually estimated from the apparent friction angle and landslide volume (Fig. 2.8; Corominas, 1996; Wong and Ho, 1996; Wong et al., 1997; Finlay et al., 1999). However, caution needs to be exercised when extrapolating these relationships to different landslide settings and environments.

4 *The impact of falling boulders or debris.* Falling rocks and boulders can present a major public safety issue in many areas, especially along road or railway cuttings or where rock cliffs back onto bathing beaches. In addition, falls can cause considerable property damage and disruption to services. On many rock slopes, boulders can roll further than the downslope extent of the talus slope. Evans and Hungr (1993) suggest an empirical minimum rockfall shadow angle of 27° (i.e. the angle between the top of the talus slope and the downslope limit of the boulder field). The travel distance of boulders can be modelled using commercially available computer programs, such as the Colorado Rockfall Simulation Program (CRSP). This program

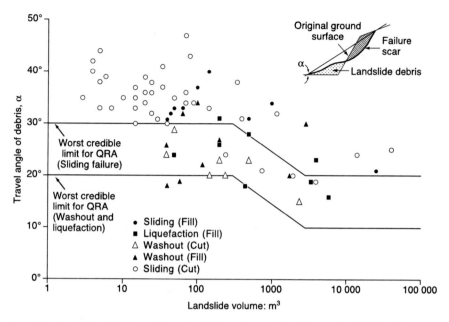

Fig. 2.8. Landslide debris mobility data, Hong Kong (after Ho et al., 2000)

uses slope and rock geometry and material properties, calculating falling rock bounce height, velocity and travel distance.

Secondary hazards

Landslides can also produce *secondary hazards*, which may result in *secondary impacts* that are even greater than those posed by the failure itself. Large landslides, such as rock avalanches and rock slides, can block narrow, steep-sided valleys and form *landslide dams* (Schuster, 1986). Table 2.6 provides an indication of the scale of such landslide dams, which tend to be a feature of seismically active, steep relief mountain areas undergoing uplift and erosion (e.g. the Indus River catchment) or deeply dissected thick sequences of weakly consolidated sediments such as loess.

Landslide dams give rise to two important flood hazards. *Upstream* or *backwater flooding* occurs as a result of the relatively slow impounding of water behind the dam. In 1983, the small town of Thistle, Utah, was inundated by a 200 m long, 50–60 m deep lake that formed behind a landslide dam on the Spanish Fork River (Kaliser and Fleming, 1986). A similar lake that formed behind a rock avalanche dam in 1513 led to the drowning of the Swiss hamlets of Malvaglia and Semione (Eisbacher and Clague, 1984). *Downstream flooding* can occur in response to failure of the landslide dam. The most frequent failure modes are overtopping because of the lack of a

61

Table 2.6. A selection of historic landslide dams (from Schuster, 1986 and Sassa, 1999)

Landslide	Year	Dammed river	Landslide volume: m^3	Dam height: m	Dam width: m	Lake length: km	Lake volume: m^3	Dam failure
Slumgullion earth flow, USA	1200–1300	Lake Fork, Gunnison River	$50–100 \times 10^6$	40	1700	3		No
Usay landslide, Tadzhikistan	1911	Murgab River	$2–2.5 \times 10^6$	300–550	1000	53		Partial
Lower Gros Ventre landslide, USA	1925	Gros Ventre River	38×10^6	70	2400	6.5	80×10^6	Yes
Deixi landslide, China	1933	Min River	150×10^6	255	1300	17	400×10^6	Yes
Tsao-Ling rockslide, Taiwan	1941–42	Chin-Shui-Chi River	250×10^6	217	2000		157×10^6	Yes
Cerro Codor Sencca rockslide, Peru	1945	Mantaro River	5.5×10^6	100	580	21	300×10^6	Yes
Madison Canyon rockslide, USA	1959	Madison River	21×10^6	60–70	1600	10		No
Tanggudong debris slide, China	1967	Yalong River	68×10^6	175	3000	53	680×10^6	Yes
Mayunmarca rock slide, Peru	1974	Mantaro River	1.6×10^9	170	3800	31	670×10^6	Yes
Gupis debris flow, Pakistan	1980	Ghizar River		30	300	5		No
Polallie Creek debris flow, USA	1980	East Fork Hood River	$70–100 \times 10^3$	11	230		105×10^3	Yes
Thistle earth slide, USA	1983	Spanish Fork River	22×10^6	60	600	5	78×10^6	No
Pisque River landslide, Ecuador	1990	Pisque River	3.6×10^6	56	60	2.6	3.6×10^6	Partial
Tunawaea landslide, New Zealand	1991	Tunawaea Stream	4×10^6	70	80	0.9	9×10^5	Yes
Rio Torro landslide, Costa Rica	1992	Rio Torro	3×10^6	100	75	1.2	0.5×10^6	Yes
La Josefina rockslide, Ecuador	1993	Paute River	$20–44 \times 10^6$	100+	500	10	177×10^6	Partial

natural spillway, or breaching due to erosion. Failure of the poorly consolidated landslide debris generally occurs within a year of dam formation. The effects of the resultant flooding can be devastating, partly because of their magnitude and partly because of their unexpected occurrence (surprise). For example, in the winter of 1840–41 a spur of the Nanga Parbat Massif, in what is now Pakistan, failed during an earthquake, completely blocking the River Indus and causing the impoundment of a 60–65 km long lake. The dam, which had been up to 200 m high, breached in early June 1841. The lake emptied in 24 hours, causing what has been called 'the Great Indus flood' during which hundred of villages and towns were swept away. A Sikh army encamped close to the river about 420 km downstream was overwhelmed by a flood of mud and water estimated to be 25 m high; about 500 soldiers were killed (Mason, 1929).

Failure of the Deixi landslide dam on the Min River, China, in 1933 resulted in a wall of water that was 60 m high 3 km downstream. The floodwaters had an average velocity of 30 km/hour and reached the town of Maowen (58 km downstream) in two hours. The total length of valley flooded was 253 km and at least 2423 people were killed (Li Tianchi *et al.*, 1986).

When fast moving landslides enter bodies of water the impact can be devastating as a consequence of the generation of *tsunamis*. In 1792, a landslide off the flanks of Mount Unzen, Japan, generated a tsunami in Ariake Bay which caused 14 500 deaths around the shoreline (McGuire, 1995). In July 1958, a major rockslide (around 30 Mm^3) occurred on the margins of Lituya Bay, Alaska, triggered by a 7.5 Magnitude earthquake on the Fairweather fault. A 30–200 m high mega-tsunami wave was generated that inundated the shoreline of the Bay, with a run-up that probably reached up to 530 m high (Miller, 1960; Mader, 2002). Recent modelling of the event has indicated that the wave was caused by the impact of the slide moving at 110 metres/second into 120 metres of water.

Although the role of landslides in generating tsunamis has been known for many years, it is only recently that the true catastrophic potential has begun to be recognised (Keating and McGuire, 2000). Moore *et al.* (1989) claim that huge submarine landslides on the Hawaiian Ridge have produced tsunamis large enough to affect most of the Pacific margin. For example, Lipman *et al.* (1988) consider that the huge Alika slide from Hawaii Island (1500–2000 Mm^3, ca 105 000 BP) produced tsunamis with run-up heights up to 325 m high on Lanai Island, more than 100 km away. Evidence of catastrophic wave erosion up to 15 m high along the New South Wales coast has also been linked to the collapse of portions of Mauna Loa volcano, over 14 000 km away (Bryant, 1991). Blong (1992) considers that the Pacific Basin may experience comparable submarine landslide generated tsunamis with a frequency of between 1 and 4 per 100 000 years, indicating a global frequency close to the higher figure. It should be noted, however, that some authorities favour an alternative explanation involving meteorite impacts (e.g. A. T. Jones, 1992; Jones and Mader, 1996).

More recently, there is a growing appreciation that at least some of the greatest tsunamis appear to be related to the collapse of volcanoes. The eruption of Mount St Helens, USA, in 1980 graphically illustrated two important points. First, that landsliding can trigger a volcanic eruption (Lipman and Mullineaux, 1981) so that the blast effects, ashfalls etc. produced in 1980 could, quite properly, be interpreted as secondary hazards generated by the landslide. Second, that volcanic edifices are prone to major instability. Structural failure of volcanic edifices is now recognised to be ubiquitous, as can be illustrated by the observation that 75% of large volcanic cones in the Andes have experienced collapse during their lifetime (Francis, 1994). The rate of collapse is unknown, with estimates ranging from 4 per century (Siebert, 1992) to more than double that number. How many collapses involve cones on coastlines or on volcanic islands is also unclear. However, about 70 gigantic slides have been identified around the Hawaiian Islands (Moore *et al.*, 1994) and there is a growing consensus that such failures occur mainly during warm Inter-glacial periods, such as the present, when global sea-levels are high (see McGuire *et al.*, 2002). For example, collapse of the western flank of the Cumbre Vieja volcano on La Palma in the Canary Islands is expected to generate a massive tsunami toward the coasts of Africa, Europe, South America, Newfoundland and possibly even the United States. It is predicted that within six hours of the failure, waves reaching 30 ft (9 m) would arrive in Newfoundland and 45 to 60 ft (14–18 m) waves would hit the shores of South America. Nine hours after the collapse, crests reaching 30 to 70 ft (9–21 m) could surge onto the East Coast of the United States, where the floodwaters might extend inland several kilometres from the coast. Accurate estimates of the scale of economic loss are yet to be made but are thought to be in the multi-trillion US dollar range (see McGuire *et al.*, 2002).

Landslides entering the more confined waters of big rivers, natural lakes or impounded reservoirs displace large volumes of water in the form of waves and surges which can have devastating local effects. The classic example of such an impact is the Vaiont disaster in the Alpine region of north-east Italy. Filling of the reservoir in 1960 resulted in the opening of a 2 km long crack on the flanks of Mount Toc. During the next three years the unstable slope moved downslope by almost 4 m until the evening of 9 October 1963 when a 250 Mm^3 rockslide occurred. The slide entered the reservoir at around 25 m per second, sending a wave around 100 m high over the crest of the concrete dam. The flood wave destroyed five downstream villages and killed around 1900 people (Kiersch, 1964; Hendron and Patton, 1985).

Landslide reactivation hazards

Landslides that have involved shearing and remain confined within the original shear surface (e.g. clay slope or mudrock failures that have not run-out beyond the shear surface) will remain prone to reactivation. The likely displacement during reactivation can be estimated as that required to change the geometry of the slide so that the average shear stress on the shear surface is not increased (Vaughan, 1995). The rate of displacement (ν) can be determined from (Vulliet, 1986; Leroueil *et al.*, 1996):

$$\nu = F(\sigma'_n, \tau) \cdot \tau$$

in which F is a function of the normal effective stress, σ'_n, and the applied shear stress, τ.

The hazards associated with the reactivation stage are similar to those experienced during the main phase of movement (i.e. the failure stage). Of particular significance is the differential ground movement and distortion as the failed mass slides along the pre-existing shear surface or surfaces. As the shear strength of the material along the shear surface is at or close to a residual value, landslide movement can be initiated by smaller triggering events than those needed to initiate first-time failures.

There is considerable evidence that there can be regular periods of land-slide reactivation, especially related to wet year sequences (e.g. Maquaire, 1994, 1997; Schrott and Pasuto, 1999). Lee and Brunsden (2000), for example, demonstrated that for the west Dorset coast, UK, there was a direct relationship between wet-year sequences and landslide activity. Wet year sequences were identified by calculating the cumulative number of years with effective rainfall (moisture balance) greater than the mean value (Fig. 2.9; the mean annual value is 319 mm). This value is set to zero every time it falls below the mean, on the assumption that the groundwater levels only become critical when they rise above the average level. On this basis a number of distinct 'wet year' sequences (of 3 years in length or more) were recognised: 1874–1877; 1914–1917; 1927–1930; 1935–1937; 1958–1960; 1965–1970; 1979–1982; 1993–1998. There have been eight 'wet year' sequences in 130 years, each with a duration of 3–6 years. As indicated on Fig. 2.9, since the late 1950s these 'wet year' sequences have broadly coincided with the timing of major reactivations within the Black Ven-Spittles mudslide complex (prior to the 1950s less attention was probably given to the recording of landslide events, other than the most dramatic failures).

In general, pre-existing landslides present only a minor threat to life, as movements, when they occur, usually involve only slow to extremely slow displacements. Even when large displacements occur, the rate of movement tends to be gentle and not dramatic, as was reported graphically for a slide near Lympne, UK, in 1725 where a farmhouse sank 10–15 m overnight,

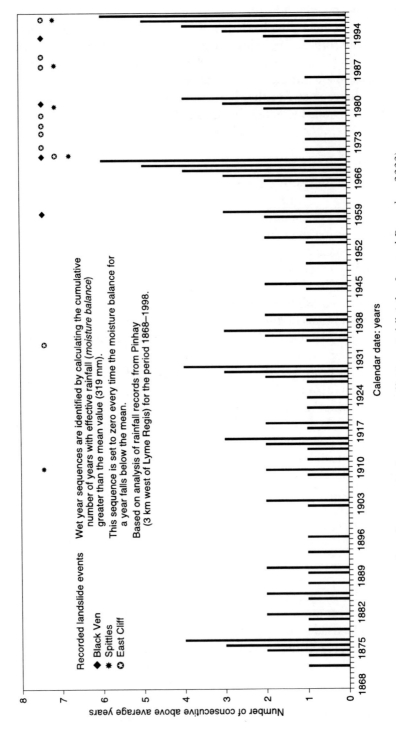

Fig. 2.9. Lyme Regis, UK: number of years with above average effective rainfall (after Lee and Brunsden, 2000)

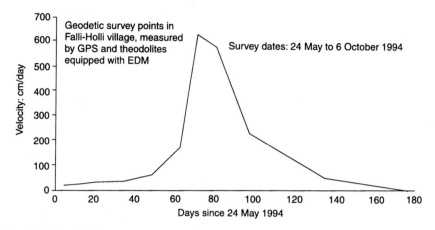

Fig. 2.10. Landslide reactivation displacements: Falli-Holli village, Switzerland (after Lateltin and Bonnard, 1999)

'so gently that the farmer's family were ignorant of it in the morning when they rose, and only discovered it by the door-eaves, which were so jammed as not to admit the door to open' (Gostling, 1756).

However, the cumulative effects of episodes of slow movement can cause considerable damage to buildings, services and infrastructure. Damage due to slope instability can lead to expensive remedial measures or, where repair is considered uneconomic, the abandonment and loss of property. In 1994, for example, reactivation of a large landslide near Freiberg, Switzerland, caused the destruction of 41 homes in the tourist village of Falli-Holli, with estimated losses of US$15 million (Lateltin and Bonnard, 1999). The maximum recorded displacements had been around 7 m per day (Fig. 2.10). Reactivation of the Portuguese Bend landslide in the Palos Verdes Hills, California, resulted in $45 million of damage (1996 prices) to roads, homes and other structures between 1956 and 1959 (Merriam, 1960). Subsequent litigation awarded $41 million (1996 prices) to property owners from the County of Los Angeles, on the grounds that the problems had been initiated by road construction works (Vonder Linden, 1989). Construction of the A21 Sevenoaks Bypass, UK, had to be halted in 1966 because bulldozers cut through innocent looking grass covered lobes of material which proved to be the remains of periglacial landslides and solifluction lobes (Skempton and Weeks, 1976). The affected portion of the route had to be realigned at a (then) cost of £2 million.

In many countries, the greatest inland landslide hazard is associated with the unanticipated reactivation of pre-existing landslides following prolonged heavy rainfall or as a direct result of human activity. Recent research has shown that many of the current instability problems in South Wales, UK, appear to be associated with ancient landslides that have been reactivated by housing development or industrial activity (Halcrow, 1986, 1988).

Acquiring information: landslide investigation

The distinction between slid and unslid areas provides a framework for investigating the nature of the landslide problems in an area (i.e. establishing landslide hazard). Landslide investigation requires an understanding of the slope processes and the relationship of those processes to geomorphology, geology, hydrogeology, climate and vegetation (Australian Geomechanics Society, 2000). From this understanding it should be possible to:

- classify the types of potential landsliding, the physical characteristics of the materials involved and the slide mechanisms. An area or a site may be affected by more than one type of landslide hazard; for example, deep seated landslides on the site, and rockfall and debris flow from above the site. It is important to recognise that hazard assessment is not merely concerned with landsliding generated from within a site or area under consideration, but must also include landslides that might intrude into the site or area from elsewhere;
- assess the physical extent of each potential landslide being considered, including the location, areal extent and volume involved;
- assess the likely triggering events;
- estimate the resulting anticipated travel distances and velocities of movement;
- address the possibility of fast acting processes, such as flows and falls, from which people will find it more difficult to escape.

There are many books that describe the range of approaches to the investigation of landslides, including Brunsden and Prior (1984), Bromhead (1986), Turner and Schuster (1996) and Clark *et al.* (1996). Many investigations will comprise a series of stages, including:

1. A *desk study* involving a thorough search of the relevant documentation, including geological, topographic and soil maps, technical papers and records, so as to develop a preliminary ground model of the area or site (e.g. Fookes, 1997; Fookes *et al.*, 2000; Charman, 2001).
2. *Historical research.* In many countries, there exist a wide range of sources that can provide useful information on the past occurrence of events, including aerial photographs, topographic maps, satellite imagery, public records, local newspapers, consultants' reports, scientific papers, journals, diaries, oral histories etc. (see Brunsden *et al.*, 1995). Among the many descriptions of landslides that can be found in historical sources, that

of the preacher John Wesley of the 1755 movements at Whitestone Cliff on the edge of the North York Moors, is particularly fascinating:

On Tuesday March 25 last, being the week before Easter, many persons observed a great noise near the ridge of mountains in Yorkshire, called Black Hamilton. It was observed chiefly in the south-west side of the mountain, about a mile from the course where the Hamilton races are run; near a ridge of rocks, commonly called Whitestone Cliffs, or Whitson-White-Mare; two miles from Sutton, about five from Thirsk.

The same noise was heard on Wednesday by all who went that way. On Thursday, about seven in the morning, Edward Abbott, weaver, and Adam Bosomworth, bleacher, both of Sutton, riding under Whitston Cliffs, heard a roaring (so they termed it), like many cannons, or loud and rolling thunder. It seemed to come from the cliffs, looking up to which they saw a large body of stone, four or five yards broad, split and fly off from the very top of the rocks. They thought it strange, but rode on....

About seven in the evening, one who was riding by observed the ground to shake exceedingly, and soon after several large stones or rocks, of some tons weight each, rose out of the ground. Others were thrown on one side, others turned upside down, and many rolled over and over. Being a little surprised, and not very curious, he hasted on his way.

On Friday and Saturday morning the ground continued to shake and the rocks to roll over one another. The earth also clave asunder in very many places, and continued so to do till Sunday morning.

While such reports are of considerable interest, it has to be recognised that journals and diaries are an unsystematic source of information heavily dependent on a diarist happening to be in the area or hearing of the event.

Over the last 150–200 years newspapers have become an important vehicle for recording dramatic events. The value of local newspaper sources has been highlighted by Lee and Moore (1991), who established the pattern of contemporary ground movement in the Ventnor landslide complex, Isle of Wight, from a systematic search of local newspapers that were dated from 1855 to 1990. The search identified over 200 individual incidents of ground movement and allowed a detailed understanding of the relationship between landslide activity and rainfall to be developed (see Examples 2.2 and 4.4). However, more subtle changes or less dramatic events in remote areas are not generally recorded in the historical archive.

When researching particular events that have been recorded in local newspapers or documents, it is necessary to make a judgement on the reliability of each data source. Potter (1978) suggests that three questions need to be borne in mind:

- What type of the event is being recorded, and with what detail?
- Who is making the report, in particular what are their qualifications to know of the event; that is, is it based on personal observation and experience, an editing of reports from other people (who themselves may have modified the facts), a plausible rumour, a complete invention or a falsification?
- In the light of knowledge of this type of event, is the report credible, as a whole, only in part or not at all?

However, as Brunsden *et al.* (1995) note, the problems with archive data can be minimised if certain 'rules of interpretation' are followed:

- never assume that the whole landslide event population is represented;
- regard surveys and diaries as a time sample and base judgement on the quality of the observations;
- use as many data sources as possible and compare trends or extremes;
- compare with other independently collected data series (e.g. instrumental climatic records);
- always assume changes in reporting quantity and quality over time. Never assume that the present standard of recording is better than in the past.

3 *Analysis of historical map sources.* Historical topographical maps, charts and aerial photographs provide a record of the former positions of various features. In many cases, historical maps and charts may provide the only evidence of cliff or slope evolution over the last 100 years or more. When compared with recent surveys or photographs, these sources can provide the basis for estimating cumulative land loss and average annual erosion rate between survey dates. However, great care is needed in their use because of the potential problems of accuracy and reliability (e.g. Carr, 1962, 1980; Hooke and Kain, 1982; Hooke and Redmond, 1989).

Where *pronounced* or *noticeable changes* have occurred in the landscape, such as rapid coastal or river cliff recession, they may have been recorded on maps or charts. Taking Britain as an example, so-called 'county' maps date from Elizabethan times to the 19th century and are of variable scale and quality. The earliest manuscript maps (estate maps, tithe or enclosure maps) date from the late 18th and early 19th centuries. They vary considerably in scale and standard of cartography. Large-scale (3–6 chains to the inch, that is between 1:2376 and 1:4752 scale) tithe maps were produced for 75% of England and

Wales during the period 1838–45, in compliance with the Tithe Commutation Act of 1836. The first Ordnance Survey maps were produced at 1 inch to the mile (1:63 360 scale) between 1805 and 1873, but are considered to be of dubious accuracy in depicting topographic detail, especially in irregular terrain well away from human settlements (Carr, 1962; Harley, 1968). Larger scale, 1:2500 and 1:10 560 scale, maps have been produced from the 1870s onwards. The number of subsequent editions of these maps depends largely on the amount of development in an area.

In Britain, therefore, the map-based record of historical landscape change consists of a number of separate measurements made, typically, five times or less over the last 100 years or so. As a consequence, it is often insufficient to identify the pattern and size of events that led to the cumulative land loss between the measurement dates or of assistance in identifying the sequence of preparatory and triggering events that generated individual landslide or recession events (Lee and Clark, 2002). The cartographic record can, at best, reveal only a partial picture of the past landslide processes.

Two main methods are often used to measure changes in the positions of landslide features, such as main scars, from historical sources (Hooke and Kain, 1982):

- *Measurement of distance changes* along evenly spaced transect lines drawn normal to the landslide or cliffline. An average annual recession rate is obtained by dividing the distance change by the time interval between surveys. A frequent problem using this method is the need to extend the transect lines from fixed points (e.g. the corner of a building, property boundary etc.) that are common to all map or chart editions.

- *Measurement of areal changes* between the landslide main scar or cliffline position at different survey dates, along segments of uniform length. The area of land loss between each successive cliff line can be measured using a planimeter or by counting squares, and is converted to an average annual recession rate by dividing by the segment length and the time interval between surveys. This method is considered to be more reliable than the transect method as it measures all mapped change and provides better spatial coverage.

4 *Aerial photograph interpretation* (API). Aerial photography (panchromatic, colour and colour infra-red), and particularly the use of stereoscopic pairs, can provide an exact and complete record of the ground surface at a given time and can represent an efficient means of recording landslide features within an area (Dumbleton, 1983; Dumbleton and West, 1970,

71

1976). The principal advantages of API in landslide investigation have been outlined by Crozier (1984) and include:

- the rapid definition of the boundaries of large landslides;
- appreciation of surrounding slope conditions;
- rapid measurement of slope form and displacement rates if aerial photographs taken at different times are available.

The optimum scale for aerial photographs is around 1:15 000, with 1:25 000 scale considered to be the smallest most useful scale for identifying landslides (Soeters and van Westen, 1996).

Information obtained from aerial photograph interpretation includes the location of drainage channels and associated drainage patterns, variability in underlying geology and soil types, slope inclination data (if stereoscopic vision is used), and areas prone to or potentially vulnerable to landsliding (Crozier, 1984; Dikau et al., 1996; Soeters and van Westen, 1996). Also, API can significantly increase the efficiency with which a site investigation is planned. The use of aerial photographs in the preliminary design of site investigation for all engineering works has become well integrated into codes of practice in Great Britain and is briefly described in the Site Investigation Code of Practice BS 5930 (BSI, 1999).

5 *Interpretation of satellite imagery.* Until recently the spatial resolution of most satellite imagery was insufficient to allow the reliable identification of landslide features smaller than 100 m (Soeters and van Westen, 1996). However, the availability of high-resolution data sets such as IKONOS and EROS (see Table 2.7) will probably make these images as versatile and useful as aerial photography. In addition to the identification of landslides, satellite images are valuable tools for indirect mapping of landslide controlling variables, such as bedrock geology and land use.

6 *Surface mapping.* Techniques such as morphological, geomorphological and engineering geological mapping can be used to establish the nature and extent of pre-existing landsliding (e.g. Brunsden and Jones, 1972, 1976; Brunsden et al., 1975a,b; Doornkamp et al., 1979; Griffiths and Marsh, 1986; Griffiths et al., 1995; Griffiths, 2001). The technique of morphological mapping is the most convenient and efficient way of recording the surface morphology of a landslide or unstable slope and allows later interpretation of form and process. Breaks of slope and slope angles are recorded using standard symbols (Fig. 2.11). In many instances this will involve the use of a tape and compass, although in some instances it may be appropriate to record slope morphology 'by eye' or to use conventional survey equipment to accurately record the detail of surface features. The degree of generalisation used in a

Table 2.7. A summary of the key features of currently available sources of satellite imagery

Data source	Comment
LANDSAT 7	Excellent, low-cost data source for regional studies (1 : 500 000 to 1 : 50 000 scales). Acquiring data since 1999 and has the majority of the Earth's surface covered with cloud-free data. Swath width of 180 km. Panchromatic band with 15 m resolution. Thermal infra-red band 6 with 60 m resolution.
SPOT	Provides medium resolution data world-wide, can be used up to 1 : 25 000 scale (Pan band). Provides stereoscopic imagery and digital elevation model creation capability. SPOT 5 has 2.5 to 5 m resolution in the panchromatic mode. Multispectral mode resolution of 10 m in the visible and infra-red ranges.
IRS	Pan band resolution of 5.8 m. Data collected as 6-bit data and requires high sun angle for good imagery. Can be output at 1 : 10 000 scale.
IKONOS	IKONOS-2, launched in 1999 and operated by Space Imaging Inc. It is the first high-resolution commercial satellite to operate successfully. Swath width of 11 km. Has both cross- and along-track viewing instruments, enabling flexible data acquisition and revisiting – 3 days at 1 m resolution, 1–2 days at 1.5 m resolution.
EROS A1	Launched December 2000. Ground resolution of 1.8 m, with a swath width of 12.5 km. Revisit time of 2–4 days.

morphological mapping survey will depend upon the scale of the base map used and the purpose of the exercise.

Engineering geological and geomorphological mapping can be regarded as the essential tool for rapid assessment of unstable terrain, portraying the ground conditions and surface form together with an interpretation of the origin of particular landforms and features; that is related to geological conditions, surface processes or human activity (e.g. Dearman and Fookes, 1974; Clark and Johnson, 1975; Brunsden et al., 1975a,b; Griffiths and Marsh, 1986).

A good example is the extensive programme of geomorphological mapping that was carried out as part of the investigations at the Channel Tunnel Portal and Terminal areas near Folkestone (Fig. 2.12; Birch and Griffiths, 1995; Griffiths et al., 1995). The Channel Tunnel terminal and portal on the UK side are located immediately below a Lower Chalk escarpment (the North Downs), which was known to be mantled by a

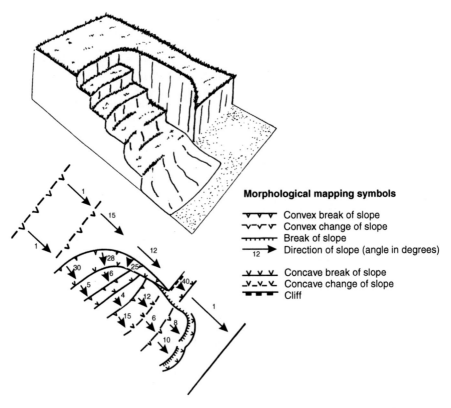

Morphological mapping symbols

▼▼▼	Convex break of slope
∨∨∨	Convex change of slope
┬┬┬┬	Break of slope
──▶ 12	Direction of slope (angle in degrees)
⋎⋎⋎	Concave break of slope
⋎⋎⋎	Concave change of slope
┳┳┳	Cliff

Fig. 2.11. Morphological mapping symbols for a landslide (after Lee and Clark, 2002)

series of large ancient landslides. The objective of the mapping exercise was to delimit and define the nature of past and contemporary landslide activity within the general area of the construction site. Field mapping was undertaken at a scale of 1:500 supported by the interpretation of 1:5000 scale, false colour infra-red aerial photographs. The mapping was carried out using standard procedures for large-scale geomorphological survey work using simple equipment (a 100 metre tape, prismatic compass and clinometer).

The large-scale geomorphological mapping refined the boundaries of the landslides shown on earlier geological maps. The mapping was also of value in showing the form and complexity of the landslide units. This complexity could then be allowed for during the interpretation of borehole data, and for the engineering design, particularly in the assessment of slope stability, as it helped define the parts of the various landslides that either needed to be, or could be, drained and also established where toe loading could be most effectively placed.

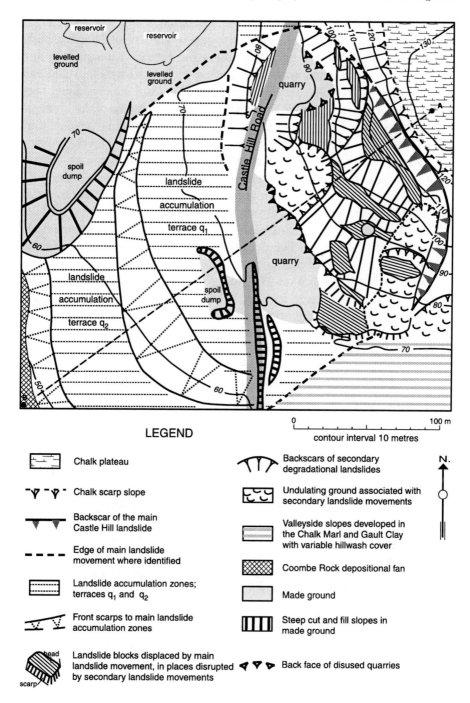

Fig. 2.12. Geomorphological map of the Castle Hill landslide, UK (from Griffiths et al., 1995)

7 *Topographic survey* can be used to obtain a detailed map of a landslide site and establish reliable cross-sections for use in stability analysis. Establishment of a network of reference points can provide a baseline against which future movements can be compared (e.g. Keaton and DeGraff, 1996). For example, detailed topographic control surveys have been carried out to monitor the movements and deformations of the Slumgullion earthflow, Colorado (Varnes and Savage, 1996; Varnes *et al.*, 1996).

8 *Measurement and monitoring of current rates of erosion.* Direct measurement is the most obvious method of obtaining information on current recession rates of coastal cliffs and river bluffs. For example, systematic surveys undertaken on a regular basis are a common approach to establishing cliff recession rates. On the Holderness Coast, Yorkshire, the local authority initiated a programme of cliff recession measurement in the early 1950s which has been continued on an annual basis ever since. A series of 71 marker posts, termed 'erosion posts' by the local authority, were installed at 500 m intervals along 40 km of the coastline, each post located at a distance of between 50 m and 100 m normal to the coast. These posts are replaced further inland from time to time if cliff retreat causes them to become too close to the cliff top. Annual measurements from each post to the cliff top – defined as the lip of the most recent failure scar – commenced in 1953 (see Example 2.10).

9 *Sub-surface investigation.* Sub-surface investigation techniques must be selected carefully on the basis of the anticipated ground conditions. Boreholes and trial pits are the most frequently used combination of investigation methods, although shafts and adits can be of value on large landslides. Investigations of pre-existing landslides should include techniques which will locate and define the three-dimensional shape of the shear surfaces or potential shear surfaces, and allow identification of the engineering properties and groundwater conditions of the slope or cliff (see Clayton *et al.*, 1982; Hutchinson, 1982; Petley, 1984).

10 *Slope monitoring* provides a means of accurately and objectively gauging the stability conditions of unstable or potentially unstable slopes (see Franklin, 1984; Dunnicliff and Green, 1988). There is a range of techniques that can be used for a variety of purposes, including extensometers and crack-meters, inclinometers and slip indicators.

11 *Groundwater monitoring* is often an essential component of site investigation, providing input data for stability analysis and an indication of a cliff's or slope's response to rainstorms etc. It is common practice to use piezometers (e.g. Franklin, 1984; Bromhead, 1986; BSI, 1999) installed in boreholes or driven directly into soft ground.

12 *Dating of events.* Dating methods often provide the only means of establishing the pattern of landsliding through time and defining links with

potential causes. For example, a clustering of events around a particular date might be associated with deteriorating climatic conditions or a specific triggering event such, as a large earthquake. *Relative dating* of landslides can be established through morphological characteristics, degree of surface weathering, re-vegetation rate (see earlier in Section 'Landslide behaviour' of this chapter) and stratigraphic relationships revealed in pits and exposures, but is of limited value in risk assessment. *Absolute dating* (i.e. obtaining a reasonably precise date for the event) is much more valuable and a number of methods are available to obtain dates for landslide events and thereby build up a picture of the event history of the area, including (Lang *et al.*, 1999):

- *Radioactive dating*, in which measurement of the ratio of stable (^{12}C) to radioactive (^{14}C) isotopes in an organic sample (e.g. charcoal, wood, peat, roots, macrofossils) allows the determination of the time elapsed since biological or inorganic fixation. The age range for radiocarbon dating is from a few centuries to around 40 000 years. Thus samples from beneath a landslide can give a maximum date and those on the surface of the landslide a minimum age. Care needs to be taken with material within the landslide as it could be significantly older than the failure.

- *Lichenometric dating*, involving the measurement of the maximum lichen size on exposed surfaces (e.g. boulders) and comparing the results with established lichen-growth curves. This is assumed to indicate the time elapsed since a particular event (e.g. rockfall, debris flow, mud floods). The same approach could be used to date failure scars.

- *Dendrochronology*, whereby the age of death of a tree (e.g. buried beneath a landslide) may be determined by cross-reference to a master tree-ring chronology for that region. The oldest trees that have colonised a now stabilised slide or debris fan can be established by counting the rings revealed by coring and will give a minimum age for the event (e.g. Fantucci and Sorriso-Valvo, 1999).

- *Luminescence dating*, based on estimating the time since a sediment was last exposed to daylight, which 'zeroes' the previously accumulated radiation damage to minerals (e.g. quartz or feldspar) in the sample. The age of a sample is derived from: Age = palaeodose/dose rate, where the palaeodose is the accumulated radiation damage and the dose rate is the rate at which the sample absorbs energy from the immediate proximity. Palaeodoses are calculated by thermoluminescence (energy is supplied by an oven) or optically stimulated luminescence. The maximum age range for these methods depends on the mineral: for quartz it is 100 000–150 000 years; for feldspars

77

it is around 800 000 years. Applications include dating alluvial fans and debris flow deposits (e.g. Lang *et al.*, 1999).

- *Amino acid dating methods* provide relative ages by measuring the extent to which certain amino acids within protein (e.g. bones, mollusc shells, eggshells) residues has transformed from one of two chemically identical forms to the other (e.g. the transformation from L-form to D-form amino acids) until equilibrium is reached. The methods can be used for timescales from a few years to hundreds of thousands of years (Lang *et al.*, 1999).

13 *Laboratory testing* of soil or rock samples is frequently undertaken to determine the composition and properties of the materials encountered during a site investigation. Two main groups of test are relevant: *classification tests* to determine the particle size distribution of the material, index property tests (liquid and plastic limits), bulk density, water content and specific gravity; *shear strength tests* for the laboratory measurement of shear strength. Further details can be found in Head (1982, 1985), Bromhead (1986) and BSI (1990).

14 *Stability analysis* is important when a judgement is needed about whether a slope is stable or not, or whether proposed stabilisation measures will be effective (see Section 'Estimating probability through use of stability analysis' in Chapter 4). Both *finite element* and *limit equilibrium* methods are available. There are a number of limit equilibrium methods of stability analysis, but they are broadly similar in concept (see, for example, Graham, 1984; Bromhead, 1986; Nash, 1987; Duncan, 1996).

Landslide investigations should be carried out by *competent investigators*, who should be able to demonstrate relevant specialist experience in the assessment and evaluation of slope stability. It is important that persons with training and experience in landsliding and slope processes are involved, because the omission or under/over estimation of the likely effects of different hazards will control the outcomes of the risk assessment.

Landslide susceptibility and hazard assessment

Landsliding is a physical process. To become a *hazard* the landslide process must pose a threat to humans and the things that humans value (see Section 'Hazard and vulnerability' in Chapter 1). The potential for adverse impact must be in the foreseeable future and relate to existing patterns of human occupance and activity, or to planned development or activities. In the case of the latter, the notion of *potential hazards* is acceptable, but the notion that all landslides are potential hazards because they may, at some stage in the future, interact adversely with human activity can lead to confusion.

These points are crucial to appreciating the distinction between *landslide susceptibility* mapping and *landslide hazard* mapping, especially as the terms susceptibility and hazard have tended to be used synonymously in the literature. Landslide susceptibility is the potential for landsliding to occur, while landslide hazard is the potential for landsliding to cause adverse consequences from a human perspective. The two are not the same because of the 'human factor'. Thus high susceptibility does not necessarily mean high hazard; indeed, landslide prone remote areas may pose no threat to humans whatsoever. To convert *susceptibility* into *hazard* requires not only an understanding of the magnitude, character and frequency of landsliding, but also the *probability and potential to cause damage*. Unfortunately this is not always attempted and far too many studies that are presented as landslide hazard assessments are, in reality, landslide susceptibility assessments, which focus on the physical phenomena and merely assume that there will be adverse consequences, or that some end user will re-interpret the results in terms of hazard.

Landslide susceptibility studies clearly form the essential basis for landslide hazard assessments. Only once the nature, magnitude, frequency and distribution of future landslides has been determined is it possible to evaluate the threat to existing and planned human activity in terms of the potential for damage and destruction of buildings, infrastructure, communication lines etc., as well as the possibility of death and injury to humans, livestock etc. The results are frequently presented in map form and the revealed spatial patterns are of value to both development planners and engineers.

The focus of landslide hazard assessment is mainly directed towards:

- *prediction*, involving establishing the general likelihood and character of landsliding in an area. This can lead to the identification of areas with different levels of threat (i.e. landslide zonation or hazard zonation with reference to landsliding), which can be used to establish land use zonation plans, development controls and patterns of building regulations;
- *forecasting*, involving the identification of the location, timing, magnitude and character of individual landslides. This is a detailed, site-specific exercise, heavily reliant on monitoring and modelling. It is the necessary basis for the development of early warning systems and the triggering of emergency action procedures.

Landslide hazard prediction generally involves the following steps (Selby, 1993; Jones, 1995b):

1 identification of areas and locations which could prove of concern;
2 definition of the extent of the areas that require study;
3 mapping and cataloguing areas of active and past landsliding in each study area;

79

4 identification of slope units, topographic situations, superficial deposits or stratigraphic units which are associated with instability and the extrapolation of these factors to locations that are potentially unstable.

The resulting hazard maps provide the spatial framework for the consideration of the threat posed by landsliding and the potential consequences, that is the risk assessment process. However, to date, landslide hazard assessments have generally been restricted to determining the nature of the hazard and where it might occur (i.e. susceptibility), rather than the probability of occurrence. This is because it is extremely difficult to determine the association between factors, such as potential triggering events, and landslide activity across a broad area. Most assessments have, therefore, concentrated on either:

• the relative susceptibility of different areas to landsliding, or
• the relative likelihood that a landslide will occur, expressed in terms such as high, medium or low hazard.

A wide range of approaches have been used to generate landslide susceptibility and hazard assessments (see Hansen, 1984; Varnes, 1984; Jones, 1992, 1995b; Hutchinson, 1992, 1995 and Hearn and Griffiths, 2001 for reviews of the various methods and their limitations). All are based on one or more of the following assumptions (Hearn and Griffiths, 2001):

1 the location of future slope failure or ground movement will largely be determined by the distribution of existing or past landslides, that is known landslide locations will continue to be a source of hazard;
2 future landslides or ground movement will occur under similar ground conditions to those pertaining at the sites of existing or past landslides, that is the conditioning or controlling factors that give rise to existing landslides can be ascertained and their distribution reliably mapped over wider areas. Together these factors provide a reasonable indication of the relative tendency for slopes to fail;
3 the distribution of existing and future landslides can be approximated by reference to conditioning factors alone, such as rock type or slope angle.

Two basic methodologies can be recognised (Carrara *et al.*, 1998):

1 *Direct mapping approaches*, in which the hazard is defined in terms of the distribution of past landslide events. This is described by Soeters and van Westen (1996) as an experience-based approach, whereby the specialist evaluates direct relationships between landslides and their geomorphological and geological settings through the use of direct observations from a survey of existing landslide sites. The main types of direct mapping are:

- *Landslide inventories.* This is the most straightforward approach to land-slide hazard mapping and involves compiling a database of pre-existing landslides within an area from aerial photographs, ground survey or historical records.
- *Geomorphological analysis*, where the hazard is determined directly in the field by the geomorphologist, drawing on individual experience and the use of reasoning by analogy with similar sites elsewhere. As this approach relies on expert judgement, reproducibility can be a major issue.
- *Qualitative map combination.* This involves the identification of key factors that appear to control the pattern of landsliding. Each of these factors is then assigned a weighted value, based on experience of the particular landslide environment. The weighted factor scores are then summed to produce hazard values that can be grouped into hazard classes. The exact weighting of factors can be difficult to deter-mine and as Soeters and van Westen (1996) noted 'insufficient field knowledge of the important factors prevents the proper establishment of the factor weights, leading to unacceptable generalisations'.

2 *Indirect mapping approaches*, involving the mapping of and statistically analysing large numbers of parameters considered to influence land-sliding to derive a predictive relationship between the terrain conditions and the occurrence of landsliding. Two different statistical approaches are used:

- *Bivariate statistical analysis*, in which each factor map (e.g. slope angle, geology, land use) is combined with the landslide distribution map and weighted values based on landslide densities are calculated for each parameter class (e.g. slope class, lithological unit, land use type). A variety of statistical methods have been applied to calculate the weighting values, including the *information value method* (Yin and Yan, 1988) and *weight-of-evidence modelling method* (Spiegelhalter, 1986).
- *Multivariate statistical analysis*, in which all relevant factors are sampled either on a large-grid basis or in morphometric units. The presence or absence of landsliding within each of the units is also recorded. The resulting matrix is analysed using multiple regression or discriminant analysis (e.g. Carrara, 1983, 1988; Carrara *et al.*, 1990, 1991, 1992). Large datasets are needed to produce reliable results.

The spatial framework for hazard assessments can be provided by units identified through terrain evaluation or geomorphological mapping (e.g. Phipps, 2001; Lee, 2001; see Section 'Acquiring information: landslide investigation' in this chapter). Each unit can be assumed to present a similar landslide susceptibility or hazard. Alternatively, relative scores for a range of controlling factors can be assigned to a series of grid squares set up to cover the area of interest. This approach lends itself to GIS-based methods of data management, analysis and output generation (e.g. Greenbaum, 1995).

Example 2.1

The combination of widespread landslide activity, dense development of a hilly terrain, highly variable seasonal rainfall regime and a large number of potentially substandard man-made slopes (mostly formed before the 1970s without proper geotechnical input and control) have created acute slope safety problems in Hong Kong. Landslides have caused over 470 deaths since 1940 (e.g. Malone, 1998; Wong and Ho, 2000).

On average, some two to three hundred slope failures are reported every year. This equates approximately to an average density of one landslide per year for every square kilometre of the developed area. Almost all of the landslides are triggered by heavy rainfall. Rainfall events that trigger landsliding at medium densities ($1-10/km^2$) occur every 2 years at the local scale (i.e. any given site), and five times per year for the region (i.e. Hong Kong as a whole), as indicated in Table 2.8.

Most of the landslides occur on man-made slopes (i.e. slopes and retaining walls) and are relatively small ($<50\,m^3$), although some can be $5000\,m^3$ or more. Until recently attention has focused on the stability of man-made slopes. However, landslides on natural terrain also present a significant hazard, as highlighted by:

- the 1.2 km long debris flow on the eastern slopes of Tsing Shan, above Tuen Mun New Town, which occurred in 1990 and reached a planned development site (King, 1996);

Table 2.8. Example 2.1: Rainfall thresholds for landslide activity in Hong Kong (Evans, 1997; Evans et al., 1997)

	Threshold I start of landsliding		Threshold II start of landsliding at medium densities ($1-10/km^2$)		Threshold III start of landsliding at high densities ($>10/km^2$)	
	Rainfall	Return period	Rainfall	Return period	Rainfall	Return period
Local	60–70 mm	2.5 times per year	180–220 mm	Every 2 years	380–450 mm	Every 20 years
Regional		25 times per year		5 times per year		Every 2 years

- numerous debris flows on the northern slopes of Lantau Island, near Tung Chung in 1992/1993 (Franks, 1996, 1999), which would have presented a threat to the North Lantau Expressway had it been opened at the time.

Concerns for the safety of new development sites led the Geotechnical Engineering Office to undertake a programme of region-wide landslide hazard mapping – the Natural Terrain Landslide Study (Evans *et al.*, 1997; King, 1997; Evans and King, 1998, 1999). As a first step, an *inventory of landslide features and areas of intense gullying* was compiled from the interpretation of high-level aerial photographs (1:20 000 to 1:40 000 scales) taken in 1945, 1964 and annually from 1972 to 1994 (excluding 1977). Most parts of Hong Kong appeared on between 20 and 23 sets of photographs.

A total of 26 870 natural terrain landslides were identified within an area of 640 km². Of this total, 8804 landslides were described as having occurred within the last 50 years (i.e. 'recent'), with evidence for 17 976 slides over 50 years old. Most failures were debris slides or debris avalanches within weathered rock or overlying colluvium. Although the majority of recorded landslides were relatively small features, a significant proportion (15%) exceeded 20 m in width.

A clear association between landslide density, geological unit and slope angle was identified and provided the basis for a regional hazard assessment. The inventory provides an indication of the relative *susceptibility* of different combinations of geological unit and slope class (terrain units), as derived from a 1:20 000 scale digital terrain model. As major landslide-triggering rainfall events occur on susceptible natural terrain, with an average frequency of once every 20 years, it was assumed that the inventory reflects the underlying susceptibility.

Five susceptibility classes were defined in terms of the average estimated frequency (landslides/km² per year):

- Very high susceptibility: >1 landslide/km² per year;
- High susceptibility: 0.5–1 landslide/km² per year;
- Moderate susceptibility: 0.25–0.5 landslide/km² per year;
- Low susceptibility: 0.1–0.25 landslide/km² per year;
- Very low susceptibility: <0.1 landslide/km² per year.

Table 2.9 and Fig. 2.13 summarise the landslide susceptibility statistics for one of the bedrocks, coarse grained granite.

The model provides a general picture of the landslide hazard across Hong Kong. However, it cannot be used to predict failure size, mobility or consequences. The landslide susceptibility figures allow a rapid assessment of the landslide potential of a given natural slope simply by:

Table 2.9. Example 2.1: Natural Terrain Landslide Study statistics for coarse-grained granite and pegmatite (from Evans and King, 1998)

Slope class (degrees)	Area: km^2	Recent slides	Density of recent slides: per km^2	Older slides	Density of older slides: per km^2	Total number of slides	Total density: per km^2
0–5	0.471	1	2.12	–	–	1	2.12
5–10	0.566	–	–	–	–	–	–
10–15	0.818	1	1.22	4	4.89	5	6.11
15–20	1.356	2	1.47	8	5.90	10	7.37
20–25	2.130	3	1.41	10	4.69	13	6.10
25–30	2.870	12	4.18	32	11.15	44	15.33
30–35	2.238	11	4.91	48	21.45	59	26.36
35–40	0.996	10	10.04	18	18.07	28	28.11
40–45	0.291	3	10.30	7	24.03	10	34.33
45–50	0.076	3	39.47	2	26.31	5	65.78
50–55	0.028	2	70.43	–	–	2	70.43
55–60	0.011	–	–	–	–	–	–
>60	0.008	–	–	–	–	–	–
Total	11.86	48	4.05	129	10.88	177	14.93

1 identifying the correct geological unit from the available 1 : 20 000 scale geological maps of Hong Kong;
2 calculating the slope angle between contours on 1 : 20 000 scale topographic maps;
3 combining both the geological unit and slope class to define a terrain unit and identify the appropriate susceptibility value from the tables presented in Evans and King (1998).

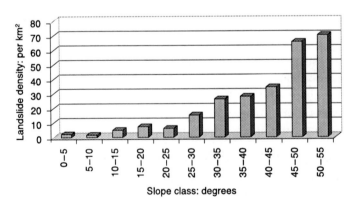

Fig. 2.13. Example 2.1: The Natural Terrain Landslide Inventory, Hong Kong: landslide susceptibility statistics for coarse-grained granite and pegmatite (after Evans and King, 1998)

Example 2.2

The Isle of Wight Undercliff, UK, is an extensive, ancient landslide complex with a permanent population of around 6500, located in the small towns of Ventnor, St Lawrence and Niton. There has been a history of problems associated with the degradation or reactivation of the landslide complex over the last 100 years or so (e.g. Hutchinson, 1991a; Hutchinson and Chandler, 1991; Hutchinson *et al.*, 1991; Hutchinson and Bromhead, 2002).

In 1988 the UK Government (Department of the Environment) commissioned an assessment of the landslide hazard in the Ventnor–Bonchurch area of the Undercliff (Lee and Moore, 1991; Lee *et al.*, 1991a,b; Moore *et al.*, 1991b). Subsequent studies for South Wight Borough Council (now the Isle of Wight Council) extended the mapping to cover the whole of the Undercliff (Moore *et al.*, 1995). The purpose of these studies was to define the nature and extent of the problems and identify ways in which landslide hazard information could be used to assist local land use planners. A suite of six 1:2500 scale map sheets were produced for the Undercliff, covering the following themes: geomorphology, ground behaviour and planning guidance.

The work undertaken involved a thorough review of available records, reports and documents, followed by a programme of detailed field investigation comprising 1:2500 scale geomorphological and geological mapping. This revealed the extent and complexity of the landslides. The following main features were distinguished (see Fig. 2.14):

- a sequence of *compound slides* which occupy a zone of similar breadth in the lower part of the Undercliff;
- *multiple rotational slides* which occupy a broad zone in the upper parts of the Undercliff, giving rise to linear benches separated by intermediate scarps. These units mainly comprise back-tilted blocks of Upper Greensand and Chalk;
- in Upper Ventnor, a *graben-like feature* occurs landward of the zone of multiple rotational slides, comprising a 20 m wide subsiding block bounded by parallel fissures, and extends parallel to the coast for over 500 m. This unit exhibits the most serious contemporary ground movements experienced in the town;
- *mudslides* which have developed on the coast where displaced Gault Clay is exposed.

The distribution, frequency and magnitude of ground movements in the Ventnor Undercliff were determined from a search through historical documents, local newspapers from 1855 to 1989, local authority records and published scientific research. This revealed nearly 200 individual incidents of ground movement over the last two centuries. The various

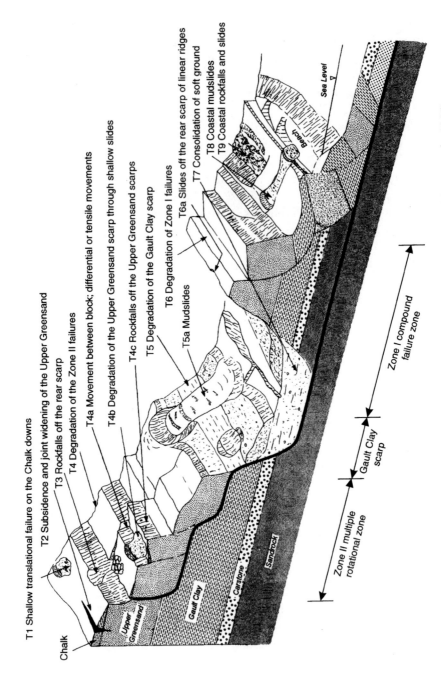

T1 Shallow translational failure on the Chalk downs

T2 Subsidence and joint widening of the Upper Greensand

T3 Rockfalls off the rear scarp

T4 Degradation of the Zone II failures

T4a Movement between block; differential or tensile movements

T4b Degradation of the Upper Greensand scarp through shallow slides

T4c Rockfalls off the Upper Greensand scarps

T5 Degradation of the Gault Clay scarp

T6 Degradation of Zone I failures

T5a Mudslides

T6a Slides off the rear scarp of linear ridges

T7 Consolidation of soft ground

T8 Coastal mudslides

T9 Coastal rockfalls and slides

Chalk

Upper Greensand

Gault Clay

Carstone

Sandrock

Sea Level

Beach

Zone II multiple rotational zone

Gault Clay scarp

Zone I compound failure zone

Fig. 2.14. Example 2.2: Landslide model of the Ventnor Undercliff (after Lee and Moore, 1991; Moore et al., 1995)

forms of contemporary instability are summarised in Fig. 2.14. These include sub-surface movements associated with the deep seated creep of the entire landslide complex and surface or superficial movements arising from the erosion or failure of steep slopes, and the differential movement and potential collapse (vents) between landslide blocks.

Photogrammetric analysis of oblique and vertical aerial photographs provided evidence of building displacements at over 100 points, between 1949, 1969 and 1988 (Moore *et al.*, 1991). The results indicated that throughout much of the area, average annual movement (i.e. long-term cumulative displacement) rates are less than 10 mm, although higher rates occur in a number of localised areas. A systematic survey of structural damage was undertaken to establish the nature and intensity of the impact of ground movement throughout the area. A five-fold classification was used to describe the intensity of damage, from negligible (class 1) to severe (class 5). Serious (class 4) and severe damage was restricted to a limited number of locations. The distribution of damage followed a similar pattern to the historical movement rate data. It was found that the patterns of damage and movement rates was related to the geomorphological setting. Serious and severe damage were mostly associated with higher movement rates at scarp or inter-block zones (see Fig. 2.15).

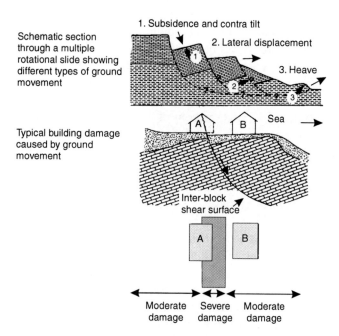

Fig. 2.15. Example 2.2: Variability of building damage caused by ground movement: Ventnor, Isle of Wight, UK (after Lee and Moore, 1991; Moore et al., 1995)

A landslide hazard map was produced at 1:2500 scale that summarises the nature, magnitude and frequency of contemporary processes and their impact on the local community, being a synthesis of the following information:

- the nature and extent of individual landslide units which together form the mosaic of landslide features;
- the different landslide processes which have operated within the town over the last 200 years;
- the location of ground movement events recorded in the last 200 years;
- the recorded rates of ground movement, over the last 30–100 years;
- the severity of damage to property caused by ground movement;
- the causes of damage to property as a result of ground movement;
- the relationship between past landslide events and antecedent rainfall.

The *geomorphological approach* was used in the production of the landslide hazard map (termed *ground behaviour* by the study). The hazard assessment involved defining the landslide activity and ground displacements within different landslide units (Lee *et al.*, 1991a). For example, examination of damage to properties within a small landslide sub-system (Fig. 2.15) revealed that these properties had been affected by a range of ground movements, including heave, subsidence, rotation and tilting. The most severely affected properties were tilted forward, as a result of heave at the toe of a small rotational landslide unit. The buildings at the crest of the landslide sub-system appear to have been affected by settlement and rotational (contra-tilt) movements. However, towards the middle of the slope, outward movements appear to have taken place, with only limited evidence of tilting. This example of contrasting movement over a short distance (300 m) illustrates the range of ground movements that may occur in the Undercliff.

The characterisation of each geomorphological unit in terms of *ground behaviour* was based on the direct evidence of measured movement rates and records of past events, together with the indirect evidence provided by the structural damage survey. A structural damage index was developed for each geomorphological unit by dividing the study area into a series of 10 m × 10 m grid cells and summing the recorded damage within each cell. An estimate of the relative *susceptibility* of different geomorphological units was developed by dividing the total structural damage score by the area of that unit that was occupied by property and other structures.

The ground behaviour map demonstrated that the problems resulting from ground movement vary from place to place according to the geomorphological setting. This formed the basis for landslide management strategies that could be applied within the context of a zoning framework. In support of the management strategy, a 1:2500 scale Planning Guidance map was produced which related categories of ground behaviour to forward planning and development control (Lee *et al.*, 1991b). The map indicated that different areas of the landslide complex need to be treated in different ways for both policy formulation and the review of planning applications. Areas were recognised which are likely to be suitable for developments, along with areas that are either subject to significant constraints or mostly unsuitable. The latter generally correspond with undeveloped areas and the designation reflects a need to be cautious about the possible effects that development may have on the stability of adjacent land. Advice was also provided on the level of stability information that should be presented with applications for development in different parts of the Undercliff.

The studies demonstrated that an understanding of ground behaviour (i.e. landslide hazard) in an urban landslide system can be achieved through a combination of geomorphological mapping, exhaustive archive research, surveys of structural damage and photogrammetric analysis. However, it was not a detailed, exhaustive study of the landslides (there were no boreholes, for example), rather it was a pragmatic assessment of the broad patterns of landslide activity to support strategic-level decision-making. Further details of ground conditions are obviously required and need to be sought from potential developers as part of the planning and development process.

Example 2.3
The valleys and escarpment slopes within the Peak District of the UK contain a large number of deep-seated landslides developed in sequences of Carboniferous limestones and Millstone Grit, together with weaker shales and mudstones (Jones and Lee, 1994). A variety of landslide forms occur, including rotational failures (e.g. Mam Tor; Skempton *et al.*, 1989), compound failures (e.g. Alport Castles; Johnson and Vaughan, 1983) and cambering. Cross (1987, 1988) described the use of a Matrix Assessment Approach (MAP; a form of *qualitative map combination*) for establishing an index of landslide susceptibility over the area.

The MAP approach relies on the premise that the landslide distribution can be explained in terms of particular combinations of *landslide attributes* (e.g. bedrock geology, slope angles, slope aspect etc.; DeGraff, 1978). The area of interest is sub-divided into a matrix of grid cells. Each grid cell has its own particular combination of *landslide attributes*. The *landslide susceptibility* of this combination of attributes is determined by calculating the frequency that the combination is associated with landsliding (i.e. the number of grid cells with landslides present where this combination occurs) compared with the overall frequency that the combination occurs across the area:

Landslide Susceptibility

$$= \frac{\text{Number of Cells with Combination C and Landslides}}{\text{Total Number of Cells with Combination C}}$$

Those grid cells without pre-existing landslides, but having a higher than average landslide susceptibility, are considered to be potential sites for failure in the future.

The computer-based method used by Cross (1987, 1988) to determine landslide susceptibility in a 928 km^2 sample area of the Peak District involved the following steps:

1   Dividing the area into northern and southern sectors. The MAP was applied to the southern sector and tested for its predictive accuracy in the northern sector.

2   Sub-dividing the southern part of the area into a series of 125 m × 125 m wide grid cells. Each cell was characterised on the basis of nine selected topographic, geological and geomorphological parameters that were believed to contribute to instability (i.e. *landslides attributes*; slope aspect (8 classes), relative relief (7 classes), height above valley floor (9 classes), height above sea level (8 classes), superficial deposits (11 soil types), soil classification (9 soil units), bedrock geology (11 lithologies), slope steepness (6 classes), bedrock combinations (9 combinations)). The landslide attribute values for each cell were derived from existing maps sources.

3   Recording the distribution of pre-existing landslides in the southern sector, from geological maps, aerial photograph interpretation and field checking. In the study area the dominant types of landslide were single or multiple rotational failures (72% and 13%, respectively), often associated with extensive debris aprons or mudsliding.

    All the grid cells with landslide features present were identified.

4   Assessing a *landslide area factor* (LAF) for each possible combination of landslide attributes (note that the number of possible combinations of attributes is equal to the product of the number of classes used for each of the nine attributes). An LAF was calculated for each possible attribute combination as the total area occupied by cells with pre-existing landslides having that combination. It is a measure of how significant the attribute combination is in controlling the distribution of landsliding.

5   Assessing a regional area factor (RAF) for each possible combination of landslide attributes. The RAF for a particular attribute combination was defined as the total area within the study area occupied by cells with that combination (i.e. cells with pre-existing landslides and non-failed slopes).

6 Calculating a landslide susceptibility index (LSI) for each combination of attributes:

$$\text{LSI (Attribute Combination C)} = \frac{\text{LAF (Attribute Combination C)}}{\text{RAF (Attribute Combination C)}}$$

It was assumed that the closer the LSI score is to unity, the greater influence that particular attribute combination has had on the occurrence of landsliding; that is the more susceptible that combination is to generating landsliding.

For example, Fig. 2.16 shows a schematic diagram of the MAP for a case using seven attributes:

$$\text{LSI} = \text{LAF/RAF}$$
$$= 98\,\text{ha}/141\,\text{ha}$$
$$= 0.695$$

7 Assigning LSI values for each grid cell in the study area, based on the calculated LSIs for the particular combination of attributes that occur in that cell. The results for the southern part of the Peak District study area are presented in Table 2.10, which indicates that:
- for *7-attribute combinations*, 642 grid cells have high and relatively high susceptibility combinations (LSI > 0.5);
- for *3-attribute combinations*, 50 grid cells have LSI values of over 0.5.
8 Calculating a mean LSI value for the mapped landslide areas:

$$\text{LSI (Mean for Cells with Landslides)} = \frac{\sum \text{LSI (Landslide Cells)}}{\text{Number of Landslide Cells}}$$

9 Identifying non-landslide cells with LSI values greater than the mean value. Table 2.11 presents the mean LSI values for mapped landslides for four different attribute combinations. For the nine-attribute case, for example, 51 cells contained LSI values greater than the mean and were located in areas not mapped as containing landslides. These locations were considered to have a high degree of susceptibility to landsliding.
10 Testing the predictive accuracy of the LSI values in the northern sector of the study area. The results are presented in Cross (1987, 1988) and indicate that the technique was successful in identifying landslide sites and high susceptibility locations. Indeed, field checking of the susceptibility assessment in the northern sector led to the identification of 15 previously unrecognised landslide sites.

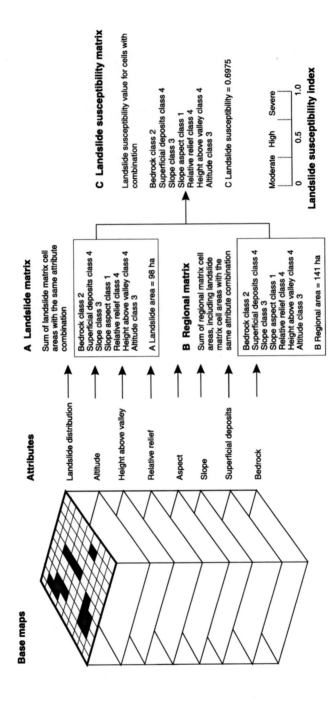

Base maps

Attributes

Landslide distribution →

Altitude →

Height above valley →

Relative relief →

Aspect →

Slope →

Superficial deposits →

Bedrock →

A Landslide matrix

Sum of landslide matrix cell areas with the same attribute combination

Bedrock class 2
Superficial deposits class 4
Slope class 3
Slope aspect class 1
Relative relief class 4
Height above valley class 4
Altitude class 3

A Landslide area = 98 ha

B Regional matrix

Sum of regional matrix cell areas, including landslide matrix cell areas with the same attribute combination

Bedrock class 2
Superficial deposits class 4
Slope class 3
Slope aspect class 1
Relative relief class 4
Height above valley class 4
Altitude class 3

B Regional area = 141 ha

C Landslide susceptibility matrix

Landslide susceptibility value for cells with combination

Bedrock class 2
Superficial deposits class 4
Slope class 3
Slope aspect class 1
Relative relief class 4
Height above valley class 4
Altitude class 3

C Landslide susceptibility = 0.6975

Moderate   High   Severe

0   0.5   1.0

Landslide susceptibility index

Fig. 2.16. Example 2.3: schematic diagram of the Matrix Assessment Approach (after Cross, 1998)

Table 2.10. Example 2.3: Frequency of grid cells with landslide susceptibility index values, using seven- and three-attribute cases (after Cross, 1988)

Landslide susceptibility classification: LSI	Seven-attribute case	Three-attribute case
Reservoirs	144	144
0–0.1	10 533	9 326
0.1–0.2	70	1 234
0.2–0.3	63	450
0.3–0.4	63	176
0.4–0.5	5	140
0.5–0.6	174	36
0.6–0.7	36	6
0.7–0.8	0	0
0.8–1.0	432	8
Number of combinations	6 560	516

Note. Seven-attribute case: bedrock geology, slope steepness, slope aspect, height above sea level, height above valley floor, relative relief, superficial deposits. Three-attribute case: bedrock geology, slope steepness, slope aspect.

Example 2.4

Landslides present a range of problems to communities within the steep-sided valleys of the South Wales Coalfield, UK. A detailed study of landslide problems in Rhondda Borough revealed 82 separate events over a 100 year period (Table 2.12; Halcrow, 1986). Notable events include: a flowslide in spoil in Pentre in 1909 which destroyed five houses and killed one person; the dumping of spoil at Pentre reactivated an old translational slide in 1916, destroying a billiard hall and several houses; houses in Pont-y-gwaith were

Table 2.11. Example 2.3: Mean LSI values calculated for mapped landslide areas and frequency of no landslide cells possessing an LSI greater than the mean value (from Cross, 1988)

Attribute combination	Mean LSI ratio for landslide areas	Number of non-landslide cells with LSI greater than the mean value
Nine-attribute case	583	51
Seven-attribute case	479	136
Four-attribute case	208	449

Note. Nine-attribute case: bedrock geology, slope steepness, slope aspect, height above sea level, height above valley floor, relative relief, superficial deposits, soils, bedrock combination. Seven-attribute case: bedrock geology, slope steepness, slope aspect, height above sea level, height above valley floor, relative relief, superficial deposits. Four-attribute case: bedrock geology, slope steepness, slope aspect, relative relief.

Table 2.12. Example 2.4: Recorded damaging landslide events in the Rhondda Valleys (from Halcrow, 1988)

Damage class	Description	Number of events
A	Failures where the estimated present-day cost of investigation, stabilisation and repair lies between £100 000 and £3.5 million. Typically, these might include landslips which damage more than two houses.	26
B	Failures where the estimated present-day cost of stabilisation and repair lies between £5000 and £100 000. Typically one or two houses are affected.	24
C	Minor failures, treated by expenditure of less than £5000. Typically these include events where only clearances of material or repairs to retaining walls are necessary.	32

Note. Costs are 1988 prices.

affected by a slide in 1961, an event which was described in the *South Wales Echo* (6 January 1961) as 'the mountains slipped forward and tons of liquid yellow clay slid into the houses'.

An assessment of the landslide potential (i.e. *landslide susceptibility*) was undertaken in order to inform the land use planning process in South Wales (Halcrow, 1986, 1988; Siddle *et al.*, 1987, 1991). The assessment involved the use of the indirect mapping approach (*univariate analysis*) and relied on the assumption that the distribution of recorded landslides reflects the occurrence of combinations of factors that have resulted in slope failure. Mapping the distribution of the relative significance of these factors formed the basis for the identification of areas with varying potential for landsliding (Fig. 2.17).

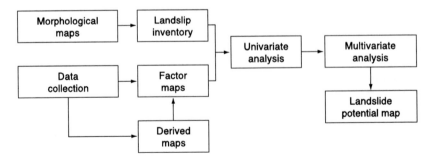

Fig. 2.17. Example 2.4: the methodology used in landslip potential mapping (from Siddle et al., 1987)

The assessment involved the following steps:

1 Creation of a landslide inventory by amalgamating the sites shown on existing geology maps and the maps produced by the South Wales Coalfield Landslip Survey (Conway et al., 1980), with information obtained from archival sources, reports held by consultants, contractors, statutory undertaking, local authorities etc., and augmented by an intensive programme of direct mapping. The results were portrayed on maps and entered into a computerised database.

2 Development of *factor maps* for the various attributes that contribute towards slope instability. A total of sixteen factors were identified as of potential importance. Maps were prepared for each factor, sub-divided into an appropriate number of *factor-zones*, each of which was characterised by a consistent set of conditions.

3 Undertaking univariate analyses to derive a rating for the degree of association between the known landslides and each factor zone. This was achieved by encoding all factor maps and landslide maps using a grid of 50 m × 50 m cells oriented parallel to the National Grid. The ratings were produced by comparing the incidence of landslides in each factor zone to that of the area as a whole:

$$\text{Rating} = \frac{\text{Proportion of a factor zone occupied by landslides} \times 100}{\text{Proportion of the entire area occupied by landslides}}$$

A rating of 100 indicated that a factor zone contained no more landslides than average, but a value above 100 indicated a possible association between the factor and landsliding, while values below 100 indicated the reverse.

4 Multivariate analysis was then undertaken to combine ratings in a simple algorithm to define the *landslip potential* of each 50 m × 50 m element. Trial and error revealed that the best results were obtained using the formula:

$$\text{LP} = \frac{\text{RP}_{n2} \times (\text{RP}_{n4} + \text{RP}_{n5} + \text{RP}_{n8})}{300}$$

where LP = landslip potential, RP_{n2} = slope angle, RP_{n4} = superficial deposit type, RP_{n5} = superficial deposit thickness, RP_{n8} = groundwater potential. The resultant landslip potential values ranged from 0 to 518. As the local planning authority has a relatively limited range of available responses to planning applications, it was decided to restrict the number of zones portrayed on the final maps to six (Fig. 2.18; Siddle et al., 1987, 1991), but grouped into four main categories:

- areas of active landsliding,
- areas of dormant landsliding,

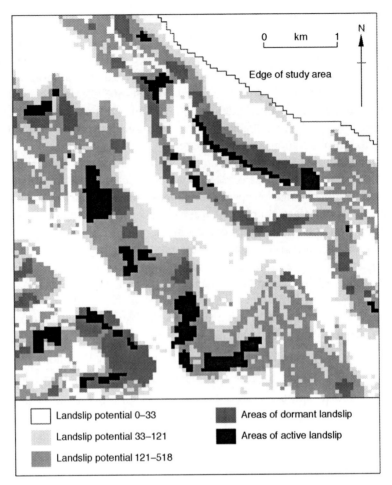

Fig. 2.18. Example 2.4: simplified landslip potential map for part of the Rhondda valleys, South Wales (after Siddle et al., 1987)

- areas of 'some landslip potential' covering two zones with LP values of 240–518 and 121–240, and
- areas with 'little landslip potential' covering two zones with LP values in ranges 33–121 and 0–33.

The resultant landslip potential maps provide the basis for defining the relative significance of landslide hazard, although they could be further refined with respect to the nature and magnitude of potential failure. They also provide a zonation scheme that can be readily communicated to the general public, for the four zones may be clearly and simply described as:

- 'areas where landsliding is occurring',
- 'areas where landsliding has occurred',

- 'areas where landsliding may occur', and
- 'areas where landsliding is unlikely to occur'.

Example 2.5

The Calhandriz area of Portugal is susceptible to a variety of landslides, developed in a combination of Jurassic marls, clays, limestones and sandstones. Typical landslide types encountered include shallow translational slides in clays and marls, deep-seated translational slides involving the limestones and sandstones, and relatively deep-seated rotational slides in clay units. The majority of the landslides appear to be triggered by rainfall, although the relationship between rainfall events and landslide activity is complex (Zêzere *et al.*, 1999a, 1999b; Zêzere, 2000):

- moderate-intense rainfall episodes (165 to 220 mm in 15 days) tend to initiate minor slides, topples and falls on the banks of rivers;
- high-intensity rainfall episodes (130 mm in one day) cause flash floods; landslides triggered by river bank erosion and shallow translational slides on steeper slopes;
- prolonged rainfall periods (495 mm in 40 days, 695 mm in 75 days) are associated with the triggering of larger, deep-seated slope movements.

Zêzere (2002) described the use of the Information Value Method (IVM; a bivariate statistical approach) to define the landslide susceptibility within an 11.3 km^2 trial area within the region. The IVM method involves determining an information value for particular slope variables that might affect instability.

The information value IV of each variable x is given by (Yin and Yan, 1988):

$$IVx = \log \frac{Tx/Nx}{T/N}$$

where Tx is the number of terrain units with landslides *and* the presence of variable x, Nx is the number of terrain units with variable x, T is the total number of terrain units with landslides, N is the total number of terrain units. Negative values of IV indicate that the presence of the slope variable is not relevant in landslide development. Positive values of IV indicate a relevant relationship between the presence of the variable and landslide distribution; the stronger the relationship, the higher the score (Yan, 1988).

The relative susceptibility of a particular *terrain unit* to the occurrence of a particular type of slope movement is given by the total information value for that unit (i.e. the sum of the information values for each slope variable).

The *landslide susceptibility* assessment of the Calhandriz area involved the following steps:

Table 2.13. Example 2.5: Slope variables considered for landslide susceptibility evaluation in the Calhandriz area of Portugal (from Zêzere, 2002)

Conditioning factors (slope variables)	Variable cases
Lithological units	LU1 – marls and clays LU2 – coralline limestone LU3 – marls, marly limestones and limestones LU4 – sandstones, marls and limestones
Slope deposits	Thickness >0.5 m Thickness <0.5 m
Slope angle	0–5° 5–15° 15–25° >25°
Relationships between slope and dip of strata	Cataclinal slopes Other slopes
Fluvial channels	Present Absent
Artificial cuts (roads)	Present Absent

1 *collection of terrain data* (i.e. slope variables), through field survey, interpretation of aerial photographs and analysis of existing topographic and geological maps (Table 2.13). The nature and distribution of landslides was mapped in the field at 1:2000 scale;
2 *definition of terrain units,* based on combinations of lithology, slope angle and slope deposits. A total of 32 *unique condition areas* were identified as separate terrain units;
3 statistical analysis of landslide susceptibility using the IVM method for four different landslide data sets: shallow translational slides (60 examples were present in the study area), translational movements (23 examples), rotational slides (19 examples) and the combined data set (102 examples).

The results of the statistical analysis are presented in Table 2.14, which indicates that the occurrence of the various landslide types are influenced to a different degree by particular slope variables. For example, rotational slides tend to be strongly associated with lithological unit 1 (lower and middle Kimmeridgian marls and clays), as indicated by the information value of 0.97. Zêzere (2002) reported that this result was in accordance with the concentration of 65% of these landslides on lithological unit 1. Clays and marls of this unit have low shear strength properties, and

Table 2.14. Example 2.5: Information values obtained for different landslide data sets in the Calhandriz area of Portugal (from Zêzere, 2002)

Variables		Total landslides	Rotational slides	Translational slides	Shallow translational slides
Lithological units	LU1	0.4079	0.9705	0.1834	−0.0551
	LU2	−0.1312	−0.3549	−0.5336	0.2237
	LU3	−0.1708	−0.8990	0.0614	−0.0328
	LU4	0.0000	0.0000	0.0000	0.0000
Slope angles	0–5°	−0.9987	−0.5287	−0.3506	−1.4542
	5–15°	−0.3072	−0.1333	−0.1784	−0.5598
	15–25°	0.0643	0.1544	0.0794	0.0066
	>25°	0.5439	0.0475	0.0977	0.8448
Slope deposits >5 m		0.3637	0.5339	0.4673	0.2126
Cataclinal slopes		0.2614	−0.5260	1.0071	−0.1878
Presence of fluvial channels		0.3701	0.6220	0.5769	0.3290
Presence of road cuts		0.6673	0.7712	0.9774	0.5388

frequently become fully saturated during wet winters; these conditions enable rotational failures to occur, even on slopes as low as 8°. The highest information values for shallow translational slides are associated with slopes higher than 25° ($IV = 0.84$). This type of landslide tends to occur on steep slopes and is caused by a decrease in the apparent cohesion of colluvium (thin, and often lying upon an impermeable bedrock), resulting from water infiltration into the soil.

Example 2.6
During the heavy rains of November 1982, several thousand landslides occurred in the Eastern Pyrenees, Spain (Corominas and Moya, 1999). Many of these slides were first-time landslides developed on steep colluvium and bedrock slopes (especially marls, sand and clay formations), including mudslides, slab slides and debris flows. It was found that most landslides were the product of a combination of factors; that is no one single factor could be used to successfully predict landslide activity. As a result, a *multivariate approach* was adopted by Baeza and Corominas (1996) to establish the *landslide susceptibility* of different geological and geomorphological settings in the Alt Berguedà region. This involved the following stages:

1 *Establish a landslide inventory* from the analysis of 1 : 15 000 and 1 : 22 000 scale aerial photography. A total of 230 failed and unfailed slopes were included in the inventory.

Table 2.15. Example 2.6: Landslide factors and discriminant functions (from Baeza and Corominas, 1996)

Variable	Function coefficients		Correctly classified (F = −0.4): %
	Standard	Unstandard	
Slope gradient (β)	0.865	0.124	General: 88.5
Watershed area (A)	0.638	1.170	General: 88.5
Land use (LU)	−0.414	−0.102	General: 88.5
Transverse section (TS)	−0.237	−0.298	Partial: Stable, 81.4; Unstable, 95.6
Mean watershed gradient (γ)	−0.213	−0.023	Partial: Stable, 81.4; Unstable, 95.6
Thickness of superficial deposits (Z)	0.204	1.280	Partial: Stable, 81.4; Unstable, 95.6
Constant		−6.219	Partial: Stable, 81.4; Unstable, 95.6
	Eigenvalue 0.9993	Wilks-λ 0.502	χ^2 189 Probability 0.000

2 *Characterise the inventory sites*, through a programme of field-based data collection. The data gathered were simple and easy to measure, including slope attributes (lithology, slope morphology, land use) and landslide characteristics (Table 2.15).

3 *Statistically analyse the field data.* Initially the most significant factors influencing slope stability were identified through the use of *factor analysis*; these were slope gradient (β), watershed area (A), land use (LU), transverse section (TS) of the failure zone (i.e. convex, rectilinear or concave), mean watershed gradient (γ) and thickness of superficial deposits (Z). *Discriminant analysis* was undertaken to establish the relative significance in terms of contribution to landslide susceptibility and weighting of these factors (see Moore, 1973; Gilbert, 1981). The discriminant function correctly classified 88% of the failed and unfailed slopes in the inventory. As shown in Table 2.15, slope gradient proved to be the most powerful discriminant variable, with a weight of 0.865, followed by watershed area (0.638) and land use (−0.414). Note that in Table 2.15 the negative values are associated with unfailed slopes, that is the large negative values of land use, transverse section and mean watershed gradient increase stability.

The analysis indicated that landsliding tends to be associated with steep slopes in large watersheds, with un-forested, concave transverse sections, thick superficial deposits and gently sloping watersheds. The interplay of these factors is important in controlling landslide activity.

4 *Test and revise the discriminant function.* A pilot area in the Eastern Pyrenees was used to test the validity of the discriminant function. The region was divided into 50 m × 50 m grid cells to which a set of attributes (discriminant function variables) was assigned. The presence or absence of landsliding was recorded for each cell. Because of the difficulty of determining the thickness of the superficial deposit cover prior to failure, only five variables were used in the testing process. The best discriminant function (*DF*) proved to be:

$$DF = 1.331 \log(A + 15) + 7.469 \sin(2\beta) - 0.120LU - 0.030\gamma$$
$$- 0.324TS - 8.499$$

Each cell provided a discriminant score derived from this function. Around 95% of the observed landslides in the test area were correctly predicted using the function.

A landslide susceptibility map was prepared using the discriminant function, with susceptibility class boundaries defined by the discriminant scores for each grid cell:

Susceptibility class	Discriminant scores
I (Very low)	<-1.8
II (Low)	-1.8 to <0.9
III (Moderate)	-0.9 to <0.1
IV (High)	0.1 to <1.4
V (Very high)	>1.4

Hazard models

As has already been shown in Chapter 1, a key preliminary stage in landslide risk assessment is the identification of the different types of hazard present or credibly envisaged for a site or within an area, and to employ this information in the development of *hazard models* (see Sections 'Hazard and vulnerability' to 'Risk assessment' in Chapter 1). The hazard models should classify the different types of hazard and quantify their future frequency and magnitude. This can involve defining the different mechanisms and scales of failure, each with a corresponding frequency and impact potential. The hazard models should cover the full range of hazards at a level of detail that is consistent with the resolution of the available data and as determined by scoping as specified in the Section 'Description of Intention' in Chapter 1. Consideration must be given to hazards originating off site or beyond the boundaries of the area under consideration, as well as within the immediate site/area, as it is possible for landslides both upslope and downslope to affect a site or location. The effects of proposed developments should also be considered, as these human-generated changes have the potential to alter the nature, scale and frequency of future landslide hazards. The adequacy and appropriateness of the hazard models will greatly affect the accuracy of subsequent frequency and consequence assessments.

Each situation will be different because of the variation in client requirements and the uniqueness of individual sites or areas. It follows that hazard models need to be individually designed to reflect site conditions and cannot be provided 'off-the-shelf'. However, the following examples provide an indication of the way in which hazard models have been developed for particular purposes. Although they might appear very different in approach, all of the examples focus on:

- *What could happen?* The nature and scale of the landslide events that might occur in the foreseeable future. Often an important issue will be the way in which hazards develop, from incubation, via the occurrence of a triggering or initiating event to the slope response and all possible outcomes.
- *Where could it happen?* The hazard model will need to provide a spatial framework for describing variations in hazard across a site or area. In many instances this framework will be provided by a geological map, although a geomorphological map is preferable.
- *Why such events might happen?* The circumstances associated with particular landslide events.

A hazard model, no matter how good it is, is not a risk assessment, merely an essential component of a risk assessment.

Example 2.7
Earthquake-triggered landslides present a major hazard in many regions. For example, in January 2001, a magnitude 7.6 earthquake off the coast of El Salvador triggered numerous landslides throughout the country, including the Las Colinas landslide, south of San Salvador. The landslide buried more than 400 homes and killed 1000 people. Keefer (1984) developed an important generic hazard model that outlined the relationship between earthquake activity and landsliding. The model was the product of a detailed study of 40 historical large earthquakes (magnitude $M = 5.2$–9.5; note these include moment magnitudes M_w as well as Richter surface-wave magnitudes M_s) from the world's major seismic regions, along with hundreds of smaller magnitude events in the USA.

The hazard model provides a preliminary indication of the zone of influence of earthquakes of different magnitudes, the materials that are

Table 2.16. *Example 2.7: Minimum earthquake magnitudes likely to trigger different types of landsliding (from Keefer, 1984)*

Landslide type*	Magnitude: M
Rock falls, rock slides, soil falls, soil slides	4.0
Soil slumps, soil block slides	4.5
Rock slumps, rock block slides, slow earthflows, soil lateral spreads, rapid soil flows	5.0
Rock avalanches	6.0
Soil avalanches	6.5

*Landslide classification based on Varnes (1978).

most susceptible to earthquake-triggered landslides and what type of landslide events might occur. Amongst the key points of the model are:

1 The smallest earthquakes likely to cause landsliding are around $M = 4.0$ (see Table 2.16).
2 The number of landslides caused by an earthquake generally increases with the event magnitude. Earthquakes with $M < 5.5$ tend to cause a few tens of landslides at most, whereas earthquakes with $M > 8.0$ usually result in many thousands of slope failures.
3 Rock falls, soil falls and soil slides appear to be especially susceptible to initiation during earthquakes (see Table 2.17), hence the huge death tolls from earthquakes in the loess regions of China.
4 Earthquake-triggered landslide activity is generally associated with the first-time failure of materials not previously involved in landsliding. The number of landslide reactivations tends to be small compared with the total number of slides. Where reactivations have occurred, they are

Table 2.17. *Example 2.7: Relative abundance of different landslide types triggered by earthquakes (from Keefer, 1984)*

Relative abundance (in the 40 historical earthquakes)	Landslide type
Very abundant >100 000	Rock falls, disrupted soil slides
Abundant 10 000–100 000	Soil lateral spreads, soil slumps, soil block slides, soil avalanches
Moderately common 1000–10 000	Soil falls, rapid soil flows, rock slumps
Uncommon 100–1000	Subaqueous landslides, slow earth flows, rock block slides, rock avalanches

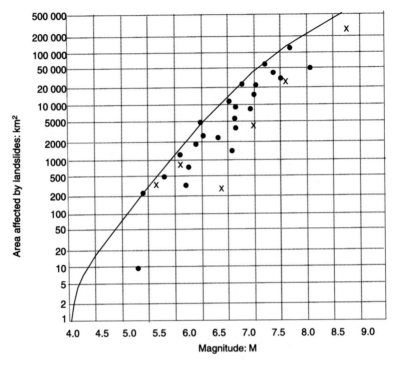

Fig. 2.19. Example 2.7: area affected by landsliding in earthquakes of different magnitude (dots for onshore earthquakes, crosses for offshore earthquakes) (after Keefer, 1984)

associated with pre-existing landslides that were already marginally stable due to other factors, such as high groundwater levels.

5 The maximum area affected by landslide activity increases from around $0\,km^2$ at $M = 4.0$ to $500\,000\,km^2$ at $M = 9.2$ (the 1964 Alaska earthquake; Fig. 2.19). The focal depth of the earthquake is an important control on the area affected. In Fig. 2.19 all onshore earthquakes with a focal depth of over $30\,km$ plot close to the curve, indicating that seismic shaking strong enough to trigger landsliding propagates over larger areas than in shallower events.

6 The maximum distance from the earthquake epicentre to the limit of landslide activity ranges from $0\,km$ at $M = 4.0$ to around $500\,km$ at $M \geq 9.0$ (Fig. 2.20a).

7 The maximum distance of landslide activity from the nearest edge of the fault surface rupture zone ranges from $0\,km$ at $M = 4.0$ to between $200\,km$ and $500\,km$ at $M \geq 9.0$ (Fig. 2.20b).

8 Landslides may occur at minimum Modified Mercalli Intensities (MMI; a measure of the severity of shaking) of IV (disrupted slides or falls) to V

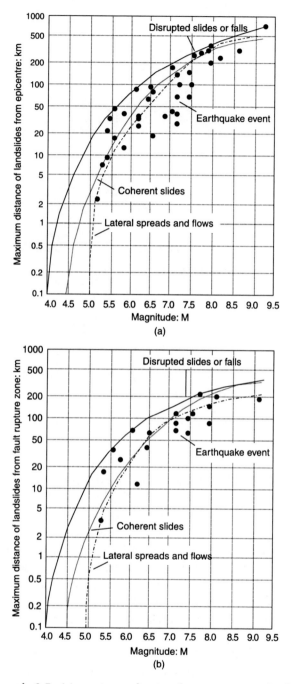

Fig. 2.20. Example 2.7: (a) maximum distance from epicentre to landslides for earthquakes of different magnitudes, (b) maximum distance from fault rupture zone to landslides for earthquakes of different magnitude (after Keefer, 1984)

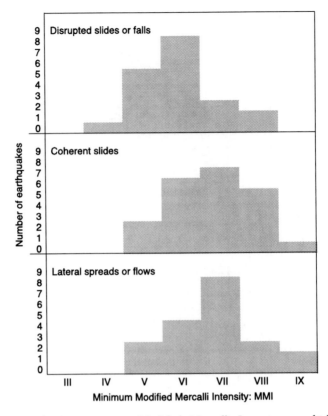

Fig. 2.21. Example 2.7: minimum Modified Mercalli Intensity at which landslides occurred in earthquakes (after Keefer, 1984)

(coherent slides and lateral spreads or flows), up to five levels lower than indicated on the standard MMI scale (Richter, 1958; Fig. 2.21).

9 The materials that are most susceptible to earthquake triggered landsliding are:
- weakly cemented, weathered, sheared, intensely fractured, or closely jointed rocks (rock falls, slides, avalanches, slumps and block slides);
- more indurated rocks with prominent discontinuities (rock falls, slides, block slides and, possibly, slumps);
- unsaturated residual or colluvial sand (disrupted soil slides and soil avalanches);
- saturated residual or colluvial sand (rapid soil flows);
- saturated volcanic soils containing sensitive clay (disrupted soil slides, soil avalanches and rapid soil flows);
- loess (rapid soil flows);
- cemented soils (soil falls);

- deltaic sediments containing little or no clay (soil lateral spreads and subaqueous landslides);
- floodplain alluvium containing little or no clay (soil slumps, block slides and lateral spreads);
- uncompacted or poorly compacted man-made fill containing little or no clay (soil slumps, block slides, lateral spreads and rapid soil flows).

This hazard model provides an initial framework for the development of site or area-specific earthquake-triggered landslide hazard models. Such models need to be based on an investigation of the local ground conditions and should be directed towards the identification of slopes that might be susceptible to landsliding during seismic activity. Keefer (1984), however, provides a note of caution:

> Not all earthquake-induced landslides are in areas with histories of landsliding or at localities where slopes are unstable under nonseismic conditions. Some materials, such as the loess of central Asia and the pumice of the Guatemalan highlands, form steep, high slopes under nonseismic conditions yet disintegrate readily in seismic shaking. In addition, few earthquake-induced landslides reactivate old landslides. Indicators of landslide susceptibility under nonseismic conditions thus should be applied with caution to earthquake-induced landslides.

Example 2.8
Tung Chung (East) is a very steep catchment (35–40° slopes) on the northern slopes of Lantau Island, Hong Kong. The catchment is susceptible to a combination of frequent shallow hillside failures and rare, large-scale channelised debris flows. An integrated programme of engineering geological and geomorphological mapping was undertaken in order to define a series of hazard models, as part of a quantitative risk assessment (Halcrow, 1999; Moore *et al.*, 2001, 2002).

Field mapping identified a series of distinct terrain units and sub-units, based on interpretation of slope morphology and surface drainage (Fig. 2.22). These include an upper catchment (Terrain Unit 1), comprising a series of convergent incised stream channels (Terrain sub-units 1.1 to 1.4), and a lower catchment (Terrain Unit 2), comprising a single incised stream channel (Terrain sub-unit 2.2). Below Terrain sub-unit 1.3 there is a narrow, sediment-filled channel storage zone (Terrain sub-unit 2.1). A large debris fan occurs at the mouth of the catchment (Terrain sub-unit 2.3) that extends onto the shoreline.

A total of 75 landslides were identified within the catchment. Most landslides occur on 35–40° slopes, are of *recent* origin (i.e. vegetated or partly vegetated failures that most likely occurred within the last 50 years; Halcrow, 1999) and have involved an initial phase of detachment as a

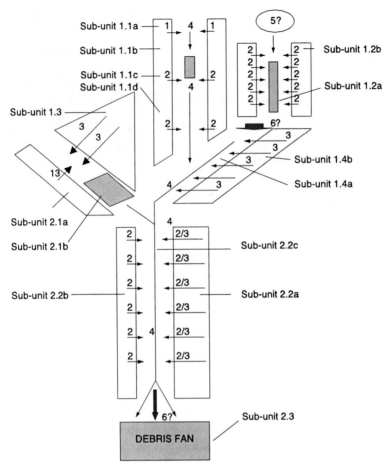

Natural hazards
1. Isolated confined hillside slides.
2. Slightly confined hillside failures supplying limited debris to stream channels and/or stores.
3. Unconfined hillside debris flows: high velocity events.
4. Channelised debris flows involving remobilisation of debris in channel stores.
5. Major first-time failure in colluvium/rock with limited secondary mobilisation of material.
6. Major first-time failure in colluvium/rock with mobilisation of material as a channelised debris flow.

Notes:
Hazards 5 and 6 represent possible historical events.
Rock and boulder falls may occur from rock faces, especially along the main stream channel.
Flash flood events may occur along stream channels.

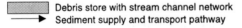   Debris store with stream channel network
———▶ Sediment supply and transport pathway

Fig. 2.22. Example 2.8: Tung Chung East, Hong Kong: a schematic summary of the landslide hazard model (from Palmer and Lee, 1999)

shallow (<3 m deep) *debris slide or avalanche* (Evans et al., 1997) within colluvium, weathered rock and/or saprolite. Evidence for *channelised debris flows*
was limited to the presence of an extensive fan of debris flow deposits at
the catchment mouth (debris fan) and the indistinct traces of debris trails
on the steep valley sides and stream channels.

Ground investigation confirmed that the debris fan deposits were probably
of debris flow origin and revealed up to 5 m of material in three distinct layers
with an estimated volume of 48 000 m^3. The evidence indicated that the fan
comprises debris from three main events of comparable magnitude of about
16 000 m^3.

A schematic summary of the distribution of terrain units and the
associated natural hazards across the catchment is presented in Fig. 2.22.
The field investigations provided the framework for deriving channelised
debris flow event scenarios for the catchment. The approach involved
estimation of a *sediment budget* for the catchment (i.e. the balance between
sediment *inputs* from landsliding and the *outputs* from the catchment). The
whole catchment was viewed as a cascading sequence of inter-related
physical systems (i.e. the terrain sub-units), so that the *outputs* from one
system provide the *inputs* to the next in line. Within the catchment, five
main stages of landslide activity were recognised: *initiation* and detachment
of material from hillslopes; *transport* and delivery of this material into the
channel system; *storage* of material within the channel system (and also, in
the short-term, on hillslopes before delivery to the channels); *entrainment*
and run-out from the catchment; and *deposition* on the debris fan.

The sediment budget for the catchment provided an indication of the
current potential sediment supply to the debris fan at the catchment
mouth. A large proportion of the potential sediment inputs remain *stored*
within the natural terrain as:

- landslide debris upon hillslopes – most landslides within the area have
  delivered less than 50% of the failed mass to the stream channels and
  valley stores. The amount of material stored within individual landslides
  was estimated visually during the field mapping;
- colluvial and alluvial deposits within channels – as ribbon-like stores along
  stream channel banks and beds (often as angular and sub-rounded
  boulders within a sandy matrix) or broader accumulations in valley floors
  (valley floor stores). The amount of material stored within stream channels was estimated as an average volume per metre run. The volume of
  material stored within valley floors was estimated from surface dimensions
  and an assumed average depth.

For relatively high frequency, shallow, open hillside landslides, debris enters
the slope or valley stores rather than being directly mobilised as channelised

debris flows. Thus, the stores act as a *buffer* between hillslope instability and stream valley transport, either as channelised debris flows or in flash floods. However, for low-frequency, high-magnitude events, debris stored upon hillslopes and within valley floors provides a *source* of generally unconsolidated sediment that can be entrained and mobilised by channelised debris flows. For example, the Tsing Shan debris flow of 1990 entrained twice as much as its initial volume as it entered a drainage line and entrained bouldery colluvium (King, 1996). It follows that the accumulation of unconsolidated debris from numerous shallow hillside landslides within valley stores over time can provide a large volume of sediment capable of being mobilised in a single episodic event.

Figure 2.23 presents a summary sediment budget for the catchment. For each terrain sub-unit the sediment budget was calculated as follows:

- the *inputs* were estimated as the volume of material supplied from open hillside landslides (i.e. landslide volume less the amount remaining on the hillslope), plus the volume of material supplied from upslope terrain sub-units;
- the hillslope and valley *storage* was estimated from field observations (i.e. length × average width × average depth);
- the *output* from each terrain sub-unit was assumed to be the balance between the inputs and storage. Where the inputs are greater than the estimated hillslope and valley floor storage, the excess is assumed to be the output from the terrain sub-unit and, hence, an input to the next sub-unit.

In this way, the contemporary catchment yield was estimated to be about $5770\,\text{m}^3$, that is the supply to the debris fan. This figure represents the difference between the inputs from the current suite of open hillside landslides and the catchment storage. The recorded landslide features have involved $23\,900\,\text{m}^3$ of sediment, of which around $13\,500\,\text{m}^3$ has been supplied to the channel system ($10\,400\,\text{m}^3$ remains in storage upon the hillsides within landslide scars), $7730\,\text{m}^3$ of which (i.e. 32% of the total) remains as channel storage and $5770\,\text{m}^3$ (24% of the total) has been transported to the debris fan by channelised debris flows and flash floods.

Landslides input $23\,900\,\text{m}^3$		Hillside storage $10\,400\,\text{m}^3$		Channel storage $7730\,\text{m}^3$		Output to debris fan $5770\,\text{m}^3$
	$\Rightarrow$		$\Rightarrow$		$\Rightarrow$	

The sediment budget calculation was used to estimate the volume of particular scenarios, one of which was the *maximum potential channelised debris flow event* that could be expected to occur within the catchment, with an

110

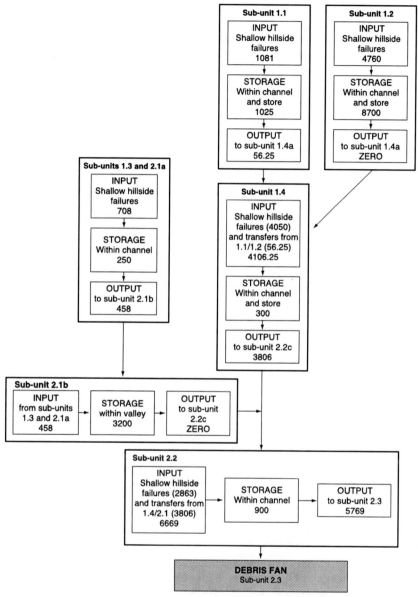

Notes:
1 The *inputs* have been estimated as the volume of material supplied from hillside failures (i.e. the landslide volume minus the amount stored on the hillside), plus the volume of material supplied from an upslope terrain sub-unit.
2 The *storage* has been estimated from field observations (i.e. length × average width × estimated average depth).
3 The *output* from the terrain sub-unit (and an input to the next sub-unit) is assumed to be the balance between inputs and storage. Where the inputs are less than the estimated channel or valley floor storage, the sediment inputs are assumed to only contribute to the storage (i.e. storage will increase over time), with zero outputs. Where the inputs are greater than the estimated channel or valley floor storage, the excess is assumed to be the outputs from the terrain sub-unit and, hence, an input to the next sub-unit.

Fig. 2.23. Example 2.8: summary sediment budget for the Tung Chung East catchment, Hong Kong (from Palmer and Lee, 1999)

111

Table 2.18. Example 2.8: Best estimate maximum potential channelised debris flow event (from Moore et al., 2002)

Terrain sub-units and landforms	Volume: m^3
Terrain sub-unit 1.2	
Maximum open hillside landslide source event	1800
Channel section 1 (35°)	
Entrainment of colluvium $= 64\,\mathrm{m} \times (2\,\mathrm{m} \times 2\,\mathrm{m/tan}\,35°)$	367
Channel section 2 (17°)	
No entrainment due to low gradient $= 135\,\mathrm{m} \times 0\,\mathrm{m^3/m}$	0
Channel section 3 (27°–30°)	
Entrainment $= 195\,\mathrm{m} \times 2.5\,\mathrm{m^3/m}$	488
Terrain sub-unit 1.1	
Maximum open hillside landslide source event	1800
Channel section 4 (28°)	
Entrainment of colluvium $= 163\,\mathrm{m} \times (1\,\mathrm{m} \times 1\,\mathrm{m/tan}\,35°) = 1.43\,\mathrm{m^3/m}$	
Since $1.43 < 2.5$, use $2.5\,\mathrm{m^3/m}$	408
Channel section 5 (25°–36°)	
Entrainment $= 157\,\mathrm{m} \times 2.5\,\mathrm{m^3/m}$	393
Terrain sub-units 1.3 and 2.1	
Maximum open hillside landslide source event	1800
Channel section 6 (30°–50°)	
Entrainment of colluvium $= 185\,\mathrm{m} \times (2\,\mathrm{m} \times 2\,\mathrm{m/tan}\,35°)$	1057
Channel section 7 (20°)	
No entrainment due to low gradient $= 65\,\mathrm{m} \times 0\,\mathrm{m^3/m}$	0
Channel section 8 (12°)	
No entrainment due to low gradient $= 75\,\mathrm{m} \times 0\,\mathrm{m^3/m}$	0
No deposition	0
Terrain sub-unit 2.2; V-shaped stream channel below confluence	
Channel section 9 (18°–20°)	
No entrainment due to low gradient $= 300\,\mathrm{m} \times 0\,\mathrm{m^3/m}$	0
Channel section 10 (25°)	
Entrainment $= 54\,\mathrm{m} \times 2.5\,\mathrm{m^3/m}$	135
Total potential output to terrain unit 2.3	8248 m^3

expected frequency of 1:50 years. This scenario was estimated by defining a maximum potential open hillside landslide for each terrain sub-unit (i.e. source area) and entrainment of colluvium from the existing channel storage areas (Table 2.18). Field evidence indicated that the largest existing open hillside landslide has an estimated volume of 1800 m^3. As rainfall events that could trigger major landslide activity are likely to cover the entire catchment, it was assumed that events of a comparable magnitude could be triggered simultaneously in each of the three sub-catchment source areas, providing an input of 5400 m^3 of sediment.

A series of channel segments were defined from each potential source of landsliding along the stream course, highlighting channel storage areas from which colluvium can be entrained by the channelised debris flow. It was assumed that all debris from channel stores could be mobilised and transported down the channel to a point where deposition takes place. Based on existing knowledge of channelised debris flow behaviour in Hong Kong, debris entrainment has been observed on channel gradients above 22° while deposition occurs where gradients reduce to 15° or less. An *initial deposition point* was therefore defined by the 15° break in slope contour, below which channel gradients are generally less than 15°.

For channel segments steeper than 22°, it was assumed that 2.5 m^3 of debris per metre length of stream channel may be entrained. This assumption accounts for the potential entrainment of loose colluvium from the steep side-slopes adjacent to the stream channel. In some locations, the potential channelised debris flow passes over colluvium previously deposited in stream channels. If the colluvium is within a channel at gradients of 22° or above, it was assumed that all material within the channel could be entrained. For example, a 5 m-deep deposit of colluvium in a 7.2 m-wide stream channel could yield 36 m^3/m length of channel (7.2 m × 5 m × 1 m = 36 m^3).

The results indicate a 'best estimate' maximum potential channelised debris flow event of about 8250 m^3. This compares with a possible 'worst-case' maximum potential channelised debris flow event of about 24 000 m^3. This latter event would involve the mobilisation and transport of all the current hillside, channel and valley floor storage in a single event. It was assumed that there would be no storage of sediment within the catchment after the event; that is for each terrain sub-unit the outputs have been estimated as the volume of material currently held in store and the volume of material supplied from an upslope terrain sub-unit. The likelihood of the 'worst-case' scenario was considered remote with an expected frequency well in excess of 1 : 200 years. Indeed, evidence from the debris fan of only three main events suggests that channelised debris flows are extremely rare. In this case, high-magnitude channelised debris flows are only likely to be triggered by the rare occurrence of multiple open hillside landslides, coupled with delivery of debris into fast flowing stream channels that increase the mobility and entrainment of landslide debris down the mountain stream course.

Example 2.9
The coastal slopes between the Cobb Gate and the Harbour in Lyme Regis, UK, form the seaward part of a landslide system composed of a variety of deep-seated failures and Pleistocene solifluction deposits, which extends around 500 m inland. These landslide mantled slopes are developed in an

interbedded sequence of Lower Jurassic (Lias) clays, mudstones, siltstones and limestones, overlain unconformably by Cretaceous sands and sandstones.

The landslides are believed to be of considerable age. As sea-level rose after the last glaciation, marine erosion would have created near-vertical seacliffs in the Pleistocene landslide features and led to their progressive reactivation. Between 1787 and 1854 the unprotected seacliffs retreated at around 0.2–0.3 m/year (High-Point Rendel, 1999). Although the construction of seawalls in the 1860s has prevented further marine erosion, the slopes have continued to be affected by instability or slope degradation because they had been oversteepened by marine erosion prior to protection. A sequence of landslide events last century resulted in considerable damage to infrastructure, amenity areas and properties, including the Cliff House landslide of 1962 (Hutchinson, 1962; Lee, 1992).

Combining surface morphological evidence and the results of borehole investigations led to the identification of five interrelated landslide systems in the area between Cobb Gate and the Harbour (see High Point Rendel, 1999; Clark *et al.*, 2000; Fort *et al.*, 2000; Sellwood *et al.*, 2000; Lee *et al.*, 2000). Each system comprises a number of inter-linked but discrete units (Table 2.19; Fig. 2.24). These units are arranged in a linear sequence, progressing inland from the seacliff to the rear scarp.

The *reactivation* of landslide systems, such as the multi-levelled systems identified at Lyme Regis, will rarely occur as a single event. It is more likely that individual units within each system will become active as a result of removal of support from downslope or localised high groundwater levels. This activity may, in turn, promote the spread of instability into adjacent landslide units, often through the effects of loading or unloading (Fig. 2.25).

Of particular significance is the sensitivity of these landslide systems to the removal of support provided by the lower landslide units (Fig. 2.25). Seawalls currently protect the lowest portions of the coastal slopes from marine erosion, preventing further oversteepening and destabilisation. Failure of these seawalls would result in renewed erosion of the lower units (Units 1A, 2A and 3A on Fig. 2.24). This would lead to the seaward movement of the translational block slides which, in turn, would unload the rear scarp. Failure of the rear scarp would cause debris to be deposited on to the back of the block slides, with the loading promoting further seaward movement of the blocks and destabilisation of the slopes above the rear scarp. In the case of the Lister Gardens and Harbour Heights systems (see Table 2.19), this unloading would ultimately lead to reactivation of the pre-existing compound failures above the sub-Cretaceous unconformity (System 5). In reality the pattern of instability is complex and interactive, so that what happens in one area directly affects the stability elsewhere. Continued removal of material from the toes of unprotected slopes by marine erosion would ensure that

Table 2.19. Example 2.9: Landslide systems on the Lyme Regis coastal slopes (from High Point Rendel, 1999)

Landslide system		Unit	Type
Name	Number		
Langmoor Gardens	System 1	A	Translational or rotational toe failures
		B	Shallow translational/debris slides
		C	Translational block slides (shallow and deep-seated)
		D	Mudslides and first-time failures of the rear scarp
Lister Gardens	System 2	A	Translational or rotational toe failures
		B	Shallow translational/debris slides
		C	Translational block slides (shallow and deep-seated)
		D	Rear scarp failures including mudslides and possibly translational slides
Harbour Heights	System 3	A	Translational or rotational toe failures
		C	Translational block slides (shallow and deep-seated)
		D	Rear scarp (degraded) failures including mudslides and possibly translational slides
Cobb Terrace	System 4	C/D	Mudslides and possibly translational slides
Coram Avenue	System 5	E	Mudslides and compound slides above sub-Cretaceous unconformity
		F	Compound slides above sub-Cretaceous unconformity

instability cycles and progressive reactivation would continue, generally similar to those observed in the naturally evolving, unprotected slopes to the east and west of Lyme Regis.

A *progressive reactivation model* was developed for the Lyme Regis coastal slopes. The model involves the following potential behaviour types:

1 almost continuous *extremely slow* creep;
2 episodes of significant *slow to very slow* ground movement;
3 distinct landslide reactivation sequences involving the re-establishment of active instability across individual units and combinations of units.

Key features of the hazard model include:

- Reactivation is driven by the chance occurrence of *initiating events* and subsequent system *responses*.
- Reactivation can occur in response to the progressive effects of repeated toe unloading after a single *initiating event* (e.g. seawall failure or a wet

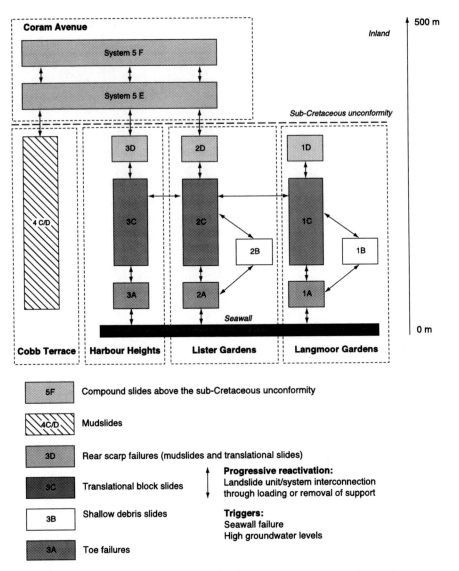

Fig. 2.24. Example 2.9: Lyme Regis, UK: schematic summary of the landslide reactivation model

year sequence). The net result is a spread upslope of active instability as higher-level slides are successively unloaded by the movements of the lower-level slides which had previously provided passive support.

- The development of reactivation sequences is *conditional* on the unloading/removal of support to units due to the movement of the adjacent, downslope units.

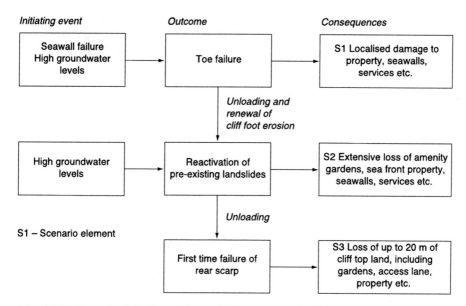

Fig. 2.25. Example 2.9: Lyme Regis, UK: summary landslide reactivation scenario (from Lee et al., 2000)

- The response to unloading/removal of support to a unit may be *delayed* or *lagged*. It is expected that reactivation at a particular level will be conditional on the occurrence of suitable triggering events (e.g. wet year sequences and high groundwater levels).
- The likelihood of a major landslide event is *increasing through time* because of progressive internal system changes (e.g. build up of stress, gradual loss of strength), the effects of climate change/sea-level rise and the cumulative impact of urban development on the slopes.

Four scenarios (i.e. plausible sequences of events) were developed, based on an understanding of the causes and mechanisms of landslide behaviour:

1 Scenario 1; almost continuous creep.
2 Scenario 2; an episode of significant ground movement in response to the exceedance of a winter rainfall threshold level.
3 Scenario 3; seawall failure causing the development of a landslide reactivation sequence that leads, ultimately, to the establishment of active landsliding throughout the coastal slopes.
4 Scenario 4; high groundwater levels initiating the development of a landslide reactivation sequence that leads, ultimately, to the establishment of active landsliding throughout the coastal slopes.

The impact of these scenarios will vary from place to place, according to the geomorphological setting. Scenarios 1 and 2 are essentially *business as usual* scenarios and, hence, the contemporary ground behaviour is considered to provide an indication of those parts of the coastal slopes that are most likely to be affected by each scenario (High Point Rendel, 1999). It was assumed that areas which have experienced the effects of ground movement in the past century will continue to be the most susceptible to the same styles of movement in the future. Scenarios 3 and 4 (development of a landslide reactivation sequence) have the potential to affect the whole area. In considering the hazard associated with these scenarios it was assumed that:

- at a particular level there will be a gradual deterioration of stability over a number of years, followed by a sudden and dramatic movement of large areas of ground. Successive phases of reactivation after an initiating event are likely to correspond with wet years and high groundwater levels;
- widespread reactivation is likely to be preceded by a period of localised ground movement (e.g. tension cracks, heave, subsidence etc.);
- peak movement rates are likely to be in the order of metres per year, and the cumulative differential horizontal and vertical movements will, over time, result in severe property and infrastructure damage.

Example 2.10
The Holderness cliffs, UK, range in height from less than 3 m to around 40 m. The cliffs are formed in a sequence of glacial tills, predominantly silty clays with chalky debris and lenses of sand and gravel. They are subject to severe marine erosion, but remain unprotected for most of their length. Long-term recession rates are in the order of 1.2–1.8 m per year (Valentin, 1954; Pethick and Leggett, 1993). At the example site, ongoing recession threatens cliff top land and property, including a caravan site.

A probabilistic hazard model was developed by Jim Hall and presented in Lee and Clark (2002). It simulated the episodic cliff recession process on the cliffline using a 2-distribution model of the type described in Chapter 4 (see Example 4.10). The model used cliff retreat data from erosion post measurements covering a 50 year period, supported by a geomorphological assessment of cliff behaviour (Lee and Clark, 2002). The data provides approximate information about the size of individual landslide events and the time interval between them. The model was fitted to the data on the assumption that in years when recession is recorded, this represents a *single landslide event*. In practice this will not always be the case, but the approximation does provide a lower bound on the number of recession events. By proceeding on the basis of this assumption, it is possible to directly derive the parameters of the prediction model using the method of moments (see Hall *et al.*, 2002).

These data were used to estimate the mean and standard deviation on the duration between recession events (Fig. 2.26) and recession event size

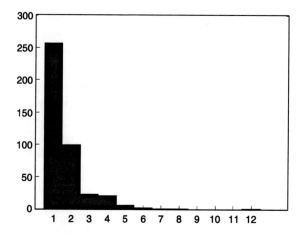

Fig. 2.26. Example 2.10: histogram of durations between recession events on the Holderness cliffs, UK (from Lee and Clark, 2002)

(Fig. 2.27) as follows: Duration between recession events: mean $= 1.64$ years, standard deviation $= 1.17$ years. Recession event size: mean $= 3.14$ m, standard deviation $= 3.44$ m. The model incorporates understanding of the cliff recession process by representing the role which storms have in destabilising cliffs (Hall *et al.*, 2002). The approach has links to renewal theory (Cox, 1962) inasmuch as the cliff is considered to be progressively weakened by the arrival of storms. The arrival of damaging storms is assumed to conform to a Poisson process (see Section 'Multiple events' of

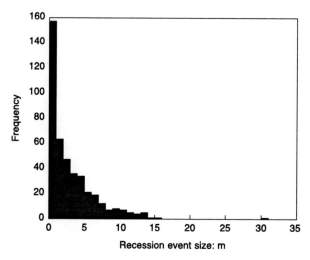

Fig. 2.27. Example 2.10: histogram of recession distance for events on the Holderness cliffs, UK (from Lee and Clark, 2002)

Chapter 4; Borgman, 1963; Shane and Lynn, 1964), that is successive storms are assumed to be independent incidents with a constant average rate of occurrence but random actual intervals between storms. This is a reasonable assumption provided the threshold for storm damage is sufficiently high. The times Ti, $i = 1, 2, \ldots, k$, between storms conform to an exponential distribution. After a number of storms of sufficient severity, a landslide occurs. Suppose that it takes k storms of given severity to cause a recession event. The time to the kth storm is the sum $T1 + T2 + \ldots + Tk$. The time between successive landslides is therefore gamma distributed with density function:

$$f_T(t|k, \lambda) = \frac{\lambda^k t^{k-1}}{\Gamma(k)} e^{-\lambda t}$$

where $\Gamma(k)$ is the gamma function, λ is a scaling parameter (the reciprocal of the return period of the significant storm event), k is a shape parameter (the average number of storms above a certain threshold which cause damage to the toe of the cliff) and t is time.

The mean of this distribution is k/λ and the variance is k/λ^2, so for the selected erosion posts, $k = 1.98$ and $\lambda = 1.21$.

The recession event size was assumed to conform to a lognormal distribution:

$$f(x \,|\, \mu, \sigma) = \frac{1}{x\sigma\sqrt{2\pi}} \exp\left(\frac{-(\ln x - \mu)^2}{2\sigma^2}\right)$$

The mean of this distribution is $\exp(\mu + 0.5\sigma^2)$ and the variance is $\exp(2\mu + \sigma^2)[\exp(\sigma^2) - 1]$, so for the selected posts, $\mu = 0.93$ and $\sigma = 0.85$.

For the hazard model, random sequences of landslide event sizes and durations between events were generated, with statistics that conform to the measured statistics at the site. From each random sequence, the annual recession distance between 1 and 49 years was extracted. Each simulation (i.e. repeated random sequences) represents a different landslide scenario. A large number of simulations (in this case 10 000) were used to generate a histogram or annual recession distance. A kernel density estimation method was used to obtain a smooth probability density estimate from the histogram. Probabilistic predictions of recession distance for selected time periods are shown in Fig. 2.28 (further consideration of the impact of this recession is presented in Example 6.5).

Example 2.11
Rockfalls can pose a significant hazard, especially in mountain regions and along coastal clifflines. A key factor in developing a hazard model for rock cliffs is developing an understanding of the magnitude/frequency distribution

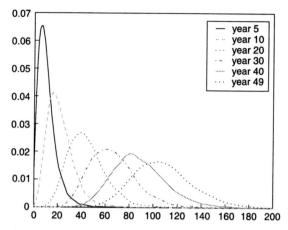

Fig. 2.28. Example 2.10: probabilistic predictions of recession distance for the Holderness cliffs, UK (from Lee and Clark, 2002)

of rockfall events. In general, the occurrence and size of events is controlled by discontinuity patterns within the rock mass. As a result, geomechanical survey of the discontinuities along a cliffline can help support the development of a potential fall frequency/volume distribution. However, this detailed approach is unlikely to be feasible or cost-effective for a lengthy cliffline. An alternative approach to developing a hazard model for rock cliffs is to derive a probabilistic recurrence rate for an event of a given size (e.g. Dussauge-Peisser *et al.*, 2002).

The observed frequency–size distribution of some natural hazards can be represented by *distribution laws*, based on the statistical analysis of historical events. For floods these can be exponential-like (e.g. the so-called Gumbel distribution; Gumbel, 1941), whereas for earthquakes the magnitude/ frequency distribution is usually described by a power law, such as the Gutenberg–Richter law (Gutenberg and Richter, 1949):

$$\log N(m) = \alpha M^{-b}$$

where $N(m)$ is the number of earthquakes with magnitude larger than M. This model is considered satisfactory for small and moderate earthquakes, but does not hold for extremely large earthquakes because it predicts an infinite energy release.

Hungr *et al.* (1999) have suggested that distributions of rockfalls from homogeneous areas could be fitted by a similar power law. The approach was used by Dussauge-Peisser *et al.* (2002) to develop simple hazard models for three cliff sites, based on the analysis of relatively long duration rockfall inventories (Fig. 2.29):

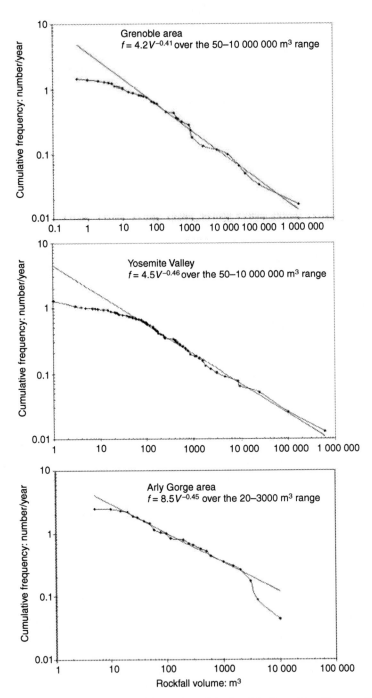

Fig. 2.29. Example 2.11: rockfall hazard models for (a) Grenoble, France, (b) Yosemite Valley, USA, (c) Val d'Arly, France (from Dussauge-Peisser et al., 2002)

1 Grenoble, French Alps; a 120 km long, 50–400 m high series of clifflines developed in upper Jurassic and lower Cretaceous limestones and marls, prone to rockfalls and topples, with volumes ranging from 0.5 to 106 m³. A rockfall record has been maintained by forest guards since 1850 (RTM, 1996), although Dussauge-Peisser *et al.* (2002) only examined events in the period 1935–95.

2 Yosemite Valley, USA; around 100 km of granite cliffs, up to 1000 m high. National Park rangers and USGS geologists have reported the occurrence of rockfalls in the Yosemite Valley since 1850, gathering a record of more than 400 events (Wieczorek *et al.*, 1995). Dussauge-Peisser *et al.* (2002) considered 101 events which occurred in the $1–6 \times 10^5$ m³ volume range, in the 78 year period between 1915 and 1992.

3 Val d'Arly Gorge, French Alps; a 7 km long cliff developed in mica schists overlain by Carboniferous sedimentary rocks. Rockfall activity threatens a minor road and daily records of events larger than 1 m³ have been kept since 1954. A total of 111 events occurred between 1954 and 1994, ranging from 1–10 000 m³ in volume, with 59 events between 1954 and 1976.

Dussauge-Peisser *et al.* (2002) concluded that, above a given volume, the occurrence of rockfall sizes corresponded to a simple power law, regardless of the period of observation (Table 2.20):

$$n(V) = \alpha V^{-b}$$

Table 2.20. Example 2.11: Rockfall hazard model characteristics (from Dussauge-Peisser et al., 2002)

	Grenoble, France	Yosemite Valley, USA	Upper Arly gorges, France
Geology	Calcareous cliffs	Granite cliffs	Metamorphic and sedimentary rocks
Number of rockfalls	87	101	59
Approximate cliff surface: km²	24	30	2.2
Time period: years	60	78	22
Range of the power law fit: m³	$50–10^6$	$50–10^6$	30–3000
Exponent b	0.41 ± 0.11	0.46 ± 0.11	0.45 ± 0.15
Exponent α	4.2	18	8.5
n_{100}: number per year	0.62	2.16	1.07
$n_{100}/10$ km²	0.26	0.72	19.45

where V is the rockfall volume, $n(V)$ is the number of events per year with a volume equal or greater than V, and α and b are constants.

They concluded that:

- the b constant (which controls the shape of the distribution) is independent of geological setting and is fairly constant at around 0.45 (with a range of 0.41 to 0.46);
- the α constant (defined as n_{100} which represents the annual number of events greater than $100\,m^3$) varies with the site. This coefficient reflects the level of activity along the cliffline and varies from 0.62 (Grenoble) to 2.16 (Yosemite Valley). When the n_{100} coefficient was normalised by dividing the surface area of each cliffline that was a potential rockfall source, values of n_{100} ranged from $0.26/10\,km^2$ (Grenoble) to $19.45/10\,km^2$ (Arly gorges).

The work of Dussauge-Peisser *et al.* (2002) provides the basis for a generic approach to developing hazard models for rockfall activity. Statistical studies of past events thus provide a tool for quantifying an overall frequency of rock falls in a given area (at the scale of a whole cliff or series of cliffs). The following stages would be involved:

1 analysis of historical records to establish a frequency/volume distribution for rockfall events;
2 calculation of the power law parameters α and b. This could involve the use of linear regression or a maximum likelihood method (Aki, 1965) in which, for example, the estimate for b would be given by

$$b = 1/\ln(10)(\langle \log V \rangle - \log V_0)$$

where $\langle \log V \rangle$ is the average value of $\log(V)$, and V_0 is the minimum volume considered, above which the inventory is assumed to be complete.

Uncertainty, assurance and defensibility

Understanding the nature and scale of potential landslide problems in an area provides the foundations for landslide risk assessment. Poor problem definition will lead to poor risk assessment. However, it is clear that some degree of uncertainty will exist in the knowledge of any site or area, because of insufficient data, natural spatial variability and the difficulties in predicting the implications of future environmental change on slope stability (see Section 'Uncertainty and risk assessment' in Chapter 1). Fookes (1997) noted that the art of geological or geotechnical assessment is 'the ability to make rational decisions in the face of imperfect knowledge'. Decisions about landslide hazard almost always incorporate uncertainty to one degree or another and there is a need to rely on judgement.

124

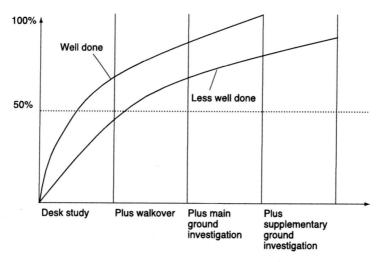

Fig. 2.30. Estimated upper and lower bounds of landslide information anticipated during the stages of a site investigation (after Fookes, 1997)

Recognition of uncertainty clearly has implications for how much effort should be spent defining the landslide hazard. There is no simple answer to this question, but a number of issues will be of relevance in reaching a decision:

1 *The need for adequate site investigation.* Experience has shown that shortfalls in data acquisition almost inevitably lead to the possibility that significant risks have been overlooked. For example, in a review of site investigation practice, the Ground Board of the Institution of Civil Engineers suggested that in any civil engineering or building projects the largest technical and financial risk lies in the ground conditions (ICE, 1991). Figure 2.30 illustrates the estimated rate of acquiring information on a typical site. To achieve a robust and reliable understanding of ground conditions that support the development of a hazard model, it is advisable not to cut corners in the investigation process.

2 *The need to provide assurance to the client, decision-maker or regulator* that the hazard model delivers a reliable basis for risk assessment. In a perfect world every landslide or potentially unstable slope would be thoroughly investigated using many of the techniques described in Section 'Acquiring information: landslide investigation' in this chapter. However, in reality, factors such as the urgency in finding a solution to a pressing problem, cost, the availability of suitable investigation equipment etc. mean that compromises have to be reached between the effort involved and the degree to which the uncertainties are reduced. In other circumstances, the client, decision-maker or regulator may only be convinced that

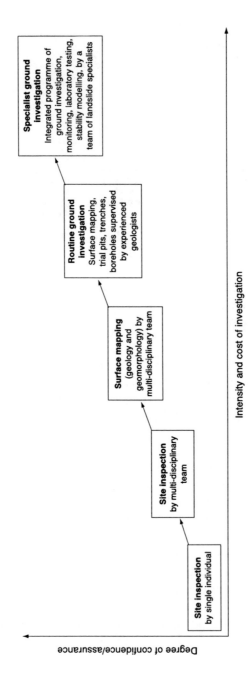

Fig. 2.31. *An illustration of how the degree of assurance in landslide hazard assessment might change with the intensity of investigation*

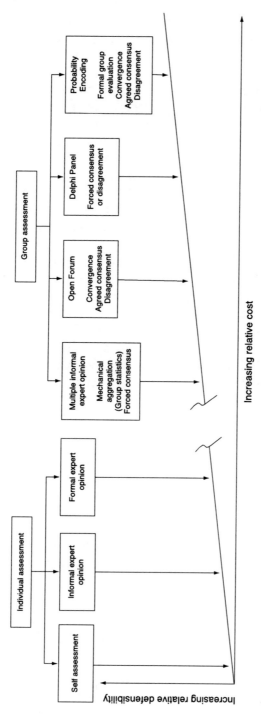

Fig. 2.32. Summary of the expected defensibility of different methods for undertaking subjective judgements (after Roberds, 1990; Fookes, 1997)

there is sufficient knowledge of the ground conditions and risks after an extensive borehole investigation. In reaching a decision as to the level of assurance that is appropriate to the particular assessment, a key factor should be the scale of consequences if the hazard model should prove to be inadequate or even wrong.

Figure 2.31 illustrates how the degree of assurance might increase with increased investigation effort. For example, a simple site inspection of an apparently stable slope may be sufficient if the assets at risk are of low value and where there would be no threat to the public should the slope fail unexpectedly. However, if an unexpected failure would cause major economic loss or loss of life, it might be appropriate to undertake a more rigorous investigation to provide a higher level of assurance that the slope is safe. More detailed investigation may not significantly reduce the uncertainties, but often will result in a higher level of confidence about the judgements.

3 *The need to ensure that the judgements are defensible.* It is important for practitioners to appreciate that their judgements about risk may provoke considerable disagreement and controversy, especially if the judgements have implications for property values or the development potential of a site. In an increasingly litigious world there will be a need to demonstrate that they have acted in a professional manner appropriate to the circumstances. Figure 2.32 introduces the notion of increasing defensibility with increasing sophistication of the approach used to overcome the problems associated with subjective judgement (see Sections 'Subjective probability' and 'Estimating probability through expert judgement' in Chapter 4 for further discussions).

3

Qualitative and semi-quantitative risk assessment

Introduction

Slope inspections and landslide studies have always involved some rudimentary form of risk assessment, although it may have seldom been recognised as such (e.g. Fell and Hartford, 1997). Informal assessments of risk have generally relied on the judgement and skill of experienced engineers, geologists and geomorphologists. Recognition of hazards, mapping of areas of current or past instability, creation of a ground model and development of an understanding of the causes and mechanisms of failure, have all been shown to be essential for making judgements about the significance of landslide problems within an area or at a particular site. As a consequence, decision-makers have often been able to act on specialists' advice without having recourse to the explicit quantification of risk. However, consistency between different specialists is difficult to achieve, both for successive inspections of the same slope or when comparing the relative significance of different landslide problems. The world is changing and both clients and consultants are becoming increasingly aware of the need to introduce ever more rigorous and systematic procedures to formalise the evaluation process and thereby enhance the 'openness', 'objectivity' and 'consistency' of such judgements.

A large proportion of this book is devoted to quantitative risk assessment methods. However, there are many instances when an estimate of risk in terms of economic impact or loss of life cannot either be realistically achieved because of constraints of time, resources and data availability, or is simply not required (see Section 'Risk estimation' in Chapter 1). For example, a decision-maker may be aware that landsliding is an issue in a particular area, but only needs advice on:

129

- which landslides present a significant or pressing threat to assets or the population. This can involve screening out those sites where landsliding is perceived to present only minor or negligible problems;
- how to prioritise those significant sites in order to ensure that the most urgent problems are dealt with first.

A variety of so-called qualitative or semi-quantitative risk assessment methods have been developed to address these types of issues, including:

- risk registers;
- relative risk scoring;
- risk ranking matrices;
- relative risk rating;
- failure modes, effects and criticality analysis (FMECA).

These approaches have the following in common:

1 *The sub-division of the area of interest into sub-units*, often on the basis of geomorphology, engineering geology or the dominant observed landslide process. This emphasises the necessity of mapping as a pre-requisite for risk assessment (see Chapter 2). In the case of coastal cliffs, the division should be made on the basis of the identification of Coastal Behaviour Units (CBUs; Lee and Clark, 2002).
2 *Estimation of the likely magnitude, frequency and impact potential* of landsliding within each identified unit during a defined period of time, using scoring or ranking schemes (see Section 'Risk assessment' in Chapter 1). Although four approaches incorporate the word 'risk' in their title, they actually focus on the boundary between 'hazard' and 'risk'.
3 *The use of expert judgement.* It is important, therefore, that effort is directed towards ensuring that all judgements can be justified through adequate documentation, allowing any reviewer to be able to trace the reasoning behind particular scores or rankings. Ideally the assessment process should involve a group of experts, rather than single individuals, as this facilitates the pooling of knowledge and experience as well as limiting bias, but this will not always be practicable (for further discussion of expert judgement, see Sections 'Subjective probability' and 'Estimating probability through expert judgement' in Chapter 4).

Risk registers

A risk register is an *active* (i.e. continuously up-datable) document that sets out *all the known risks (threats)* in an area or at a particular site and the decisions taken about how to monitor and manage them. The structure of the register should reflect the overall risk assessment and management process, concentrating on both the nature of the potential problems arising from different types/scales of slope failure and the recommended mitigation actions. It should be reviewed and updated as and when new risks (threats) are identified, or following the implementation of further studies or mitigation measures.

Risk registers, despite being a very basic or rudimentary form of risk assessment, can prove a useful tool in the screening and prioritisation of landslide problems at the early stages of a project. They can also provide a means of tracking landslide management decisions through a project and thereby help to ensure that low priority issues are not ignored.

Example 3.1

The coastal cliffs of Scarborough's South Bay, UK, are 2 km long and developed in a sequence of Jurassic rocks and glacial tills. The cliffs were purchased by Scarborough Borough Council in the 1890s. At that time, the Council undertook a programme of coast protection, landscaping and drainage as part of a range of slope treatment measures. The landscaped slopes have since become a major recreational resource of the town and are covered with an elaborate network of paths. Despite the stabilisation works, the cliffs (often sloping at 30–40° and over 50 m high) are regularly affected by slow, minor landslides that result in disrupted footpaths or small losses of cliff top land. Of greater concern, however, were the results of geomorphological mapping and historical research undertaken following the unexpected and dramatic 1993 Holbeck Hall landslide, which revealed that other large and damaging landslides have occurred in historic times (Lee *et al.*, 1998a; Lee, 1999; Lee and Clark, 2000).

As a consequence of the Holbeck Hall landslide, a preliminary assessment of the risks posed by coastal instability was undertaken, based on geomorphological assessment and a thorough review of historical sources (Rendel Geotechnics, 1994). The South Bay cliffline was sub-divided into eight separate cliff sections, following the recognition of *four previously failed major landslides*, separated by *four intact coastal slopes*, that is cliffs unaffected by major failures (Fig. 3.1). A range of failure scenarios were identified for each section and recorded in a risk register, two examples of which are shown in Table 3.1. Each failure type was evaluated in terms of likelihood, based on the historical frequency, and the expected scale of potential consequences to the current assets along and adjacent to the cliffline:

- *small-scale shallow failures* of the coastal slopes may lead to slight to moderate damage to footpaths and other structures, but would not lead to cliff top loss. This type of failure can be expected to occur somewhere within South Bay, on average, every year;

131

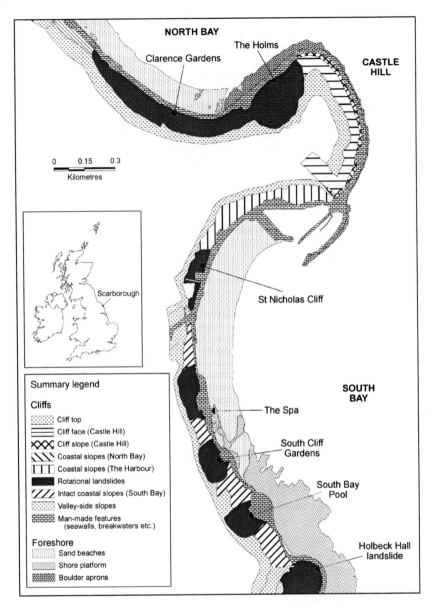

Fig. 3.1. *Example 3.1: summary geomorphological map of the Scarborough urban coast (after Lee, 1999)*

- *large failure* involving rapid cliff top recession and run-out of debris. Such an event could lead to total loss of the seawalls, coastal slope structures and cliff top property within the affected area. The historical frequency of failure of the intact steep slopes was estimated to be four events in 256 years (see Example 4.2);

Table 3.1. Example 3.1: Part of the landslide risk register developed for Scarborough's South Bay (adapted from Rendel Geotechnics, 1994) (note that the Action Plan has been periodically reviewed and amended in the succeeding years)

Cliff section	Setting	Elements at risk	Landslide hazard	Likelihood of failure	Consequences	Recommended Action Plan (1994)	Actions (1995)
Holbeck Gardens	Intact, protected coastal cliff	Cliff footpaths Putting green Cliff top property Seawall and promenade	Small-scale shallow translational failures on the cliff face	Failures could occur most years	Slight to moderate footpath damage Slide could expand and trigger a major landslide (unloading)	Council to monitor slopes and record failures; all failures to be treated ASAP	Consultant has been commissioned to implement action plan
			Large landslide involving rapid run-out and loss of up to 100 m of cliff top land	This type of failure could be expected in South Bay, once every 100 years	Loss of cliff top property and putting green, footpaths and seawall/ promenade Potential for loss of life and injury because of rapid run-out	Council to install inclinometers and piezometers; monitor slopes; establish contingency plan for emergency works; establish planning controls	Consultant has been commissioned to implement action plan
Spa Cliff	Previously failed (1737) major landslide	Footpaths The Spa Pavilions and Ocean Ballroom Cliff top property Seawall and promenade	Rockfalls and small slides off the rear cliff face	Failures could be expected to occur during wet periods	Moderate to serious damage to cliff top road and infrastructure	Council to install crack monitoring devices; monitor slopes and record failures; treat all localised failures ASAP; establish planning controls	Consultant has been commissioned to implement action plan
			Small-scale shallow translational failures on the coastal slope	Failures could occur most years	Slight to moderate footpath damage		
			Reactivation would be associated with slow, intermittent ground movement	Reactivations could be expected to occur during wet periods (say, 1 year in 10)	Slight to moderate and severe damage to the Pavilions, Ocean Ballroom and seawalls		

133

- *reactivation of the pre-existing landslides* involving episodes of significant ground movement during wet winters.

Although this approach represents only a basic form of risk assessment (i.e. risk identification), it does provide a mechanism for ensuring that all instability risks on the Scarborough urban coastline were identified at an early stage of the landslide management process. Once the risks had been identified they could be screened and prioritised. The development of the risk register greatly facilitated the formulation of an Action Plan that included:

1 A *strategic coastal defence study*. The preliminary risk assessment recommended the preparation of an integrated cliff and foreshore management plan; this was subsequently developed as part of a strategic coastal defence study of the Scarborough urban frontage (High Point Rendel, 2000) and involved:
   - a review of the condition, performance and residual life of all the existing defences;
   - detailed sub-surface investigation of identified priority slopes in South Bay, including the installation of monitoring instrumentation;
   - a quantitative risk assessment of the threat to the seawalls from landslides;
   - an assessment of the consequences of seawall failure;
   - an assessment of wave overtopping problems;
   - the identification of coastal defence options for the next 60 years;
   - an environmental assessment of the defence options;
   - the development of a prioritised and costed programme of works for monitoring, maintenance and improvements to allow progress within budget constraints.
2 *Land use planning*. A planning guidance map was produced for the local planning authority which related the landslip potential and risk for forward planning and development control purposes.
3 *Monitoring and early warning*. A systematic observation and monitoring strategy was developed to provide the basis for early warning of potential problems in priority areas. The equipment employed includes simple crack monitoring studs on buildings, inclinometers, piezometers and tiltmeters.

Relative risk scoring

In many instances it is inappropriate to evaluate risk in absolute terms because of the difficulties in assigning meaningful values for the hazard, the assets or elements at risk and the possible adverse consequences. In such circumstances, it can be useful to assess the relative levels of the threat, or *relative risk*, to different sites posed by particular hazards, based on both factual data and subjective appraisal (see Sections 'Hazard assessment' and 'Consequence assessment' of Chapter 1). The value of relative risk assessment is that it can enable sites to be compared quickly and thereby allow early decisions to be made about where limited financial resources should be directed (e.g. Clark *et al.*, 1993; Hearn 1995).

The *relative risk scoring* approach utilises the basic definition of risk outlined in Chapter 1:

Risk = Probability (Hazard) × Adverse Consequences

However, the hazard (i.e. landsliding), probability and adverse consequence elements in this equation are all represented by relative scores or rank values, with the risk being the product of these scores.

The probability of landsliding of a particular magnitude can be represented by a *hazard number*:

Hazard Number = Hazard Score × Probability Score

Similarly, the adverse consequences can be represented by a risk value (i.e. the relative value of the assets or elements at risk) and the vulnerability of the assets or elements at risk:

Adverse Consequences = Risk Value × Vulnerability

The risk, expressed as a *risk number*, is calculated as follows:

Risk Number = Hazard Score × Probability Score × Risk Value

× Vulnerability

The risk numbers produced can then be used to place each site within an arbitrarily defined scale of *risk classes* that allow some comparison between sites and, thereby, provide a basis for management decisions.

Example 3.2
A series of recent rockfalls and rockslides off the 50 m-high South Shore Cliffs, Whitehaven, UK, raised concerns over the safety of foreshore users. Boggett *et al.* (2000) describe the use of a qualitative risk scoring scheme to evaluate the problems and identify where emergency remedial works were required.

A detailed geomorphological and geotechnical mapping exercise was undertaken to collect information on the cliff conditions and failure mechanisms, which facilitated the division of the cliffline into five distinct sections sub-divided into morphological zones. Then scoring scales were developed for the four main components of the risk assessment; hazard magnitude or

135

Table 3.2. Example 3.2: South Shore Cliff, Whitehaven: Qualitative risk assessment – hazard and risk scoring scheme (from Boggett et al., 2000). Each column is a separate ranking scheme

Number	Hazard H	Probability P (chance of occurrence in ten years)	Risk value R	Vulnerability V
1	Small failure/ erosion	Unlikely	Hardstanding areas not in use	Little or no effect
2	Moderate failure and occasional small falling blocks	Possible	Unoccupied building/public right of way (beach)	Nuisance or minor damage
3	Substantial failure and occasional large falling blocks	Likely	Roads/footpath	Major damage
4	Deep failure (>30 m) and large rockfall		Major structure/ mine buildings	Loss of life
5	Major failure		Residential area	

hazard (H), *probability* of occurrence within the specified ten year time frame (P), the significance of the *elements at risk* or *risk value* (R) and the likely scale of adverse impact or *vulnerability* (V), as shown in Table 3.2.

Relevant scores for each component were then allocated to each section of the cliff and multiplied together so as to yield a *risk number* (R_N), as shown in

Table 3.3. Example 3.2: South Shore Cliff, Whitehaven: Examples of cliff section risk assessment (see Fig. 3.2 for locations; from Boggett et al., 2000)

Cliff sub-section	Hazard score H (1–5)	Probability score P (1–3)	Risk value score R (1–5)	Vulnerability score V (1–4)	Risk R_N ($H \times P \times R \times V$) (4–300)	Risk class
1	4	2	3	3	72	IV
2	3	3	2	3	54	III
3	3	2	2	2	24	II
4	4	3	3	3	108	V
5	2	2	2	2	16	II

Risk classes:

R_N >100	V	Highest	
R_N 60–100	IV		
R_N 30–60	III	Moderate	
R_N 10–30	II		
R_N 0–10	I	Lowest	

Note: Each cliff section contains a number of different morphological zones
1. Shore platform
2. Lower cliff
3. Mid cliff
4. Upper cliff
5. Cliff-top

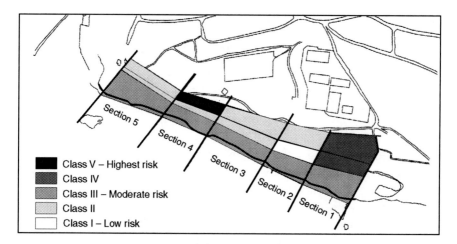

Fig. 3.2. Example 3.2: Whitehaven cliffs: risk zonation plan (after Boggett et al., 2000)

Table 3.3, with a theoretical maximum of 300:

Risk Number = Hazard × Probability × Risk Value × Vulnerability

The risk numbers for each cliff section are presented in Table 3.3. These scores were used to classify each section into one of five arbitrarily defined risk *classes*. A risk zonation plan was then produced (Fig. 3.2) that defines the key problem areas. The high-risk zone (section 4) coincides with the site of a large tension crack that is proceeding as a slow slide. As the slide develops, the likelihood of rockfall or topple increases, resulting in a high risk.

This is a classic example of the risk scoring technique. Nevertheless, it may be criticised for only having three divisions of probability and four of vulnerability. There is also no indication of 'degree of certainty' which often makes for a useful addition.

Example 3.3
Concerns have been raised about the integrity of an existing oil pipeline, in the near East region, that crosses a number of pre-existing landslides. A preliminary risk assessment was required to determine the relative levels of risk at a number of selected landslide sites along the route and to suggest a way forward for managing the potential problems associated with future movements. This was undertaken by an international team, based on a short field visit on which they were accompanied by local specialists.

The landslide hazard at each site was expressed in terms of a *hazard score* for each of the three mechanisms that could affect the integrity of the pipe-line; namely lateral or vertical displacement, loading, or the creation of an

Table 3.4. *Example 3.3: Typical hazard scores*

	Displacement	Loading	Span	Sub-total	Multiplier	Total
Site 1	3	2	1	6	1	6
Site 2	5	1	4	10	5	50

unsupported span. All of the judgements were made with reference to a scoring scale of 1 to 5 (very low to high) and assumed a 30 year design life for the pipeline. Scores were agreed by the five-man field team.

The three *hazard scores* were then summed (maximum score = 15; Table 3.4) and then adjusted by a *multiplier*. The multiplier was on a 1–5 scale, reflecting the length of pipe that crosses, or was at risk from, a particular landslide feature (small slides, less than 10 m wide, crossed by only a limited length of pipeline were given a multiplier of 1; large slides, greater than 100 m wide, with considerable lengths of pipe at risk were given a multiplier of 5). Thus, the maximum adjusted hazard score = 75.

The likelihood of a landslide event occurring sufficient to threaten the integrity of the pipeline over the assumed 30 year design life was judged on a scale of 1–5 (unlikely to extremely likely). The *consequences* in terms of pipe damage were assessed, with scores ranging from 1 (minor buckling) to 5 (full-bore rupture).

The estimated risk was expressed as a *risk score* and calculated as follows (Table 3.5):

Risk Score = Hazard Score × Multiplier × Likelihood × Consequence

so that the theoretical maximum value is 1875.

The risk score for each site was then translated into one of six arbitrarily defined *risk classes* (Table 3.6). A suggested programme of further work was developed for landslides in each different risk class.

This represents an interesting development of the risk scoring technique which places great emphasis on hazard. It could be modified to include repeated failure and recalibrated to place greater emphasis on consequences.

Table 3.5. *Example 3.3: Typical likelihood scores and total risk scores*

	Hazard	Multiplier	Likelihood	Consequence	Total risk
Site 1	6	1	2	1	12
Site 2	10	5	5	5	1250

Table 3.6. Example 3.3: Risk classes and suggested action plan

Risk class	Risk scores	Action plan
A	<10	Do nothing
B	10–25	Consideration given to further studies to investigate the landslide hazard
C	25–50	Landslide stabilisation works may possibly be required
D	50–100	Landslide stabilisation works may be required, but further studies required to refine judgements
E	100–500	Landslide stabilisation works likely to be required. Further investigations will be required, including a comprehensive assessment of risks
F	>500	Large-scale mitigation works will be required. Urgent requirements for further investigations, including a comprehensive assessment of risks

Example 3.4

Visitors to the Giant's Causeway World Heritage site, Northern Ireland, are at risk from slope failures from 100 m-plus-high basalt cliffs. In 1994 a substantial portion of the lower cliff footpath was closed following a large failure that destroyed key sections of the path. As some 400 000 people visit the site each year, there was a need for an assessment of the risks to which they are exposed.

McDonnell (2002) describes a scoring approach that was developed specifically for the site. The cliffline was sub-divided into a series of separate sections, based on the geomorphological and geotechnical conditions. Slope stability was modelled at a number of generalised cliff sections, using the Universal Discrete Element Code (UDEC). This information was extrapolated and applied to other cliff sections. Each section of the cliffline was allocated a *stability number* from 1 (very low hazard potential) to 5 (very high hazard potential).

The risk within each section was calculated as a *relative risk score* as follows:

Relative Risk Score = Hazard Score + Visitor Concentration Score

+ Visitor Perception Score

Note that in this instance risk has been defined as the sum of these scores, and not the product as was the case in the previous examples.

The hazard score was achieved by the summation of four separate components: the stability number, slope angle, presence or absence of percolines

139

Table 3.7. Example 3.4: Hazard scores for the Giant's Causeway cliffs (adapted from McDonnell, 2002)

Hazard classi- fication	Geotechnical stability		Slope angle		Pore pressure		Slope loading	
	Stability number	Hazard number	Slope class	Hazard number	Percolines	Hazard number	Dumped material	Hazard number
Very low	1	2	0–5°	1	None	0	None	0
Low	2	4	5–15°	2				
Moderate	3	6	16–20°	3				
High	4	8	21–29°	4				
Very high	5	10	>30°	5	Present	5	Present	5

(i.e. springs and water seeps), and the presence or absence of material dumped by the site staff after previous failures – this dumping has the potential to exacerbate stability problems at the site (Table 3.7). Potential hazard scores range from 3 to 25.

Consideration of the elements at risk was confined to the visitors, and expressed as a *visitor concentration score* and their vulnerability was represented by a *visitor perception score* (Table 3.8). The latter score was intended to reflect the average visitor's ability to recognise and avoid a hazard. The risk scores (the sums of the hazard, visitor concentration and perception scores) were used to assign each section into a risk class (Table 3.9), allowing a *relative risk map* to be produced for each path along the cliffline (Fig. 3.3).

Table 3.8. Example 3.4: Visitor concentration and perception scores (adapted from McDonnell, 2002)

Classification	Visitor concentration		Visitor perception		
	Class	Score	Non-recognition: %	Landslide type	Score
Very low	Very low	1	10	Complex, translational slides	1
Low	Low	2	10–20	Undercutting, slump collapse	2
Moderate	Moderate	3	20–40	Toppling	3
High	High	4	40–80	Mass block release, spalling	4
Very high	Very high	5	80–100	Isolated block release, relict landslide	5

Table 3.9. Example 3.4: Risk scores and classes (adapted from McDonnell, 2002)

Risk score	Risk class	Risk description
3–4	1	Very low
5–6	2	Low
7–8	3	Moderate
9–11	4	High
12–13	5	Very high

This is an interesting and innovative example of a risk scoring technique, because it actually attempts to embrace the true meaning of risk despite giving hazard a 70% weighting. The use of the inverse of hazard perception as a measure of 'awareness' or 'surprise' is particularly interesting. Whether or not the 'Very High Risk' category should begin with a risk score of 12 (out of a potential 35) is a matter of debate.

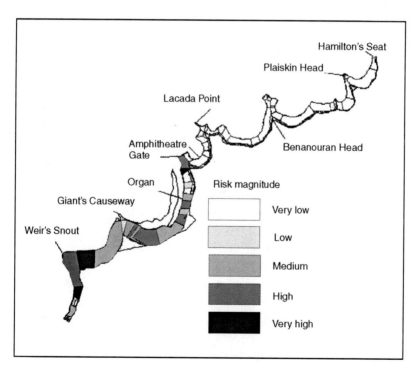

Fig. 3.3. Example 3.4: the Antrim coast, UK: magnitude of risk to visitors on the upper coast path (after McDonnell, 2002)

Risk ranking matrices

An alternative approach to risk scoring involves the development of a risk matrix, in which a measure of the likelihood of a hazard occurring is matched against the increasing severity of consequences to provide a ranking of risk levels (see Fig. 1.6, Section 'Risk estimation' in Chapter 1). Although rankings are value judgements, experienced landslide specialists should be able to make realistic assessments of the likelihood of events and consequences, based on an appreciation of the landslide environment, together with knowledge of the particular site.

Tables 3.10 and 3.11 present typical scales of hazard likelihood and consequences that could be adapted to particular circumstances. *Relative risk levels* can then be assigned to different combinations of hazard and consequences (e.g. Table 3.12). Each relative risk level should mark a step-up in the degree of threat and a change in the acceptability or tolerability of the risk (e.g. Table 3.13). Although the designation of risk levels can appear somewhat arbitrary, they do provide a framework for making comparisons between different sites within an area.

Example 3.5

Uncontrolled logging of forest land is known to result in an increase in landslide activity (e.g. Sidle *et al.*, 1985). Van Dine *et al.* (2002) describe the use of risk matrices to assist the British Columbia Ministry of Forests plan timber

Table 3.10. *Indicative measures of landslide likelihood (from Australian Geomechanics Society, 2000)*

Level	Descriptor	Description	Indicative annual probability
A	Almost certain	The event is expected to occur	$>\approx 10^{-1}$
B	Likely	The event will probably occur under adverse conditions	$\approx 10^{-2}$
C	Possible	The event could occur under adverse conditions	$\approx 10^{-3}$
D	Unlikely	The event might occur under very adverse circumstances	$\approx 10^{-4}$
E	Rare	The event is conceivable but only under exceptional circumstances	$\approx 10^{-5}$
F	Not credible	The event is inconceivable or fanciful	$<10^{-6}$

Note. '$\approx$' means that the indicative value may vary by say $\pm$ half of an order of magnitude, or more.

Table 3.11. Indicative measures of consequence (from Australian Geomechanics Society, 2000)

Level	Descriptor	Description
1	Catastrophic	Structure completely destroyed or large-scale damage requiring major engineering works for stabilisation
2	Major	Extensive damage to most of structure, or extending beyond site boundaries, requiring significant stabilisation works
3	Medium	Moderate damage to some of structure, or significant part of site, requiring large stabilisation works
4	Minor	Limited damage to part of structure, or part of site, requiring some reinstatement/stabilisation works
5	Insignificant	Little damage

Table 3.12. Qualitative risk assessment matrix: levels of risk to property (from Australian Geomechanics Society, 2000)

Likelihood	Consequences to property				
	1 Catastrophic	2 Major	3 Medium	4 Minor	5 Insignificant
A (almost certain)	VH	VH	H	H	M
B (likely)	VH	H	H	M	L–M
C (possible)	H	H	M	L–M	VL–L
D (unlikely)	M–H	M	L	VL–L	VL
E (rare)	L–M	L–M	VL–L	VL	VL
F (not credible)	VL	VL	VL	VL	VL

Table 3.13. Indicative risk level implications (from Australian Geomechanics Society, 2000)

Risk level		Example implications
VH	Very high risk	Extensive detailed investigation and research, planning and implementation of treatment options essential to reduce risk to acceptable levels; may be too expensive and not practical
H	High risk	Detailed investigation, planning and implementation of treatment options required to reduce risk to acceptable levels
M	Moderate risk	Tolerable provided that treatment plan is implemented to maintain or reduce risks. May be accepted. May require investigation and planning of treatment options
L	Low risk	Usually accepted. Treatment requirements and responsibility to be defined to maintain or reduce risk
VL	Very low risk	Acceptable. Manage by normal slope maintenance procedures

Table 3.14. Example 3.5: Ratings of potential consequences (from Van Dine et al., 2002)

	People	Property	Water supply
High	Death	Destruction of multiple residences	Destruction of multiple water intakes or very high increase in turbidity
Moderate	Serious injury	Destruction of single residence, or damage to multiple residences	Destruction of a single water intake or high increase in turbidity
Low	Minor injury	Damage to single residence	Damage to a single water intake or moderate to low increase in turbidity

harvesting operations at Perry Ridge, where local residents were concerned about the resulting threats to people, property and water supply. The risk assessment undertaken was described as a 'consensual qualitative risk assessment', based on available information and empirical evidence, together with the experience and judgement of the three geohazard specialists involved in the project.

The 76 km^2 area was subdivided into 32 catchments or hydrological units. A range of potential hazards were identified from previous studies, including flash floods, debris flows and landslides. The existing probability of each type of event was rated 'high', 'moderate', 'low', 'very low' or 'none', based on the past occurrence of the event, independent of its magnitude.

Consequences were rated 'high', 'moderate' and 'low', based on the elements at risk (people, property and water supply) and the inferred severity of the events which could affect those assets. Examples of potential consequences are presented in Table 3.14. The ratings for the three groupings of the elements at risk were not intended to be directly comparable.

Three risk matrices were developed, one for each of the elements at risk (people, property and water supply; Table 3.15).

Using a risk matrix for each of the 32 hydrological units, the position of the established past hazard–consequence relationship for each of the three recognised hazards was plotted, that is between 3 and 9 of the 33 available cells contained information. The geohazard experts then assessed the likely relationships between hazard and adverse consequences that would exist following the commencement of logging activity, and the position of these *with logging* relationships were also plotted on the matrix. In those instances where the *with logging* case revealed a significant increase in estimated risk involving the identification of very high (VH), high (H) or moderate (M)

Table 3.15. Example 3.5: Risk matrices developed for different elements at risk (from Van Dine et al., 2002)

Hazard	Consequence to people			Consequence to property			Consequence to water supply		
	High	Moderate	Low	High	Moderate	Low	High	Moderate	Low
High	VH	H	M	VH	H	M	VH	H	M
Moderate	H	M	L	H	M	L	H	M	L
Low	M	L	VL	M	L	VL	M	L	VL
Very low	L	VL			L	VL		L	VL

risks, alternatives to logging or road building were suggested and landslide management recommendations were made.

This represents a good example of a fairly standard application of the risk matrix technique in the context of landsliding.

Example 3.6

The Sarno landslide disaster in Campania Region, Italy, in August 1998, killed 153 people (Del Prete *et al.*, 1998; Guadagno, 1999). As a result the Italian Government passed new legislation on landslide and flood risk assessment and mitigation – the Landslide Risk-Assessment and Reduction Act (Gazzetta Ufficiale della Repubblica Italiana, 1998). This requires Regional Governments and National River Basin Authorities to identify and map areas where landslide risk is most severe, and to take action to reduce economic damage and societal risk. The law was accompanied by a 'technical document' providing a general framework and guidelines for the assessment of landslide hazard and risk (Gazzetta Ufficiale della Repubblica Italiana, 1999).

Cardinali *et al.* (2002) describe the risk assessment of parts of Umbria, central Italy, where landslides cover around 14% of the entire land area (8456 km^2). The approach used was a hybrid combination of both risk scoring and risk matrices. The study area was divided into a series of 'landslide hazard' zones (LHZ), defined as the area of possible (or probable) short-term evolution of an existing landslide, or a group of landslides, of similar characteristics (i.e. of type, volume, depth and velocity). The LHZs were identified and delimited from aerial photographs or field observation and determined for each type of failure observed in an area (e.g. fast-moving rock falls, rapid-moving debris flows, slow-moving earth-flow slumps or compound failures); note that each different landslide type within an LHZ represents a separate landslide scenario.

Table 3.16. Example 3.6: Types of elements at risk (from Cardinali et al., 2002)

Code	Description
HD	Built-up areas with high population density
LD	Built-up areas with low population density and scattered houses
IN	Industries
FA	Animal farms
SP	Sports facilities
Q	Quarries
MR	Main roads, motorways, highways
SR	Secondary roads
FR	Farm and minor roads
RW	Railway lines
C	Cemeteries

The risk within each LHZ was determined as follows:

Risk $= f$ (Hazard $\times$ Vulnerability)

The measure of risk used was termed the *specific risk*, that is the risk to a specific type of asset (e.g. building, road etc.) when a landslide occurs (Einstein, 1988). The specific risk was defined separately for each different type of asset (Table 3.16), within each LHZ.

Landslide hazard was defined for each LHZ in terms of:

1 *Landslide frequency*. Four classes were used:
   - Low frequency (1), when only one landslide event was observed in the 60 year period 1941–2001 that was covered by the available aerial photography of the area;
   - Medium frequency (2), when two events were observed;
   - High frequency (3), when three events were observed; and
   - Very high frequency (4), when more than three events were observed.
2 *Landslide intensity*. Defined in four classes (slight, medium, high and very high), based on the estimated volume (v) and the expected velocity (s; see Chapter 5):

   Intensity $= f(v, s)$

Levels of hazard were defined using a 2-digit code (Table 3.17) that identified separately the estimated frequency and intensity. Cardinali *et al.* (2002) suggest that such a coding system allows users to determine whether the hazard is due to a high frequency of landslides (i.e. high recurrence), great intensity (i.e. large volume and high velocity), or both.

Estimates of the vulnerability of each type of asset at risk were based on the inferred relationship between the intensity and type of the expected

146

Table 3.17. Example 3.6: Landslide hazard classes used for each landslide hazard zone (from Cardinali et al., 2002)

Expected landslide frequency	Landslide intensity			
	Slight (1)	Medium (2)	High (3)	Very High (4)
Low (1)	1 1	1 2	1 3	1 4
Medium (2)	2 1	2 2	2 3	2 4
High (3)	3 1	3 2	3 3	3 4
Very high (4)	4 1	4 2	4 3	4 4

landslide, and the likely damage the landslide would cause. Three levels of damage were envisaged – superficial or aesthetic damage (A), functional damage (F) and structural damage or destruction (S).

A risk matrix was prepared to define the specific landslide risk to each type of asset. As in the previous examples, the matrix relates landslide hazard classes to damage classes. However, instead of describing the risk in qualitative terms such as 'low', 'medium' and 'high', a unique coding value (the *specific risk index*) was used in each of the matrix cells (Table 3.18). The specific risk index shows the landslide frequency, the landslide intensity,

Table 3.18. Example 3.6: Risk matrix presenting levels of specific landslide risk (from Cardinali et al., 2002)

Hazard (see Table 3.17)	Vulnerability (expected damage)		
	Minor damage	Major damage	Total damage
1 1	A 11	F 11	S 11
1 2	A 12	F 12	S 12
1 3	A 13	F 13	S 13
2 1	A 21	F 21	S 21
1 4	A 14	F 14	S 14
2 2	A 22	F 22	S 22
2 3	A 23	F 23	S 23
3 1	A 31	F 31	S 31
3 2	A 32	F 32	S 32
2 4	A 24	F 24	S 24
3 3	A 33	F 33	S 33
4 1	A 41	F 41	S 41
4 2	A 42	F 42	S 42
3 4	A 34	F 34	S 34
4 3	A 43	F 43	S 43
4 4	A 44	F 44	S 44

and the expected damage caused by the specific type of landslide. If more than a single class of elements at risk is present in an LHZ, a different value of specific risk was computed for each class.

The Landslide Risk-Assessment and Reduction Act requires the ranking of the most dangerous landslide areas according to the expected (total) land-slide risk. In order to provide a measure of the total risk, Cardinali *et al.* (2002) combined the detailed information given by the specific landslide risk index for each LHZ into one of the four classes of landslide risk required by the law:

- *Very high landslide risk* (severe risk, R4) was assigned where rapid and fast-moving landslides could cause severe damage to the structures and infrastructures, as well as pose a direct threat to the population. These were the areas where debris flows and rock falls could cause casualties or homelessness.
- *High landslide risk* (R3) was assigned to the areas where slow-moving landslides could cause structural and functional damage to structures and infrastructure. No casualties are expected.
- *Moderate landslide risk* (R2) was attributed where aesthetic damage to vulnerable elements is to be expected, caused by slow-moving slope failures or by fast or rapid moving landslides of slight intensity.
- *Low landslide risk* (R1) was assigned to areas within an LHZ where no element of risk is currently present.

Relative risk rating

This approach adopts similar principles to those used in risk scoring and risk matrices, although the hazard and its consequences are only considered implicitly rather than explicitly through specific scoring or ranking systems. It is a descriptive approach in which a range of risk categories are defined, each with a characteristic degree of hazard and level of consequence. The approach has proved useful in situations where the elements at risk are uniform, or broadly similar, throughout an area, such as a pipeline, highway or cliff foot walkway (e.g. Palmer *et al.*, 2002), but are exposed to spatial variation in the degree of hazard. The method provides a means of identifying the *relative risk* through the area.

The study area is sub-divided into units, generally on the basis of geomorphology and geology. Information is then gathered about the distribution, nature and frequency of landsliding, the assets at risk and the likely levels of adverse consequences within each unit. Risk categories are then developed that summarise the range of hazard and consequence conditions within the area, that is the categories are developed specifically for the particular area of interest. Each unit is then assigned a risk category.

Class: Geology	Class: Classification class	Geometry: α, h (m)	Instability features: Rockfall	Spalling	Debris flows	Slumps	Wash out	Creep	Position: Flat topped ridge	Valley side-slope	Hazard: Hazard class
Chalk <20% profile superficials	A1	>53, >10	√	√	×	×	√	×	√	×	High
	A2	45–60°, 8–18	√	√	×	√	√	√	√	×	Moderate
									×	√	Low
	A3	30–45°, 5–12	×	√	×	√	√	√	√	×	Low
	A4	30–40°, <6	×	×	×	×	×	√	√	×	Very low
									×	√	Negligible
Composite 60–20%	B1	40–47°, 5–15	×	×	√	√	√	√	√	×	Moderate
	B2	30–45°, 4–12	×	×	×	√	√	√	√	×	Low
									×	√	Very low
	B3	30–40°, <5	×	×	×	×	×	√	×	√	Very low
Superficials >60%	C1	30–40°, 2–6	×	×	×	√	×	√	√	√	Low
	C2	25–35°, 1–4	×	×	×	×	×	√	√	√	Very low
Retained	D		×	×	×	×	×	√	×	√	Very low

Fig. 3.4. Example 3.7: classification scheme used for the Chalk cuttings on the Metropolitan Line (from Phipps and McGinnity, 2001)

Example 3.7

Although it is known as the 'underground', the London Underground network also includes some 270 km of surface track, including 20 km of Chalk cuttings on the Metropolitan Line. In May 1992 a large debris flow from one such cutting blocked the track between Chalfont Latimer and Chorleywood, disrupted services for 33 hours and led to £400 000 worth of mitigation measures. As a consequence, London Underground Limited commissioned an earth structures condition survey in 1993 for this stretch of line in order to provide a systematic, centralised record of the condition of cuttings and embankments.

Phipps and McGinnity (2001) describe the assessment of the condition of the Chalk cuttings on the Metropolitan Line, including an assessment of the risk to London Underground infrastructure and customers, adjacent property and the general public. A programme of geomorphological and geological mapping led to the development of a Chalk cutting classification scheme, based on slope geometry and materials (Fig. 3.4). A range of instability types and associated hazard classes (Table 3.19) were defined for each cutting type. Each individual cutting was then assigned into a cutting class and hazard class.

The variety of different cutting types presented a number of distinct hazards to the various assets along the route. For the majority of cuttings,

Table 3.19. Example 3.7: Landslide hazard classes on Chalk cuttings (from Phipps and McGinnity, 2001)

Hazard class	Description
High hazard	Rockfall events may occur annually which can reach the cut slope base, triggered mainly by freeze/thaw activity. Washout and debris flows from deep superficial deposits at slope crest may occur after heavy but foreseeable local rainfall.
Moderate hazard	Rockfalls infrequent, replaced by frequent minor slumping. Slumping and creep of accumulated debris at the slope base. Debris flows only occur after most exceptional rainfall. Conditions for washouts may be present.
Low hazard	Potential for spalling reduced in Chalk cut-slopes by reduced inclination of cutting. Some slumping has occurred in the past of superficial dominant cuttings and present movements are ongoing locally. Creep widespread. Local conditions suitable for the possibility of washouts.
Very low hazard	Cut slopes are inherently stable although ongoing creep is in evidence locally.
Negligible hazard	Cut slopes are inherently stable, with rare creep.

Table 3.20. Example 3.7: Risk categories for Chalk cuttings on the Metropolitan Line (from Phipps and McGinnity, 2001)

Risk category	Description
1	Significant failures could recur at any time, the magnitude of which would provide a direct risk to the line. Remedial measures should be implemented immediately in order to reduce the risk. Ground investigations would be required for detailed design of remedial works.
2	Annual rockfall events are possible from Chalk cuttings that could locally cover parts of the track. In the long-term, remedial measures would negate the possible impacts of rockfalls and reduce maintenance commitment. Continued degradation of the cuttings is ongoing and slow movement of debris would impinge on the tracks if maintenance is not ongoing to address the debris build up.
3	Only extreme, rare rockfall events from Chalk cuttings may impinge on the track. Flows and slumps have occurred previously from cuttings with significant superficial cover, but their inception is dependent on adverse triggering factors that are infrequent but not easy to predict from existing historical records. Remedial measures would reduce the occurrence of possible major failure events in superficial soils, although the decision to implement such measures would be dependent on a ground investigation. Continued degradation of the cuttings is ongoing and very slow movement of debris may impinge on the tracks if maintenance is not ongoing to address the debris build up.
4	Rockfall from chalk cuttings is not expected to constitute a threat to service disruption. Slumps have been identified that under extreme conditions may impinge on the track, although this is expected to be highly unlikely. Slow creep of debris may over a number of years approach the tracks.
5	Any movements from cuttings that could approach the tracks are thought to be extremely unlikely and not considered to constitute a risk to the railway.

the assets at risk were lineside services, rolling stock and track. However, a lack of historical records of past incidents led to the view that it would be difficult to develop a quantitative statement of the risks. Instead, it was decided that risk categories could be developed, reflecting the behaviour of the cuttings and the characteristic instability types and hazards, together with the proximity of the track to the cutting slopes. The risk categories are presented in Table 3.20, each of which points towards a suggested management response to the risks.

The approach classifies the cuttings according to evidence of actual and potential instability. By considering the impact of cutting failures, the

151

hazard assessment has been broadened into providing an indication of the risk to the infrastructure and public. In doing so, it provides a tool for evaluating the risk presented by a particular cutting relative to all the other cuttings along the route and, hence, can be used to support the prioritisation of maintenance and remedial works.

The FMECA approach

Failure modes, effects and criticality analysis (FMECA) is a systematic approach to analysing how parts of a system, such as an engineered slope, might fail (the failure mode). For example, possible failure modes on a protected coastal cliff might include collapse of the seawall, leading to a renewal of marine erosion at the cliff foot, or blocked drains leading to locally high pore water pressures. Each failure mode can be assessed to determine the effects of that failure and to identify how critical it might be to the stability of the overall system. For example, seawall failure will result in an almost immediate renewal of cliff foot erosion and the re-establishment of active instability on the coastal slope. By contrast, failure of the drainage system is likely to have a less immediate or direct impact on the stability of the slopes. Drainage failure could lead to small-scale landslides that, under certain circumstances, may lead progressively to the decline in overall stability, resulting in an increase in the likelihood of a larger event.

The FMECA approach provides a structured framework for the qualitative analysis of various components of a system, using engineering judgement to generate scores or rankings, rather than probabilities. The FMECA approach has been used as a risk assessment tool in the dam industry (e.g. Sandilands *et al.*, 1998; Hughes *et al.*, 2000). Lee (2003) has suggested that the approach may also be useful in the strategic monitoring of coastal slopes that have been stabilised by a combination of structural elements, including, for example, a concrete seawall at the cliff foot and slope drainage measures.

The approach involves the development and analysis of an *LCI diagram* (location, cause, indicator) for each slope (Fig. 3.5). An LCI diagram sets out the individual constructed components of each man-made slope (e.g. seawall, drainage network, retaining structures etc.) and how their lack of integrity might contribute to the overall failure of the slope. Failure through a range of possible causes (e.g. undercutting or high pore water pressures) and with different indicators (e.g. blocked drains, seepage) is considered by means of *indicator-cause pathways*. The level of detail presented in an LCI diagram should reflect the available knowledge about any potential indicator-cause pathway.

The assessment procedure involves scoring three key factors on a range of 1 to 5, for each *indicator-cause pathway*:

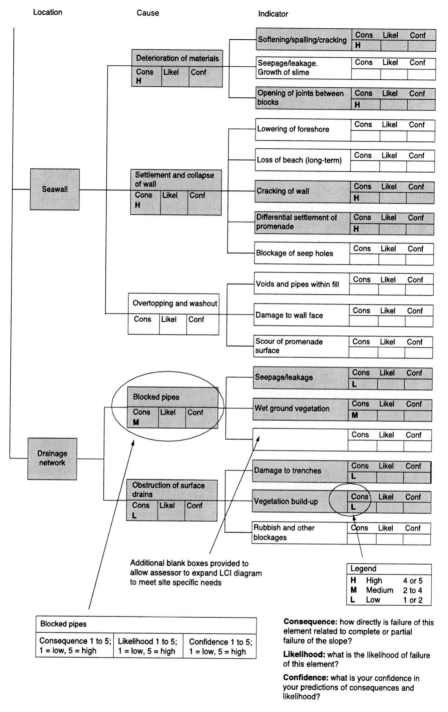

Fig. 3.5. An LCI diagram developed for a stabilised coastal slope (after Lee, 2003)

1 The *consequence* expressed in terms of how directly is failure of an element related to complete failure of the slope:

> 1 = failure of element is unlikely to lead to failure of the slope;

> 5 = failure of element is highly likely to lead to failure of the slope.

For example, failure of a seawall will almost inevitably lead to failure of the engineered slope, because marine erosion will begin to undercut the cliff foot. However, small-scale failures of engineered slopes resulting from the collapse of a gabion-basket retaining structure behind a footpath are a common feature on many coastal slopes, but rarely lead to failure of the whole slope.

2 The *likelihood* of failure of an element, ranging from 1 (low) to 5 (high).

3 The practitioner's *confidence* in the reliability of his/her predictions of the consequence and likelihood factors. The confidence score ranges from 1 (very confident) to 5 (no or little confidence). This score allows a measure of uncertainty to be included within the assessment. Table 3.21 presents

Table 3.21. The FMECA approach: key considerations for defining a confidence score in an LCI diagram (after Hughes et al., 2000)

Issue	Comment
Detectability	The ease with which potential failure mechanisms can be detected prior to failure occurring, through the use of instrumentation, that is a function of the cost/resources required to monitor signs of pre-failure movement within different components.
Construction quality	The quality of construction materials and the workmanship will vary between engineered slopes and between individual components of a slope. This can sometimes be readily identified and incorporated into the *likelihood* score. Sometimes, evidence of poor quality or bad workmanship may not be readily apparent. The *confidence* score should take account of any uncertainty regarding construction quality.
Operational maintenance	Maintenance is essential for ensuring the continued integrity of the structures. The *confidence* score should take account of any uncertainty regarding the future maintenance programme actually being undertaken. For example, poorly funded or ad hoc programmes may be subject to significant change and are likely to be unreliable.
Quality of records	A full record of the 'as-built' construction and operational maintenance is essential for a reliable assessment of structural performance. Good records do not reduce the likelihood of failure, but they increase the confidence in the allocated likelihood score.
Incompleteness of knowledge	The *confidence* score should take account of any significant gaps in knowledge about the condition, behaviour and performance of the structures.

a range of factors that should be considered when determining the confidence score. In some situations many of these issues may prove insignificant and the term effectively will represent *detectability*, that is the degree to which the potential failure mode can be detected prior to its occurrence. Increased slope monitoring can, therefore, result in a decrease in the confidence score by reducing the uncertainty about the ground conditions.

Considerable experience is required to develop and use an LCI diagram. It is important that the scores are the product of careful scrutiny, ideally by a group or panel of experts. Hughes *et al.* (2000) recommend that the process should be 'transparent' and the reasoning behind the allocation of each value should be clearly documented.

The results of the LCI diagram analysis are used to identify those structural elements that contribute most to the overall risk. A number of measures can be defined, including:

Element Score = Consequence of Failure × Likelihood of Failure

This provides a measure of the degree of risk associated with a particular element of the slope, such as a seawall. High scores indicate those elements where remedial measures may be needed to reduce the risk:

Criticality Score = Element Score × Confidence

This gives a measure of the hazard that a particular indicator-cause pathway creates for the slope. High criticality scores can reflect uncertainty in consequence and likelihood scores, highlighting the need for further investigation.

A measure of the relative risk associated with failure of particular elements of the slope can be established from the product of the criticality score and an impact score:

Relative Risk = Criticality Score × Impact Score

An *impact score* can be determined through the use of the types of scoring or ranking systems described earlier in the chapter. Hughes *et al.* (2000) present an expanded scoring framework for assessing the impact of dam failure that has been modified here to suit high-velocity landslide events (Table 3.22; note that further adjustments would be needed to make it appropriate for slow moving landslides). The scores for each type of *economic impact* are combined to provide a single measure of impact for the site or area. This is achieved by adjusting each impact score by a weighting factor, and adding the adjusted scores; these factors are based on expert judgement and should be reviewed and modified accordingly, depending on the local circumstances.

155

Table 3.22. FMECA approach: impact scoring system (from Hughes et al., 2000)

	Score	Population at risk
Residential properties affected		
0	0	0
0–15	1	30
15–50	2	100
50–250	3	500
Estimate (>250)	4	2 × estimate
Non-residential: number of people affected		
0	0	0
0–150	1	150
150–500	2	500
500–1000	3	1000
Estimate (>1000)	4	Estimate
Infrastructure affected		
None	0	0
Minor roads	1	25
Major regional infrastructure	2	50
Major national infrastructure	3	100
Major international infrastructure	4	Estimate
Recreational sites: number of people affected		
0	0	0
0–10	1	10
10–50	2	50
50–100	3	100
Estimate (>100)	4	Estimate
Industrial sites		
None	0	N/A
Light industrial	1	N/A
Public health industries	2	N/A
Heavy industrial	3	N/A
Nuclear, petrochemical	4	N/A
Utilities		
None	0	N/A
Local loss of distribution	1	N/A
Local loss of distribution/supply	2	N/A
Regional loss of distribution/supply	3	N/A
Significant impact on national services	4	N/A
Agriculture/habitat site		
Uncultivated/grassland	0	N/A
Pasture	1	N/A
Widespread farming	2	N/A
Intensive farming/vulnerable habitat/monument	3	N/A
Loss of international habitat/monument	4	N/A

Note. N/A = not applicable.

Table 3.23. FMECA approach: standard tables for calculating impact scores (from Hughes et al., 2000)

Impact	Population at risk (PAR)	Exposure factor*	Total (PAR × Exposure)
Residential property		0.5	
Non-residential property		0.5	
Infrastructure		0.5	
Recreation		0.5	
		Total loss of life	

Impact	Score	Weight	Total (Score × Weight)
Residential property		0.15	
Non-residential property		0.15	
Infrastructure		0.1	
Recreation		0.05	
Industrial		0.25	
Utilities		0.25	
Agriculture/habitats		0.05	
		Total score	

Impact	Score	Factor	Total (Score × Weight)
Economic impact		100	
Potential loss of life		1	
		Total impact score	

Note. *Exposure varies with forewarning.

Loss of life is estimated from the total number of people at risk:

Loss of Life = Population at Risk × Exposure

For high-velocity landslides, the *exposure factor* will vary with the length of forewarning time and the ability of people to escape or be evacuated. Hughes et al. (2000) suggest that the vulnerability factor may range from 0.5 if there is little or no forewarning to only 0.0002 for a warning time of 90 minutes.

The economic impact scores are combined with the estimated loss of life to give an overall impact score (Table 3.23).

The relative risk also allows comparison between the same elements on different slopes.

Example 3.8
As described in Example 3.1, the coastal cliffs of Scarborough's South Bay consist of a sequence of previously failed major landslides, separated by

intact coastal slopes (Fig. 3.1). It has been recognised that the intact slopes have the potential for large-scale failure involving rapid cliff top recession and run-out of debris. The FMECA approach has been used at this site to provide a repeatable approach to the strategic monitoring of the condition of the slopes. This example considers one of the intact slopes (Site A).

The cliff under consideration is around 2 km long, 55–60 m high and developed in a sequence of glacial tills (around 25–30 m thick) overlying Jurassic sedimentary rocks (predominantly sandstones and mudstones). The cliff face slopes at 30–32°, becoming steeper towards the base, where a near vertical mass concrete seawall, built around 1889–93, together with a concrete promenade, provides protection from wave attack. The cliffs were landscaped and partially drained around the same time to improve the stability of the slopes and allow them to be opened for public use. A cliff top road runs parallel with the cliffline, set-back 80–140 m from the cliff edge. A row of large private houses and hotels lines the landward side of the road.

It has been recognised that there is a possibility of a major landslide at the site which could lead to the loss of cliff top property (High Point Rendel, 2000). Following the development of a hazard model for the site, three scenarios were proposed that might lead to major landsliding:

- Scenario 1: the development of a major landslide caused by the expansion of the shallow landslides which occur on the cliff face, probably in response to prolonged heavy rainfall and blocked drainage.
- Scenario 2: the development of a major landslide caused by a combination of exceptionally high groundwater levels, progressive failure of the mudstones at the base of the cliff and the gradual deterioration of the slope drainage system.
- Scenario 3: the development of a major landslide due to failure of the seawall and renewed marine erosion.

The site was visited during 2002 and a number of features recorded that relate to the condition and performance of the slopes and coastal defences. Specific features of note were:

- the presence of tension cracks crossing a number of pavements, probably indicative of the early stages of development of shallow landslides;
- the poor condition of the seawalls which are considered to have a residual life of around ten years (defined as the period over which the structure could be expected to perform as an acceptable coastal defence with routine repairs and maintenance);
- the high probability of seawall failure during storm events. The probability of major structural failure of the seawall was assessed for a range of failure

Table 3.24. Example 3.8: Site A, South Bay, Scarborough – estimated annual probability of structural failure of the seawall (from High Point Rendel, 2000)

Principal seawall failure mechanisms					
Loss of apron	Undermining of toe	Block plucking	Break up of wall face	Overtopping and washout	Combined annual probability
0	0.0025	0.015	0.03	0.008	0.056

mechanisms (Table 3.24), as part of a condition survey (visual inspection and review of damage records etc.). The combined annual probability of failure is the sum of the probabilities of each of these mechanisms, and is estimated to be 0.056 (1 in 18).

The documentation of the current condition and risk was based on the FMECA approach and involved the following stages:

1 *Completion of an LCI diagram.* The scores entered into each of the boxes were assigned on the basis of observations and knowledge of the site conditions (Fig. 3.6).
2 *Calculation of the impact score.* The estimated risk-free market values of the properties within the cliff top area that might be affected by a major landslide were compiled with the assistance of a local estate agent. This yielded a value for potential direct losses of £1.8 million and an impact score of 4 (Table 3.25).
3 *Calculation of criticality and risk scores.* Criticality scores were calculated from the scoring attributed to elements in the LCI diagram. Rankings were attributed to the *criticality score, consequence × likelihood* product and the *confidence score.*

The results of the FMECA analysis are presented in Table 3.26 which indicates the following:

- The highest risk scores were associated with differential settlement and cracking of the seawall and the possible presence of voids within the fill materials.
- The highest scores for *consequence × likelihood* are associated with the potential for differential settlement, cracking and opening up of joints within the seawall. These elements pose the greatest risk to the safety of the slope and should be the focus of prompt remedial action.
- High *confidence scores* (i.e. high uncertainty) are associated with the potential for beach loss and the occurrence of voids within the seawall fill materials. These are priority areas for further investigation. Resolution

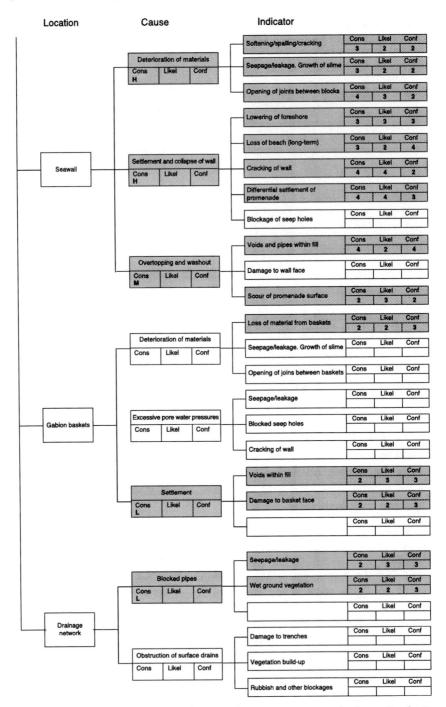

Fig. 3.6. Example 3.8: the LCI diagram for an intact coastal slope, South Bay, Scarborough (after Lee, 2003)

Table 3.25. Example 3.8: Direct losses and impact scores (from Lee, 2003)

Direct losses	Impact score
<£10 000	1
£10 000 to £100 000	2
£100 000 to £1 million	3
£1 million to £10 million	4
>£10 million	5

of the uncertainty could result in the consequence and likelihood scores going up or down.

- Calculation of risk scores for individual elements of this slope and others along the cliffline allows a comparison between the different sites. For example, similar elements from adjacent slopes might have identical *consequence × likelihood* or *criticality* scores. However, a higher impact score would indicate that the risk generated by failure of the element at one site would be greater than at another. Priority needs to be given to addressing the higher-risk elements first.

Qualitative risk assessment: an easy option?

Qualitative methods are of value where the available resources or data dictate that more formalised quantitative assessment would be inappropriate or even impractical. Nevertheless, it remains essential that any qualitative risk assessment be based on a sound hazard model (see Section 'Hazard models' in Chapter 2) and a good appreciation of the full range of possible outcomes (see Chapter 5). The use of qualitative methods should not be seen as an *easier option* than the quantitative methods described in the following chapters. Indeed, there are a number of significant issues that can limit the reliability and/or usefulness of the general approach, including the following:

- The use of subjective scales to rank hazards and adverse consequences can be problematic, as perceptions of what actually constitutes a *high* or a *low* risk will vary considerably. This can lead to misunderstandings between professionals and unnecessary alarm among those the assessment is intended to inform.
- The need to ensure that the uncertainties associated with the identification of hazards and adverse consequences are fully documented and clearly conveyed to the users.

161

Table 3.26. Example 3.8: Site A, South Bay, Scarborough – risk summary table (from Lee, 2003)

Location	Cause/indicator	Criticality score	Criticality rank	Consequence × likelihood	C × L rank	Confidence score	Confidence rank	Risk score (Impact* × Criticality)
Seawall	Differential settlement	48	1	4 × 4 = 16	1 =	3	3 =	192
	Cracking of wall	32	2 =	4 × 4 = 16	1 =	2	8 =	128
	Voids and pipes within fill	32	2 =	4 × 2 = 8	5	4	1 =	128
	Lowering of foreshore	27	4	3 × 3 = 9	4	3	3 =	108
	Opening of joints	24	5 =	4 × 3 = 12	3	2	8 =	96
	Loss of beach	24	5 =	3 × 2 = 6	6 =	4	1 =	96
	Softening/spalling/cracking of materials; seepage/leakage: materials	12	9 =	3 × 2 = 6	6 =	2	8 =	48
	Scour of promenade surface	12	9 =	2 × 3 = 6	6 =	2	8 =	48
Gabion baskets	Voids within fill	18	7 =	2 × 3 = 6	6 =	3	3 =	72
	Damage to basket face	12	9 =	2 × 2 = 4	11 =	3	3 =	48
	Loss of material from basket	8	12	2 × 2 = 4	11 =	2	8 =	32
Drainage	Blocked pipes: wet ground; seepage/leakage	18	7 =	2 × 3 = 6	6 =	3	3 =	72

Note. *Impact score = 4; see Table 3.25.

- The difficulty in establishing whether the risk levels identified at a site are acceptable. This can be particularly important where there is potential for loss of life or where there is a dispute between parties about the best way forward in managing the landslide issues.
- The problems in establishing an overall risk factor for a site where there might be multiple landslide hazards (e.g. rockfalls, debris flows and rotational slides), each with a different magnitude, frequency and range of potential adverse consequences.

As a final word, however, it is worth emphasising that 'the quality of a landslide risk assessment is related to the extent that the hazards are recognised, understood and explained which is not necessarily related to the extent to which they are quantified' (Powell, 2002). Attempts to quantify what are in effect qualitative judgements can have the effect of assigning a spurious degree of accuracy to the assessment (de Ambrosis, 2002).

4

Estimating the probability of landsliding

Introduction

What is the chance that a landslide will occur? To answer this question it is necessary to appreciate the circumstances that might lead to a landslide and to know something of the history of landsliding in the area. The ultimate cause of all landsliding is the downward pull of gravity. The stress imposed by gravity is resisted by the strength of the material. A stable slope is one where the resisting forces are greater than the destabilising forces and, therefore, can be considered to have a *margin of stability*. By contrast, a slope at the point of failure has no margin of stability, for the resisting and destabilising forces are approximately equal. The quantitative comparison of these opposing forces gives rise to a ratio known as the 'Factor of Safety' (F):

$$\text{Factor of Safety} = \frac{\text{Resisting forces}}{\text{Destabilising forces}} = \frac{\text{Shear strength}}{\text{Shear stress}}$$

The Factor of Safety of a slope at the point of failure is assumed to be 1.0.

When a slope fails, the displaced material moves to a new position so that equilibrium can be re-established between the destabilising forces and the strength of the material. Landsliding, therefore, helps change a slope from a less stable to a more stable state with a *margin of stability*. No subsequent movement will occur unless the slope is subject to processes that, once again, affect the balance of opposing forces. In many inland settings landslides can remain inactive or dormant for thousands of years. However, on the coast and along river cliffs, marine erosion removes material from the cliff foot, reducing the margin of stability, and promotes further recession.

As slope movements are the result of changes which upset the balance of forces so that those offering resistance to movement are exceeded by those producing destabilisation, the stability of a slope can be viewed in terms of its ability to withstand potential changes:

- *stable*, when the margin of stability is sufficiently high to withstand all transient forces in the short to medium term (i.e. hundreds of years), excluding excessive alteration by human activity;
- *marginally stable*, where the balance of forces is such that the slope will fail at some time in the future in response to transient forces attaining a certain magnitude;
- *actively unstable slopes*, where transient forces produce continuous or intermittent movement.

On any slope the margin of stability will vary through time in response to weathering, basal erosion and fluctuations in groundwater levels. It will usually rise to a peak immediately after a landslide event and then decline progressively to lower levels as basal erosion or other slope processes (e.g. weathering) affect the slope stability (Fig. 4.1; Brunsden and Lee, 2000). This perspective makes it possible to recognise two categories of *causal factors* that are active in promoting failure:

1 *preparatory factors*, which work to make the slope increasingly susceptible to failure without actually initiating landsliding (e.g. the long-term effect of erosion at the base of a slope, weathering etc.);

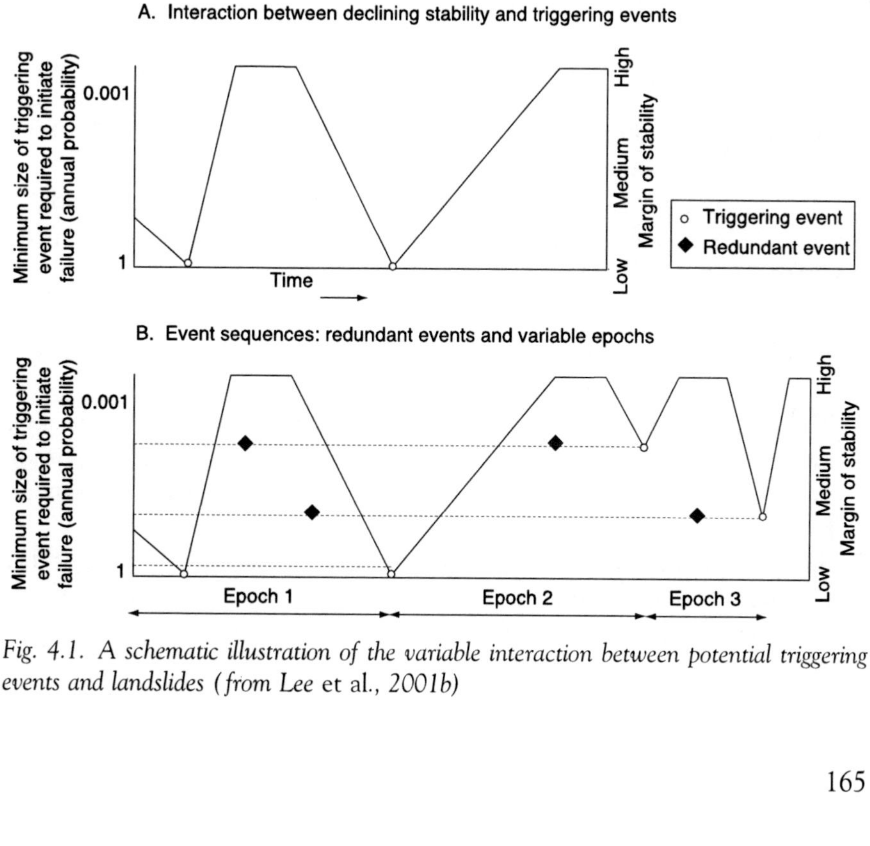

Fig. 4.1. *A schematic illustration of the variable interaction between potential triggering events and landslides (from Lee et al., 2001b)*

165

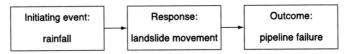

Fig. 4.2. *An example event sequence involved in the generation of landslide risk to a pipeline*

2 *triggering factors*, which actually initiate landslide events (e.g. rainstorm events).

Figure 4.1 highlights the complex relationship between preparatory and triggering factors. On coastal cliffs, for example, there are rapid temporal changes in the margin of stability as short-term fluctuations in causal factors are superimposed on the trends imposed by relatively steady erosion at the base of the cliff (Lee *et al.*, 2001b). As the margin of stability is progressively reduced over time by the operation of preparatory factors, so the size of event required to generate a fluctuation sufficient to trigger failure or initiate recession becomes smaller. Events of a particular magnitude fail to initiate instability until one chances to occur at a time when preparatory factors have lowered the margin of stability to such an extent that its influence becomes critical. Thus a particular size of event may not necessarily always result in failure or recession. Potential trigger events that do not actually result in failure or recession may be described as *redundant* (Lee *et al.*, 2001b) or *ineffective*. As Fig. 4.1 indicates, this can mean variable time periods (*epochs*) between failures or recession events, depending on the sequences of differing magnitudes of storm or rainfall events. In addition, the same size triggering events may not necessarily lead to the same scale of failure. Thus the response of a cliff or slope to rainstorms of a particular size is controlled by the recent history.

The combination of triggering and preparatory factors also provides a framework for modelling landslide scenarios (see Section 'Hazard assessment' in Chapter 1; Fig. 1.4). The occurrence of landslides involves an inter-related sequence of events driven by (Fig. 4.2):

1 An *initiating* or *triggering event* (e.g. an earthquake, high groundwater levels or construction activity).
2 The slope *response*, controlled by the nature and stability state of the slope as determined by the preparatory factors, together with propagating conditions such as high groundwater levels, progressive removal of toe support by stream or gully undercutting etc.
3 Subsequent *outcomes* determined by the style of landslide behaviour, the topography, the locations of human valued assets relative to the landslide movement and their vulnerability.

As discussed in Section 'Hazard models' of Chapter 2, the development of a hazard model provides a sound framework for estimating the probability of landsliding. Investigation effort (see Section 'Acquiring information: landslide investigation' of Chapter 2) should be directed towards addressing the following questions:

1 What is the condition of the slopes within the area of interest?
   - is there evidence of active instability (e.g. tension cracks, settlement, heave etc.) and associated damage to structures?
   - are there pre-existing weaknesses within the slope (e.g. pre-existing shears of non-landslide origin, or thin clay bands etc.)?
   - are existing slope stabilisation or erosion control measures showing signs of distress and could their failure initiate landsliding?
2 What type of landslide events can be expected to occur in the area of interest?
   - are these events associated with pre-failure movements, first-time failure, post failure movement or reactivation?
   - what are the likely mechanisms of failure (e.g. falling, toppling, spreading, flowing or sliding) and landslide types?
   - what is the range of event sizes that could occur (e.g. maximum credible event size; see Example 2.8)?
   - what velocities and travel distances could be expected (see Section 'Nature of landslide hazards' in Chapter 2)?
3 How frequently have landslides occurred in the past?
   - how many events have occurred over a particular time period (see Examples 2.1 and 2.2)?
   - what is the rate of cliff recession, both over the short-term or the longer-term (see Example 2.10)?
   - what is the magnitude/frequency distribution of past events (see Example 2.11)?
   - is there any evidence that the frequency of landsliding has been increasing (e.g. as a result of human influence on slopes or climate change) or decreasing (e.g. climate change, sea-level fall, reforestation) over a particular time period?
4 When did past landslide events take place?
   - is there evidence for the timing of past landslide activity?
   - can dates be determined for all recorded landslide events or only a sample?
5 Can past landslide events be associated with a particular triggering factor, such as rainfall or earthquakes?
   - are there reliable records of rainfall or earthquakes that cover the time period represented by the landslide inventory?
   - is there a trend in the frequency or timing of the triggering factors, or can the records be considered to be stationary?

The probability of landsliding can be estimated using *semi-quantitative methods* that rely on historical records, geomorphology and expert judgement, or *quantitative* methods such as probabilistic stability analysis and probabilistic simulation models. However, before describing these methods it is useful to set out a basic introduction to those aspects of probability that are relevant to landslide risk assessment.

Traditional approaches to analysing slope stability are essentially *deterministic*, in that parameters are represented by single values (e.g. a shear strength value or year when the event will occur). Confidence in the results can be improved by varying the values, that is by sensitivity testing. But sensitivity tests do not incorporate the likelihood of a particular value, beyond the engineer's judgement about the range within which the true value is expected to lie.

Probabilities and probabilistic methods are useful because of our ignorance of the true future frequency of events or values of different parameters, such as shear strength. A common problem is that while it is possible to say with certainty that a landslide event will occur, it is not possible to say exactly when it will occur. It may be possible to estimate the annual probability of a particular event (known as *prediction*), but not what year it will occur in (known as *forecasting*). Probabilistic methods (predictions) are essentially sophisticated sensitivity tests in which single data values are replaced by probability distributions that cover all possible values or outcomes.

Discrete events

Life is full of uncertainties, but it would be intolerable if we could *never* predict the behaviour of the objects that surround us (Belsom *et al.*, 1991). However, experience is the basis for making judgements about the likelihood of outcomes in different situations, such as the chance of being struck by lightening in a rain storm.

As mentioned in Chapter 1, the probability of an event occurring ranges from 0 (impossible) to 1.0 (certain) and is a function of the number of favourable outcomes, such as throwing a 6 with a single die, compared with the total number of *possible* outcomes (there are 6 sides to the die):

$$\text{Prob. (Event)} = \frac{\text{Number of Favourable Outcomes}}{\text{Total Number of Possible Outcomes}} = 1/6 \text{ or } 0.1667$$

In most situations it is not possible to define probability in such precise terms, because the chance of each outcome is not equal and may be dependent on a complex range of other factors. An alternative way of expressing probability is in terms of the frequency that a particular outcome occurs during a particular number of trials or experiments:

$$\text{Prob. (Event)} = \frac{\text{Number of Favourable Outcomes}}{\text{Total Number of Trials}}$$

168

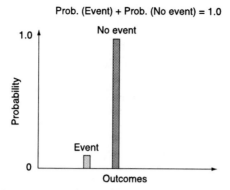

Fig. 4.3. Discrete events

For potentially unstable slopes there are two possible outcomes – *failure* or *no failure*. These two outcomes are *mutually exclusive* (Fig. 4.3), so the probability of an *event* together with the probability of *no event* must equal unity:

Prob. (Event) + Prob. (No event) = 1

which can also be expressed as

Prob. (Event) = 1 − Prob. (No event)

or

Prob. (Event) = 1 − (1 − Prob. (Event))

It is common practice to use discrete units of time, such as years, as individual trials. The chance of a landslide occurring in a particular year is, therefore, expressed in terms of an *annual probability*. Thus the probability of a landslide is:

$$\text{Annual Probability (Landslide)} = \frac{\text{Number of Recorded Landslides}}{\text{Length of Record Period in Years}}$$

For example, if records for a particular area or stretch of cliffline reveal three landslides in the past 300 years, then the annual probability is:

$$\text{Annual Probability} = \frac{3}{300} = 0.01 \quad \text{or} \quad 1\%$$

The resultant probability value, based as it is on experience or continuous records, is known as an *a posteriori* probability. It is important to appreciate that a 1% annual probability of occurrence also means a 99% probability of non-occurrence in any year.

However, an annual probability of 0.01 (1 in 100 or 1%) does not mean that an event will occur once in a 100 years or every 100th year, or even

Probability distributions for input parameters

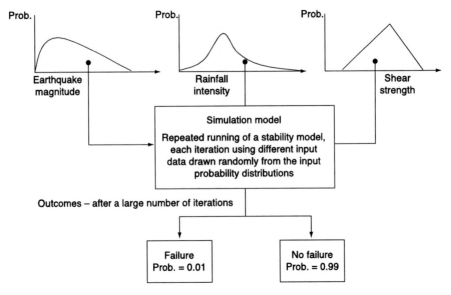

Fig. 4.4. A schematic diagram of a simulation model to illustrate the concept of annual probability

100 years apart, but rather that if a particular year were repeated a large number of times there would, on average, be 1 with a 'failure' for every 99 with 'no failures'. If a simulation model were to be developed that involved conducting a very large number of trials of the same slope, each one with randomly sampled input parameters drawn from probability distributions for the variables that set the environmental and internal slope controls on slope stability (e.g. rain storms or earthquakes of a particular magnitude, shear strength values), then on average there would be 1 failure for every 100 trials (Fig. 4.4).

It is often useful to define the probability of an event over a longer time period, such as the design life of a structure. Over time the probability accumulates in a predictable manner. If the annual probability of an event is 0.01, then the probability for the event *not* to occur in year 1 is 0.99, calculated as follows:

Prob. (No event) $= 1 -$ Annual Prob. (Event) $= 1 - 0.01 = 0.99$

The probability for the event *not* to occur in n years, taking $n = 10$ years as an example, is:

Prob. (No event, n years) $= (1 -$ Annual Prob. (Event)$)^n$

$$= (1 - 0.01)^{10} = (0.99)^{10} = 0.90 \quad \text{or} \quad 90\%$$

Thus the probability of an event occurring in n years, where $n = 10$, is 0.1 or 10%.

In hazard and risk probability studies it is normal to use the reverse of the above equation. Thus to calculate the probability of an event, whose annual probability is known, occurring in a defined period of time (years), it is usual to calculate the probability of it not occurring and then to take the result from 1. Thus using an annual probability of 0.01 and a period of 'n', where $n = 20$ years:

Prob. (Event) in 'n' years $= 1 - $ (Annual Probability of no Event)^n

Prob. (Event) in 'n' years $= 1 - (1 - $ Annual Prob. Event)^n

Prob. (Event) in 20 years $= 1 - (1 - 0.01)^{20} = 1 - (0.99)^{20}$

$$= 1 - 0.8179 = 0.1821 \quad \text{or} \quad 18\%$$

which indicates an 82% probability that such an event will not occur in the 20 year period. But note that the probability of occurrence has risen from 0.01 (1%) for a single year to 0.10 (10%) for 10 years and 0.18 (18%) for a 20 year period, even though the annual probability has remained the same. Thus, the longer the time period the greater the likelihood of an event occurring, although there is clearly not a linear relationship between the two. Recalculating the above equation for periods of 50 years and 100 years reveals that the probability of a 0.01 annual probability event occurring has risen to 39.5% and 63.4% respectively.

The accumulation of probability over time is known as the *cumulative probability* (Fig. 4.5). This can help define an annual probability by working

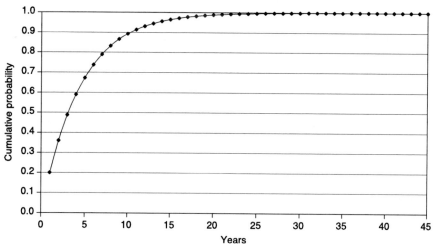

Fig. 4.5. *Cumulative probability over 45 years for an event with an annual probability of 0.2*

back from a judgement of the expected time period over which a single failure would be expected to have occurred. For example, if a slope is judged to have, approximately, an estimated 95% chance of failure in the next 14 years or a 100% chance of failure within the next 45 years, it is possible to 'track back' to yield an estimated annual probability of approximately 0.2, as demonstrated in Table 4.1, using the following equation:

$$\text{Annual Prob. (Event)} = 1 - (1 - \text{Probability Event in } n \text{ years})^{1/n}$$

$$= 1 - (1 - 0.95)^{1/14} = 0.2$$

In Table 4.1, the probability of occurrence in Year n gets progressively smaller as the years pass, despite the annual probability remaining constant at 0.2. This is because the probability of an event occurring in any year needs to take into account the possibility that the event actually occurred in the previous years, that is it is a 'one-off' event that may have already happened. Thus, the annual probability for a one-off event occurring in year 2 (and subsequent years) needs to be modified as follows:

Probability of failure in Year n

= Annual Probability of failure

× (Cumulative Prob. failure not occurred by Year n)

Some landslide events, such as debris flows, rockfalls or landslide reactivations, can be regarded as part of a random series of events with a distinct magnitude/frequency relationship. The average time between two events that equal or exceed a particular size (magnitude) is known as the return period or recurrence interval and is normally expressed as T_r. Thus an event that is expected to be equalled or exceeded on average every n years has a return period of n years. The return period is usually computed from a continuous series of recorded events as follows:

Return Period (T_r)

$$= \frac{\text{Number of Events exceeding a certain defined magnitude}}{\text{Number of years of continuous records}}$$

This allows the annual probability to be calculated as follows:

$$\text{Annual Probability } (P) = \frac{1}{T_r}$$

For events that occur on a regular basis, such as spring thaw debris flows within a defined catchment, it may also be possible to use a simple form of *extreme event analysis* similar to that used for flood predictions. This requires

Table 4.1. The probability of an event with an annual probability of 0.2 (1 in 5) occurring over a 45-year time period

Year n	Annual probability	Cumulative probability[1]	Probability of occurrence in year n[2]	Probability not occurred in n years[3]
1	0.2	0.2000	0.2000	0.8000
2	0.2	0.3600	0.1600	0.6400
3	0.2	0.4880	0.1280	0.5120
4	0.2	0.5904	0.1024	0.4096
5	0.2	0.6723	0.0819	0.3277
6	0.2	0.7379	0.0655	0.2621
7	0.2	0.7903	0.0524	0.2097
8	0.2	0.8322	0.0419	0.1678
9	0.2	0.8658	0.0336	0.1342
10	0.2	0.8926	0.0268	0.1074
11	0.2	0.9141	0.0215	0.0859
12	0.2	0.9313	0.0172	0.0687
13	0.2	0.9450	0.0137	0.0550
14	0.2	0.9560	0.0110	0.0440
15	0.2	0.9648	0.0088	0.0352
16	0.2	0.9719	0.0070	0.0281
17	0.2	0.9775	0.0056	0.0225
18	0.2	0.9820	0.0045	0.0180
19	0.2	0.9856	0.0036	0.0144
20	0.2	0.9885	0.0029	0.0115
21	0.2	0.9908	0.0023	0.0092
22	0.2	0.9926	0.0018	0.0074
23	0.2	0.9941	0.0015	0.0059
24	0.2	0.9953	0.0012	0.0047
25	0.2	0.9962	0.0009	0.0038
26	0.2	0.9970	0.0008	0.0030
27	0.2	0.9976	0.0006	0.0024
28	0.2	0.9981	0.0005	0.0019
29	0.2	0.9985	0.0004	0.0015
30	0.2	0.9988	0.0003	0.0012
31	0.2	0.9990	0.0002	0.0010
32	0.2	0.9992	0.0002	0.0008
33	0.2	0.9994	0.0002	0.0006
34	0.2	0.9995	0.0001	0.0005
35	0.2	0.9996	0.0001	0.0004
36	0.2	0.9997	0.0001	0.0003
37	0.2	0.9997	0.0001	0.0003
38	0.2	0.9998	0.0001	0.0002
39	0.2	0.9998	0.000 04	0.0002
40	0.2	0.9999	0.000 03	0.000 13
41	0.2	0.9999	0.000 03	0.000 11
42	0.2	0.9999	0.000 02	0.000 09
43	0.2	0.9999	0.000 02	0.000 07
44	0.2	0.9999	0.000 01	0.000 05
45	0.2	1.0000	0.0000	0.0000

Notes. 1 Cumulative probability calculated as follows: Probability (n years) $= 1 - (1 - 0.2)^n$.
2 Probability of occurrence in Year n calculated as follows: Probability (Year n) $=$ Cumulative probability (Year n) $-$ Cumulative probability (Year $n - 1$). 3 Probability not occurred in n years calculated as follows: Probability (no event, n years) $= 1 -$ Cumulative probability (n years).

Table 4.2. Percentage probability of the N-Year event occurring in a particular period

Number of years in period	N = Average return period in years							
	5	10	20	50	100	200	500	1000
1	20	10	5	2	1	0.5	0.2	0.1
5	67	41	23	10	4	2	1	0.5
10	89	65	40	18	10	5	2	1
30	99	95	79	45	26	14	6	3
60	–	98	95	70	31	26	11	6
100	–	99.9	99.4	87	65	39	18	9
300	–	–	–	99.8	95	78	45	26
600	–	–	–	–	99.8	95	70	45
1000	–	–	–	–	–	99.3	87	64

Note. Where no figure is inserted the percentage probability >99.9.

continuous records of events over a number of years, so that the size of the maximum magnitude event for each year is known. These extreme events are then ranked and the return period calculated using the so-called Weibull formula (Riggs, 1968), as follows:

$$\text{Return Period } (T_r) = \frac{\text{Number of years in series} + 1}{\text{Rank number of event}}$$

Thus, if there are 49 years of records, then the largest event (rank number 1) is the 50-year event (annual probability 0.02) and the second largest event is the 25-year event and so on.

The relationship between probability (expressed as a percentage), return period and the length of period under consideration is shown in Table 4.2. This indicates that an event with a return period of a 100 years (the 100 year event) *will not* have a 100% probability of occurring in a period of 100 years; the true figure is 65%. It also follows that 100-year events do not occur 100 years apart or once every 100 years; indeed, it is possible for there to be more than one occurrence in any period of a 100 years (as discussed in Section 'Multiple events' of this chapter) and even for there to be occurrences in successive years. Note should also be taken of the fact that very long-return-period events (rare events) have only a small likelihood of occurring during the design life of a building or structure, or within the planning horizons of most organisations. Thus the 500-year event has only an 11% chance of occurring during the lifetime of a building (taken here as 60 years) while the probability of a 1000-year event occurring is 6% (Table 4.2), thereby highlighting the problem of planning for extreme events.

Table 4.3. *The length of historical record required to esti-mate return period events with 95% and 80% reliability (after Benson, 1960)*

Return period	Length of record in years required to deliver reliability of return period estimate	
	95% reliable	80% reliable
2.33	40	25
10	90	38
25	105	75
50	110	90
100	115	100

It is important to stress that return period statistics are extremely sensitive to the length and quality of the historical record. For example, Benson (1960) demonstrated that to achieve 95% reliability on the return period of a 50-year event required 110 years of records; such lengthy data sets are not common (Table 4.3). It must also be stressed that the computation of return periods assumes uniformitarianism, that is past conditions are the same as those of the present and the future. As landsliding is particularly sensitive to human activity and environmental change, it is necessary to treat computed return periods with great scepticism. Indeed, all *a posteriori* probabilities need to be used with great care.

Multiple events

The preceding section focused on the probability of a single 'one-off' event occurring over a particular time period, using the equation:

$$\text{Prob. (Event, } n \text{ years)} = 1 - (1 - \text{Annual Prob. (Event)})^n$$

This equation can be expanded into a *binomial series* to predict the number of favourable outcomes x (i.e. landslides) in n trials in which the probability p remains constant in each trial. For given values of n and p the set of probabilities of this form for $x = 0, 1, 2, \ldots, n - 1$ is called a binomial probability distribution. Figure 4.6 presents a chart that can be used to predict the probability of an event occurring at least x times in n years (Gretener, 1967). For example, there is a 95% chance ($P_{n,x}$) that an event with an estimated annual probability (Po) of 0.01 (1/100) will occur at least once in 300 years ($nPo = 3$), at least five times in 900 years ($nPo = 9$) and at least 10 times in 1600 years ($nPo = 16$).

175

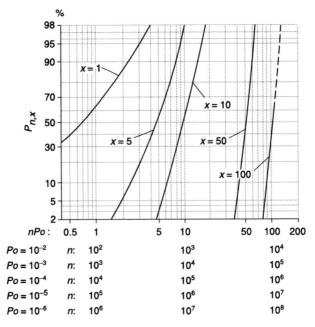

Fig. 4.6. *Probability of a rare event* ($P_{n,x}$) *to occur at least x times in n years (after Gretener, 1967). See text for explanation*

The binomial distribution applies when the total possibilities are known, that is how often an event occurs and how often it does not occur. But for so-called natural phenomena such as landslides, while it may be possible to count the number of landslides that have occurred in a particular area during a particular period of time, it is not possible to count how many land-slides did not occur. Similarly, while the binomial distribution deals with individual and discrete trials or lengths of time, it may be better to consider landsliding as a phenomena occurring in continuous time during which there could be any number of events. Both of these difficulties point to the need to employ the *Poisson probability distribution*; a distribution widely used in predicting earthquakes and volcanic eruptions, hence it being sometimes called the *model of catastrophic events*.

The general formula for the Poisson distribution is:

$$P(n) = \frac{(\lambda t)^n \, e^{-\lambda t}}{n!}$$

where t = exposure period in the same units as λ, n = number of predicted occurrences, e = constant (inverse of natural logarithm = 2.7183), λ = a *posteriori* probability of occurrence (equivalent to 'p' in the binomial formula). Thus so long as the *a posteriori* probability is known and conditions in the future are considered to be the same as those of the past, then the probability of an event occurring within differing time periods into the future can be calculated. For example, if studies of a stretch of cliff have revealed nine rockfalls in the past 80 years ($\lambda = 0.1125$), then the probability of one rockfall occurring in the next 25 years is

$$P_{(1)} = \frac{(0.1125 \times 25)^1 \times 2.7183^{-0.1125 \times 25}}{1}$$

$$= 2.8125 \times 0.060\,054 = 0.168\,901 \quad \text{or} \quad 17\%$$

and the probability of two rockfalls is

$$P_{(2)} = \frac{(0.1125 \times 25)^2 \times 2.718^{-0.1125 \times 25}}{2}$$

$$= \frac{7.910\,16 \times 0.060\,054}{2} = 0.237\,52 \quad \text{or} \quad 24\%$$

The Poisson distribution can be used in another way if the average number of events per unit time is known, that is per month, per season, per year, per decade, per century etc. In this case the Poisson expansion is used:

$$e^{-z}, \quad z \cdot e^{-z}, \quad \frac{z^2 \cdot e^{-z}}{2!}, \quad \frac{z^3 \cdot e^{-z}}{3!}, \quad \frac{z^4 \cdot e^{-z}}{4!}, \cdots$$

where z is the average value per unit time.

For example, if records for a stretch of coastal cliffs reveal the occurrence of 140 rockfalls in 100 years, then substituting the average of 1.4 per annum for 'z' in the above expansion reveals:

$$e^{-1.4} = 0.2466 \text{ probability of no rockfalls in any one year}$$

$$1.4e^{-1.4} = 0.3452 \text{ probability of one rockfall in any one year}$$

$$\frac{(1.4)^2 e^{-1.4}}{2} = 0.2417 \text{ probability of two rockfalls in any one year}$$

$$\frac{(1.4)^3 e^{-1.4}}{6} = 0.1127 \text{ probability of three rockfalls in any one year}$$

$$\frac{(1.4)^4 e^{-1.4}}{24} = 0.0395 \text{ probability of four rockfalls in any one year}$$

$$\frac{(1.4)^5 e^{-1.4}}{120} = 0.0110 \text{ probability of five rockfalls in any one year}$$

Continuous probability distributions and sampling

A number of parameters, such as recession rates or soil shear strength, can be regarded as random variables with continuous probability distributions (Fig. 4.7). These parameters can have an almost infinite number of possible values as compared with the binomial case where there are only two possible outcomes – failure or no failure.

The most widely used distribution is the so-called *normal distribution* (or Gaussian distribution) that takes the form of a symmetrical bell-shaped curve. The sum of the probabilities beneath the curve is 1. The shape of the distribution is defined by the mean and the standard deviation; the mean determines the centre of the probability distribution and the standard deviation determines the dispersion around the mean. Figure 4.8 highlights the relationship between the mean and the standard deviation; 68% of the values will lie within one standard deviation of the mean, 95% within two standard deviations and 99.7% within three standard deviations. Other

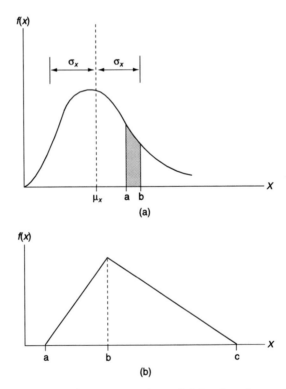

Fig. 4.7. *Probability density functions: (a) the probability that the variable x is between the values a and b is given by the shaded area; (b) a triangular distribution in which the value b represents the 'best estimate' and a and c represent the upper and lower limits of the range (after Wu et al., 1996)*

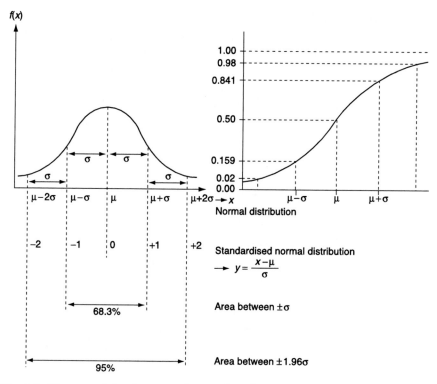

Fig. 4.8. The normal distribution, indicating the relationship between the mean (μ) and the standard deviation (σ)

common distributions include the exponential, log-normal, Weibull, gamma, Gumbel and beta (see, for example, Hahn and Shapiro, 1967; Benjamin and Cornell, 1970; Ang and Tang, 1984; Montgomery and Runger, 1994).

The use of continuous probability distributions is a key feature of fully quantitative hazard and risk assessments as they allow a full range of input parameters to be tested, each one weighted according to its likelihood of occurrence. A widely used method of achieving this is through repeated trials involving the random sampling of, for example, shear strength or recession rate, from an input probability distribution. This process is known as *Monte Carlo* simulation (see Example 4.10). The approach involves establishing a quantitative model of the slope or cliff and proceeding through numerous analyses, each one using randomly sampled input parameters drawn from the probability distribution. The higher the probability the more likely the value will be sampled. Each analysis is essentially deterministic, but repeated simulation allows a probability distribution to be produced for the results (i.e. an output distribution).

179

Subjective probability

Historical records of landslide activity and quantitative stability analyses allow *objective* estimates to be made about event probabilities. However, there are many situations where the lack of available 'hard' information does not allow this type of approach. An alternative strategy involves making judgements about the expected event probability, based on available knowledge plus experience gained from other projects and sites (see Chapter 1). This is known as *subjective* probability assessment and relies on the practitioner's *degree of belief* about the likelihood of a particular outcome, that is failure or no failure.

Subjective assessment of uncertainty is common in engineering practice, where this uncertainty is often expressed by such vague phrases as, 'possibly failing in the next 10 years', 'very likely to reactivate' etc. Experienced practitioners can replace this vagueness with estimates of likelihood or quantitative subjective probabilities. Thus, rather than saying a slope will 'probably fail next year', one could state that there is a '50% probability that it will fail by the end of next year'.

It is important to stress that there are undoubted problems associated with the use of the subjective probability approach, especially where it is undertaken by single individuals. The main potential problems have been identified by Roberds (1990):

- *Poor quantification of uncertainty*, where uncertainty concerning landslide processes is ignored or not expressed in a consistent fashion. For example, if it is estimated that there is a 90% probability of failure, it should be expected to happen, on average, 9 times out of 10; often the probability will be lower in reality.
- *Poor problem definition*, where the focus of the investigation is directed towards one element of a slope at the expense of another, because of the assessor's experience and background.
- *Motivational bias*, where the assessor's perception is influenced by non-technical factors. For example, if the objective is to demonstrate a benefit:cost ratio of greater than 1.0 for a scheme, then the landslide scenarios, timing of losses and consequences may all be overstated. In other circumstances an assessor may be over-cautious and consciously overstate the likelihood of failure so as to avoid underestimating the chance of a major event, that is *conservative* judgement.
- *Cognitive bias*, where the assessor's judgement does not match the available facts, thereby introducing bias. For example, greater weight may be given to recent laboratory test results and stability analyses than to the known historical record of events and the performance history of the slopes. There may also be a tendency for an assessor to underestimate

Table 4.4. A summary of subjective assessment techniques (after Roberds, 1990)

Technique	Potential problems			
	Poor quantification of uncertainty	Poor problem definition	Motivational biases	Cognitive biases
Individual				
Self assessment	O	O	O	O
Independent review	O	O	O	O
Calibrated assessment	O	O	O	O
Probability encoding	●	●	●	●
Group				
Open forum	O	O	O	O
Delphi panel	O	O	●	●

O Technique partially mitigates potential problem.
● Technique effectively mitigates potential problem.

the uncertainty associated with values of parameters produced by laboratory testing, such as shear strength.

A range of techniques are available for eliminating or reducing the effects of these potential problems, involving more rigorous individual assessments or group consensus. These techniques will help ensure that the judgements are defensible should it become necessary to resolve controversy at a later date (Table 4.4; Fig. 2.32).

The simplest approach is *self assessment*, where the rationale behind every judgement has to be well documented. This should include a description of the available information and the methods of analysis and interpretation used, so as to enhance the defensibility of the judgement. The method may not, however, resolve any problems as both the original judgement and later self-assessment are subject to the same operator bias arising from an individual's perception.

A common method of improving a judgement is to seek a second opinion from an expert or colleague (i.e. *independent review*). As was the case for self assessment, the independent review should be well documented. It should also be open for review. Although this represents an important improvement over self assessment, similar problems may remain, especially if the expert undertaking the independent review is influenced by the same biases as the original assessor.

A systematic approach to developing a judgement is through the use of *calibrated assessments* (e.g. Agnew, 1985), where the assessor's biases are identified and calibrated, with the assessments adjusted accordingly. This method involves an independent review of the original assessment, together

181

with an assessment of the individual's biases by, for example, a peer group review or objectively through a set of experiments or questionnaires. The increased cost and difficulties involved in identifying and objectively quantifying the biases are the main drawbacks of this approach.

The most systematic and defensible approach to developing subjective probability assessments, but also the most expensive, is *probability encoding* (e.g. von Holstein, 1972; Merkhofer and McNamee, 1982). This involves the training of staff to produce reliable assessments of the probability of various events in a formal manner. This involves six stages:

1 training the team member to properly quantify uncertainty;
2 identifying and minimising the team member's bias tendencies;
3 defining and documenting the item to be assessed in an unambiguous manner;
4 eliciting and documenting the team member's rationale for the assessment;
5 eliciting, directly or indirectly, the team member's quantitative assessment of uncertainty and checking for self-consistency. The subject's uncertainty can be established by determining the probability of various states through comparison with reference situations, such as poker hands, or by choosing between two lotteries (e.g. probability wheels or intervals) until indifference is achieved;
6 verifying the assessment with the team member and repeating the process if necessary.

Group consensus about a judgement is clearly desirable despite the increased cost. However, there may be significant differences of opinion between team members. It is necessary to attempt to resolve these differences of opinion, but they may persist, so that the following outcomes can result:

- *Convergence.* A single assessment is determined that expresses the common belief of all individuals in the group as expressly agreed to by every group member.
- *Consensus.* A single assessment is determined, although the assessment may not reflect the exact views of each individual. The consensus assessment may be a compromise derived from the individual assessments of group members but without the express agreement of the individuals concerned (forced), or the group may expressly agree to it for a particular purpose (agreed).
- *Disagreement.* Multiple assessments are determined where convergence or consensus on a single assessment is not possible (e.g. owing to major differences of opinion).

In general, convergence is the most desirable outcome because it is defensible, but may be difficult to achieve. Agreed consensus (i.e. with the concurrence of the group) is slightly less defensible but also less difficult to achieve. Forced consensus, without concurrence of the group, may be difficult to defend but is easy to achieve. Disagreement may be the most difficult outcome to use because it involves more than one view, but is defensible.

The most widely used approaches to develop convergence or agreed consensus are *open forum* and the so-called *Delphi panel*. Open forum relies on the open discussion between group members to identify and resolve the key issues related to the landslide problems. The results can, however, be significantly distorted by the dynamics of the group, such as domination by an individual because of status or personality.

Delphi panel is a systematic and iterative approach to achieving consensus and has been shown to generally produce reasonably reproducible results across independent groups (e.g. Linstone and Truoff, 1975; Foster, 1980). Each individual in the group is provided with the same set of background information and is asked to conduct and document (in writing) a self-assessment. These assessments can then be used in one of three ways. They can be given to an independent individual who produces an average interpretation, which is then returned to each individual within the group for further comment and so on until consensus is achieved. Alternatively the independent individual analyses and compares responses and goes back to the individuals concerned with comments, requesting revision or explanation as to why extreme or unusual opinions are held; a process of controlled feedback that is continued until there is agreement. The third approach is for assessments to be simply provided anonymously to the other assessors, who are encouraged to adjust their assessment in light of the peer assessment. Irrespective of the process adopted, it is typical for individual assessments to converge and for iterations to be continued until consensus is achieved. As the Delphi technique maintains anonymity and independence of thought, it precludes the possibility that any one member of the panel may unduly influence any other.

Estimating probability from historical landslide frequency

It is increasingly recognised that there exist serious problems in using historical data to predict future landslide activity because of the non-linear relationship that exists between triggering factors such as rainfall and landslide activity, the variable effects of human development on vegetation cover, land use and water balance and the uncertain consequences of climate change (D. K. C. Jones, 1992). Nevertheless, it is a commonly held assumption that the historical frequency of landsliding in an area can provide an indication of the future probability of such events. The approach relies on the assumption that the proportion of times any particular event has occurred in a large number of trials (i.e. its *relative frequency*) converges to a limit as the number of repetitions increases. This limit is called the probability of the event. There is no need for the number of different possible outcomes to be finite. This is essentially an empirical approach depending on a long period of reliable and continuous record.

Probability and frequency are fundamentally different measures and should not be used interchangeably. For example, in a game of chance (e.g. rolling dice) there are a finite number of different possible outcomes, which are assumed to be equally likely. The frequency of any event is then defined as the proportion of the total number of possible outcomes for which that event does occur. Thus, the probability of throwing a 6 with a single die is 1 in 6 (0.167). However, an individual may roll the die 100 times and record 25 successful throws with a 6 on one occasion, or 10 on another and 5 on another etc. In each case the probability based on historical frequency would be different from the classical probability of 1 in 6. Indeed, it is possible for 100 throws of the die to yield no 6 at all at one extreme and a hundred 6s at the other, although the probability of the latter is *vanishingly small*, thereby drawing attention to the fact that *improbable* (i.e. extremely low probability) does not mean *impossible* (cannot happen). The reliability of the historical frequency to provide an estimate of the annual probability will only tend to increase in the following circumstances. If the number of separate 100 throws of a die is greatly increased, then the mean value for the number of 6s thrown will progressively approach 1 in 6 (i.e. 16.67 per 100). Alternatively, greatly increasing the number of throws (trials) will have the same effect.

In the context of landsliding, therefore, obtaining annual probability data from historical records requires the following:

1 Ascertaining the extent to which environmental change during the period under consideration (the past as determined by historical records and the future as determined by risk assessment scoping) can be dismissed as insufficiently important to invalidate the basic assumption that 'the past is a valid guide to the future'.
2 Ensuring that only that portion of the historical record which is complete and accurate is used, while employing every means available to make the usable record as long as possible.
3 Undertaking steps to overcome the problems of sample size, as discussed above, by using every available means of increasing the sample population.

The three following examples illustrate how different types of historical records of landsliding, together with observational information, can provide a basis for estimating the probability of failure. In the first example, the approach adopted to overcome the problem of short-term records was to increase the sample population size. In the second and third example, various approaches were used to try to extend the historical record, with varying degrees of success.

Example 4.1

Failure of cut slopes presents a significant hazard in Hong Kong and, occasionally, leads to fatalities. For example, a 27 m high cut on the Fei Tsui Road failed in August, 1993. The landslide involved an estimated 14 000 m^3 of debris which ran out across the road, killing a young boy and injuring his father who had been walking along the pavement (GEO, 1996; see Example 6.13).

An approach for assessing the probability of failure of individual cut slopes has been developed that is based on the historical frequency of failure throughout Hong Kong, with appropriate adjustment factors for local site conditions (Finlay and Fell, 1995; Fell *et al.*, 1996a). Systematic recording of landsliding in Hong Kong has been undertaken since 1978, with landslide incident reports prepared for all recorded slides since 1983 (this includes incidents as small as a single boulder or 1 m^3 of debris). Table 4.5 summarises the landslide statistics for the period 1984 to 1993 and provides a performance database for the estimated 20 500 cut slopes in the area. During this 10-year period a total of 2177 cut slope failures were reported. The average historical frequency of failure of 217.7 events/year was used to provide an estimate of the average annual probability of failure for every one of the entire population of cut slopes:

Annual Probability of Failure (Hong Kong-wide)

$$= \frac{\text{Number of Historical Events}}{\text{Total Number of Cut Slopes}} = \frac{217.7}{20\,500} = 0.0106$$

This Hong Kong-wide estimate does not differentiate between cut slopes that are more or less likely to fail. However, by using a combination of site-related factors it is possible to adjust the estimate for a particular cut, as follows:

Probability of Failure (Cut c) = Probability of Failure (Hong Kong-wide)

$$\times \text{Adjustment Factor } (F)$$

If the slope is considered less likely than average to fail, then the adjustment factor (F) will be less than 1.0, whereas the factor would be over 1.0 if the slope is considered more likely than average to fail.

The adjustment factor (F) comprises a *primary site factor* (F′) and a factor that takes account of the *history of instability* at the site (Fe).

185

Table 4.5. *Example 4.1: Landslides in Hong Kong recorded by the Geotechnical Engineering Office (from Fell et al., 1996a)*

Year	Cut slopes		Retaining walls		Fill slopes		Natural slopes		Rock falls		Others		Total	
	All slides	Major slides	All slides	Major slides	All slides	Major slides	All slides	Major slides	All slides	Major slides	All slides	Major slides	All slides	Major slides
1984	70	4	11	0	14	1	4	1	7	0	14	0	120	6
1985	145	4	25	3	13	1	10	1	17	0	44	1	254	10
1986	115	6	26	0	18	3	9	2	29	0	36	0	233	11
1987	193	7	27	0	14	1	14	0	28	0	31	1	307	9
1988	73	3	14	0	15	1	7	1	22	0	26	0	157	5
1989	435	39	41	9	22	5	25	1	30	2	67	0	620	56
1990	41	1	5	0	12	2	8	2	9	0	24	1	99	6
1991	49	2	5	0	9	1	9	0	13	0	3	1	88	4
1992	439	14	44	3	55	6	51	2	40	1	12	0	641	26
1993	617	71	25	0	38	10	94	11	41	0	12	1	827	93
Total	2177	151	223	15	210	31	231	21	236	3	269	5	3346	226
Average	217.7	15.1	22.3	1.5	21	3.1	23.1	2.1	23.6	0.3	26.9	0.5	334.6	22.6

Table 4.6. Example 4.1: Listing of factors influencing cut slope failure probabilities (from Fell et al., 1996a)

Factor	Maximum value	Mininimum value	Independent factor	Factor component	Relative importance (maximum)
F_1'	1.25	0.25	Age	Age	High
F_2'	4	0.9	Geology	Unfavourable joints	Very high
				Recent colluvium	High
F_3'	2	0.1	Slope geometry	Slope angle	Very high
				Slope height	High–average
F_4'	4	0.1	Geomorphology	Angle above slope	Very high
F_5'	4	0.5	Groundwater	Groundwater	Very high
				Percentage chunam cover	Very high
				Drain condition	High–average
				Drain blockage	Very high
				Vegetation upslope	High

The primary site factor (F') is the product of five independent factors (Table 4.6):

Primary Site Factor $(F') = F_1' \times F_2' \times F_3' \times F_4' \times F_5'$

The independent factors are:

1 *slope age* (F_1'); For those slopes that were constructed prior to the establishment of geotechnical control procedures in 1977, a factor of 1.25 is used. For post-1977 slopes the factor is 0.25,
2 *geology* (F_2'); The key conditions assumed to influence failure are unfavourable discontinuity orientations and the presence of colluvium within the cutting (see Table 4.7a),
3 *slope geometry* (F_3'); Although the potential for failure tends to increase with slope angle, experience has shown that the higher cuts have generally been better designed. Therefore, the slope geometry factor (Table 4.7b) decreases with slope height,
4 *geomorphology* (F_4'); The most significant geomorphological condition is considered to be the gradient of the terrain above the cut (Table 4.7c),
5 *groundwater* (F_5'); Table 4.7d presents the groundwater factors which reflect a combination of evidence of groundwater seepage and the percentage *chunam* cover (a concrete coating sprayed over the cut face). The factor is further adjusted by taking into account drainage conditions and the presence of vegetation upslope (Table 4.7e). For example, a factor of 2.0 should be used for a slope with 10% chunam cover and no visible

Table 4.7a. Example 4.1: Values of geology factor (F_2') (from Fell et al., 1996a)

Condition	F_2' value
Continuous adversely oriented joints, sufficient to cause a landslide of significant magnitude	4
Extensive discontinuities adversely oriented joints, sufficient to cause a landslide of significant magnitude	3
Some discontinuities adversely oriented joints, sufficient to cause a landslide of significant magnitude	2
Recent colluvium of greater than 1 m depth present, sufficient to cause a landslide of significant magnitude	2
Otherwise use	0.9

Table 4.7b. Example 4.1: Values of slope geometry factor (F_3') (from Fell et al., 1996a)

Slope height: m	Multiplying factor F_3'		
	Slope angle <50°	Slope angle 50–60°	Slope angle >60°
<5	0.7	1.1	2.0
5–10	0.5	1.0	1.8
10–20	0.4	0.85	1.6
>20	0.3	0.7	1.2

Table 4.7c. Example 4.1: Values of geomorphology factor (F_4') (from Fell et al., 1996a)

Terrain gradient: degrees	F_4' value
0–5	0.1
5–15	1.0
15–30	1.1
30–40	1.3
40–60	2.0
>60	4.0

seepage; a further 0.25 and 1.0 should be added if there are grass and leaking pipes upslope, respectively.

Finlay and Fell (1995) recommend that the product of these five site factors is then *sense-checked* against previous slope inspections and a number of empirical cut slope assessment methods that have previously been developed in Hong Kong (e.g. the CHASE approach, Brand and Hudson, 1982; the Ranking System, Koirala and Watkins, 1988). Adjustments should be

Table 4.7d. Example 4.1: Values of groundwater factor (F_5') (from Fell et al., 1996a)

Percentage chunam cover	Groundwater factor F_5'		
	Water existing within slope in the upper two-thirds of the cut height	Water existing within slope in the lower third of the cut height	No visible seepage
0–25	4.0	3.0	2.0
25–50	3.7	2.7	1.7
50–80	3.4	2.4	1.4
80–100	3.0	2.0	1.0

Table 4.7e. Example 4.1: Additional adjustment values for the groundwater factor (F_5') (from Fell et al., 1996a)

Factor	Description	Value to adjust F_5' (add or subtract)
Drain condition and discharge capacity	Poor	Add 0.25
	Fair	No adjustment
	Good	Subtract 0.25
Drain blockage	Yes	Add 0.25
	No	No adjustment
Vegetation upslope	None/grass	Add 0.25
	Shrubs/trees	No adjustment
	Paved	Subtract 0.25
Service pipes upslope	Present, leaking	Add 1.0
	Present, not leaking	Add 0.5
	Present	Add 0.25
	Not present	No adjustment

made to the primary site factor if there is disagreement between the results.

The previous history of instability at the site, together with observations about the current presence or absence of signs of landslide activity, provides a final adjustment factor (Fe; Table 4.8).

For an individual cut slope, therefore, the estimated annual probability of failure is

Probability of Failure (Cut c) = Probability of Failure (Hong Kong-wide)

× Primary Site Factor (F')

× History of Instability Factor (Fe)

A worked example for a hypothetical slope is shown in Table 4.9.

Table 4.8. Example 4.1: Multiplying factor Fe for evidence of instability and history of instability (from Fell et al., 1996a)

		Multiplying factor Fe		
		Major distress, e.g. slumping, large cracks	Some signs of distress, e.g. minor cracking	No evidence of instability
History of instability	Yes	10	3	1.5
	No	6	2	0.5
Limits to probability values	P_{max}	1.0	0.1–1.0	0.1–1.0
	P_{min}	0.1	No limit	No limit

Table 4.9. Example 4.1: Assessment of the probability of failure of a hypothetical cut slope (based on Finlay and Fell, 1995)

Factor	Site factor	Site comment	Value
Primary site factor (F')	Slope age	Pre-GEO slope	1.25
	Geology	Colluvium present	2.0
	Geometry	10–15 m high cut, 50–60° slope	0.85
	Geomorphology	35–40° slope above the cut	1.3
	Groundwater	0% chunam and seepage in the lower third of the cut	3.0
		Trees upslope	+0.0
		No drainage upslope	+0.0
		No service pipes upslope	+0.0
	Total ($F' = F'_1 \times F'_2 \times F'_3 \times F'_4 \times F'_5$)		8.28
History of instability factor (Fe)	No history of instability Signs of minor cracks		2
Total Adjustment Factor ($F = F' \times Fe$)			16.56
Probability of failure (Hong Kong-wide)			0.0106
Estimated annual probability of cut failure = Hong Kong-wide $\times F$			0.176

Although this method is specific to man-made slopes in Hong Kong the general approach has broader applicability. Of particular note is the establishment of the probability of failure for an average slope in an area, then using a series of adjustment factors to estimate the probability of failure of specific slopes relative to the average probability.

Example 4.2
As described in Example 3.1, the 2 km long cliffline in South Bay, Scarborough, UK can be sub-divided into eight separate sections dominated by

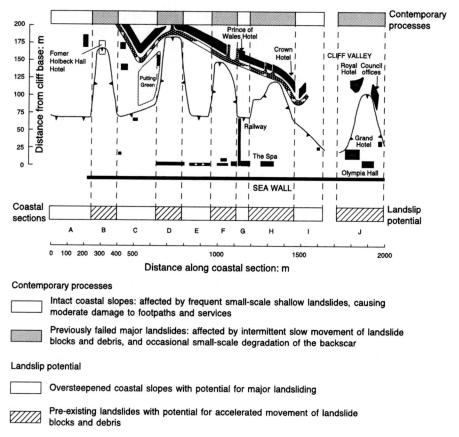

Fig. 4.9. Example 4.2: summary of cliff instability hazards, South Bay, Scarborough (after Rendel Geotechnics, 1994; Lee, 1999)

two contrasting geomorphological units: large, pre-existing landslides and intervening intact (i.e. unfailed) steep slopes. The recent history of landsliding in both of these settings was established through a search of journals, prints, reports, records and local newspapers (held on micro-fiche) archived at the Scarborough local library, and admiralty charts held at the Hydrographic Office, Taunton (Lee and Clark, 2000).

The earliest reported major landslide in South Bay is the 1737/38 failure at the site of the present day Spa (Fig. 4.9). Descriptions of the slide can be found in Schofield (1787) and Whittaker (1984), among others, along with a number of artists' illustrations. During this event an acre of cliff top land (205 m by 33 m) sunk 15.5 m, complete with cattle grazing on it. This was accompanied by 5.5–6.4 m of toe heave on the beach and at the cliff foot, creating a bulge around 25 m broad and 90 m in length. The following extract is from a long account by Schofield (1787), based on an eyewitness account published in the Philosophical Transactions (No. 461):

> On Wednesday December 28, in the morning a great crack was heard from the cellar of the spaw-house, and upon search, the cellar was found rent; but, at the time, no further notice was taken of it. The night following another crack was heard, and in the morning the inhabitants were surprised to see the strange posture it stood in, and got several gentlemen to view it, who being of opinion the house could not stand long, advised them to get out their goods, but they continued in it. On Thursday following (5 January 1738), between two and three in the morning, another crack was heard, and the top of the cliff behind it rent 224 yards in length, and 36 in breadth, and was all in motion, slowly descending, and so continued, till dark. The ground thus rent, contained about an acre of pasture land, had cattle feeding upon it, and was on a level with the main land, but sunk near 17 yards perpendicular. The sides of the cliff nearest the spaw stood as before, but were rent and broken in many places, and forced towards the sea. The ground, when sunk, lay upon a level, and the cattle next morning were still feeding on it.

In contrast to this vivid description, little has been found out about the timing of the two other major pre-existing slides on South Cliff (the South Bay Pool and South Cliff Gardens landslides; sections D and F on Fig. 4.9), other than that they both appear to have fresher morphology (and, hence, are probably younger) than the Spa landslide. As they are shown on the earliest reliable map of this coastline (an Admiralty Chart of 1843), they probably occurred in the period 1738–1842. The most recent event was the 1993 Holbeck Hall landslide which led to the destruction of a large cliff top hotel (Clements, 1994; Clark and Guest, 1994; Lee, 1999).

The historical frequency of failure of the intact steep slopes was estimated to be four events in 256 years (i.e. one event in 64 years). Thus, the annual probability of failure (*Pf*) of any one of the eight original intact slopes was estimated to be

Annual Prob. Failure (Section s) $= 4/(8 \times 256) = 0.00195$ (1 in 512)

The historical frequency of landslides can be used to provide an indication of the future probability of such events and the basis for modelling the 'survival probability' of the cliffline. The survival probability of the remaining four intact coastal slopes was modelled by Meadowcroft *et al.* (1999) as a series of repeated statistical trials involving only two possible outcomes: success (i.e. survival) or failure (i.e. landslide). The binomial distribution (see Section 'Multiple events' in this chapter) was used to estimate the probability

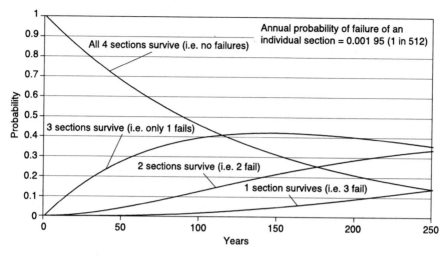

Fig. 4.10. Example 4.2: modelled survival probability of cliff sections, South Bay, Scarborough (from Lee et al., 2001)

that one or more cliff sections will survive in a particular time period. This distribution can be used for problems when:

- there is a fixed number of trials (i.e. cliff sections);
- the trials (cliff sections) are independent;
- the outcome (i.e. recession scenario) of any trial is either success or failure;
- the probability of failure is constant for each trial (i.e. each cliff section).

Figure 4.10 presents the results of the simple binomial experiment undertaken to model the future behaviour of the four remaining intact slopes, over the next 250 years. It was assumed that once one of the sections fails, it does not fail again. The experiment gave the probability of the number of successes (i.e. the number of cliff sections surviving) in a given number of trials (i.e. cliff sections) for each year. For example, the experiment suggested that, at 250 years into the future, there is a 35% chance that three cliff sections will have survived intact (i.e. only one failure) and a 33% chance that only two sections will survive. The method also predicts that there is a 2% chance that all four sections will have failed over this time period.

This example highlights a number of important limitations in the use of historical data. While the archive is clearly biased towards large, dramatic events, especially the older sources, it would be wrong to assume that all large events over a particular timescale have been recorded. As was noted in the introduction to this chapter, chance plays an important role in determining which events are recorded and thereby greatly influences the completeness of the record. Thus often the only evidence for some major

193

failures is the degraded surface form. These can be compared with failures of known date, but deducing an age for such features is problematic unless the materials and situation are very similar. In such cases absolute dating is the only true determinant.

The time series generated through the archive search allowed an estimate to be made on the frequency of particular types of major landslide events along the cliffline. A degree of caution is needed here, as the construction of seawalls along the foot of the entire cliffline (around the 1890s) has significantly altered the environmental controls on landslide activity. The cliffs are no longer subject to marine erosion which would promote failures by causing under-cutting of the cliffline and removing debris from the landslide toes. The 1993 Holbeck Hall landslide demonstrated that, despite these defences, major landslides can and still do occur (Clark and Guest, 1994; Lee, 1999). The estimation of landslide probability from historical frequency needs to consider whether the conditions that generated the pre-1842 landslides have remained valid following the construction of the seawalls and will continue to be valid in the future. If they are, then the time series and subsequent modelling provide reasonable predictions for decision making. However, if changes in environmental conditions have been profound, then great uncertainty surrounds the validity and usefulness of the results.

Example 4.3
In 1982 a rockfall landed on a car that was stuck in traffic in a large cutting (the 'Argillite Cut') on British Columbia Highway 99, killing a woman and disabling her father. The history of rockfall activity was used by Bunce *et al.* (1997) to provide an estimate of the probability of potentially hazardous incidents within the 476 m long cutting. Records of rockfall incidents have been collected by the British Columbia Ministry of Transportation and Highways since 1952. Prior to 1988, however, incidents were only recorded during inspections by Ministry geotechnical and engineering staff. Systematic recording began in 1988 with the appointment of Capilano Highway Services Ltd as maintenance contractors. Between November 1988 and December 1992 there were nine incidents when a rock was found on the road, indicating that the average number of rockfalls in the 4.12 year period was about 2.2 per year.

Only falls larger than 0.15 m diameter appear to have been recorded and Bunce *et al.* (1997) suggest that there may have been occasions when private citizens, police and the maintenance contractor's employees removed rockfall debris without any record being made. The assessment of rockfall frequency was improved by the mapping of rockfall marks on the highway surface. These marks tend to be distinctive features and are often isolated, sharply defined, circular to angular, concave depressions. A total of 60 rockfall impact marks were identified along a 393 m stretch of road that had been re-surfaced 4.75 years prior to the study. Some of these marks were interpreted as multiple impacts from a single event, suggesting that only 35 events had occurred that might have posed a threat to motorists. The number of rockfalls within the whole of the cutting was estimated as follows:

Number of rockfalls

$$= \frac{\text{Length of Argillite Cut}}{\text{Length of re-surfaced section}} \times \text{Number of rockfall incidents}$$

$$= \frac{476}{393} \times 35 = 42.4 = 8.93 \text{ incidents/year (over the 4.75 year period)}$$

The historical records and impact marks suggest a frequency of rockfalls of between 2.2 and 8.9 incidents a year. These figures can be used to provide an estimate of the probability of rockfall activity. However, as the probability cannot exceed 1.0, it is more appropriate to estimate the probability for a shorter unit of time than a year, such as a month. The risk can then be calculated on a monthly basis and then multiplied by 12 to yield an annual figure (see Chapter 6):

Risk = (Monthly Probability × Consequences) × 12

If the annual mean number of events is 2.2, then the monthly mean (x) is 2.2/12 (i.e. 0.18). Therefore, assuming a Poisson distribution (see earlier), the probability of at least one event in a month is

Probability = 1 − Prob. (no event in the month)

$$= 1 - \exp(-x) = 1 - e^{-2.2/12} = 0.167$$

This is a minimum figure. If the estimated frequency of rockfalls based on impact marks is used (8.925 or 0.7438 per month), then the probability of one or more events per month rises to 0.525.

In the Argillite Cut example, Bunce *et al.* (1997) used the annual frequency of rockfalls as the *number of trials* in a binomial model to determine the probability that rockfalls would hit one or more vehicles in a year (see Example 6.11).

Estimating probability from landslide-triggering events

Most landslides are associated with a particular triggering event, such as a heavy rainstorm or large earthquake. An indication of the probability of land-sliding can be obtained through establishing initiating thresholds between parameters, such as rainfall or seismic activity, and landsliding. The most

195

readily defined threshold is one that identifies the minimum conditions (or *envelope*) for landslide activity; above this, the conditions are necessary, *but not always sufficient*, to trigger landslides and, below this, there is insufficient impetus for failure.

If the frequency of these triggering thresholds can be determined from analysis of climatic (e.g. Schrott and Pasuto, 1999) or earthquake records (e.g. Keefer, 1984; see Example 2.7), this can be used as a basis for estimating landslide probability. In doing so, however, it is important to recognise that the occurrence of an event that exceeds the triggering threshold may not necessarily lead to slope failure, as some events will be redundant or ineffective because of the recent history of slope development (see Fig. 4.1).

The probability of a landslide event is the product of the annual probability of the trigger (e.g. earthquake or high groundwater) and the *conditional probabilities* of the subsequent slope response. For example, suppose a triggering event (E) has a probability $P(E)$. *Given that this event occurs*, the failure outcome O has the probability $P(O|E$; note that the sign '$|$' denotes *given*). The probability of this sequence of events occurring is

Probability $= P(E) \times P(O|E)$

Triggering thresholds provide a measure of the average likelihood of landsliding in an area. However, the stability of individual slopes and, hence, the response to a triggering event will vary from place to place. Some slopes will be very susceptible to a triggering event, whereas others may be able to withstand much higher magnitude events.

In each of the following three examples the approach relies on the identification of triggering thresholds from historical records of rainfall or earthquakes and landslide activity. The first is concerned with the reactivation of a large, deep-seated landslide complex. This type of event is generally associated with prolonged heavy rainfall. In general, the deeper the slide the longer the period of antecedent heavy rainfall needed to initiate failure. The period may vary from several days (e.g. Reid, 1994) to many months (e.g. Van Asch *et al.*, 1999). In many areas, it may be that the pattern of wet years appears to control the occurrence of landslides (e.g. Bromhead *et al.*, 1998 at the Roughs, Kent, UK).

The second example considers the impact of particular rainstorms on landslide activity. Whereas weekly or monthly rainfall patterns may be adequate to explain the pattern of deep-seated landslides, it is often the high intensity events of limited duration (e.g. hours) that are critical for controlling shallower failures. Shallow landslides may be associated with either a critical pore water pressure threshold being exceeded (e.g. Terlien, 1996; Terlien *et al.*, 1996; Corominas and Moya, 1996) or as a result of the increased weight of the saturated soil (e.g. Van Asch *et al.*, 1999). Harp (1997) reported several hundred landslides triggered in the hours and days following an exceptional rainstorm in Washington State, USA (over 50 cm in 7 days). Casale and Margottini (1995) describe how widespread cata-strophic landslide activity in Northern Italy during 1994 was associated with exceptional one- and two-day rainfall totals that exceeded all previous historical maxima.

The third example considers the combined probability of a landslide triggered by an earthquake from three different sources, in this case separate fault systems in California. The approach combines an assessment of the like-lihood of the maximum-magnitude earthquake on each fault over a particular period with the use of expert judgement (see Section 'Estimating probability through expert judgement' in this chapter) to estimate the probability of this event actually triggering a landslide at the study site.

Example 4.4

Many pre-existing landslide systems are sensitive to variations in ground-water levels and, hence, sequences of wet and dry years. An assessment of the climatic influence on landslide activity can, therefore, be used to assess the probability of reactivation. Lee *et al.* (1998b) describe how a combination of landslide systems mapping, historical records and rainfall analysis provided a pragmatic tool for assessing the annual probability of significant ground movement events in different parts of the 12 km long landslide (the Under-cliff) on the south coast of the Isle of Wight, UK (see Example 2.2). Land-slide activity tends to occur in the winter when rainfall totals are higher and evaporation rates are lower, and consequently much more of the rainfall is effective in raising groundwater levels. The relationship between landslide activity and winter rainfall is not a simple one. Some landslide systems within the Undercliff are more sensitive to rainfall events while others appear only to show signs of movement during extremely rare conditions. Indeed, some systems only show signs of significant movement (i.e. resulting in the development of tension cracks, subsidence features etc.) during extreme conditions. An assessment of the probability of significant move-ment has formed the basis for a pragmatic approach to landslide forecasting by the Isle of Wight Council.

The relationship between landslide reactivation and rainfall was established as follows:

1 *Identification of landslide systems.* Detailed geomorphological mapping, at 1:2500 scale, of the Undercliff delimited a series of discrete landslide units within broader landslide systems (Lee and Moore, 1991; Moore *et al.*, 1995).

2 *Analysis of historical records.* Reports of past landslide events were consolidated by a systematic review of available records, including local newspapers (from 1855 to the present day). Over 200 reported incidents have occurred over the last two centuries. Each record was classified according to the nature and scale of event: major landslide events; localised ground movement and minor landslide events; events not directly related to rainfall such as coastal cliff falls (T9; Fig. 2.14) or subsidence and joint widening of the Upper Greensand (T2; Fig. 2.14). Events were matched to the relevant landslide system (identified by the geomorphological mapping), revealing marked concentrations of past activity at the extreme eastern and western ends of the Undercliff.

3 *Analysis of rainfall records.* A composite data set was derived from the three rain gauges that have operated within the Undercliff since 1839. The antecedent effective rainfall was calculated for four-month moving periods between August and March (the wet period of the year), from 1839/40 to 1996 – this having previously been shown to be a sufficiently long time series to allow the identification of the prolonged periods of heavy rainfall that appear to control landslide activity in the Undercliff (Lee and Moore, 1991). Two methods were used to calculate the effective rainfall for the Undercliff:

- 1839/40 to 1992. The effective rainfall was calculated using Thornthwaite's formula from monthly data from the Ventnor area and temperature data for Southampton (1855–1989) and Ventnor (1959–92);

- 1992 to 1996. An automatic weather station was installed in Ventnor in 1992 which records both daily rainfall and potential evapotranspiration (using the Penman–Montieth method). Monthly effective rainfall was calculated by subtracting potential evapotranspiration from the rainfall total. Although this approach provided similar results to the effective rainfall derived from the long-term data set, there was not an exact match; it was recognised that this introduced a degree of uncertainty into the subsequent stages of the analysis.

The combined data series was used to calculate the likelihood of different four-month antecedent effective rainfall totals (4AER) occurring within the whole Undercliff in any single year (i.e. the return period) as follows:

$$\text{Return Period } (T_r) = \frac{\text{Number of Years } (156) + 1}{\text{Ranking in the Sequence}}$$

It should be noted that the average winter rainfall (September to January) increased by about 75 mm or 22% over the period 1839–2000 (Fig. 4.11). It is important to appreciate that this trend constrains the validity of deriving return period statistics from a non-stationary data set. In this

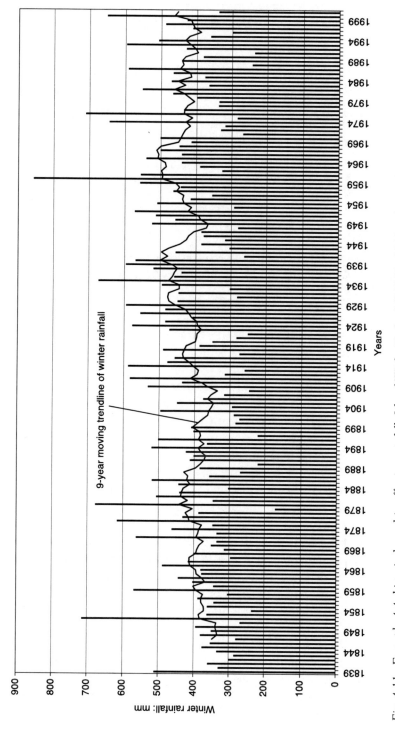

Fig. 4.11. Example 4.4: historical trend in effective rainfall, Isle of Wight Undercliff 1839–2000 (after Halcrow, 2001)

199

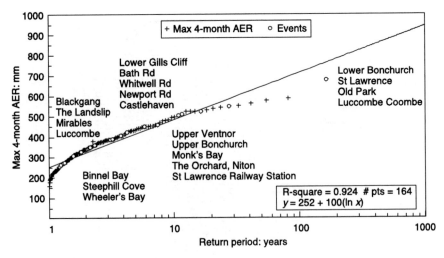

Fig. 4.12. Example 4.4: relationship between antecedent effective rainfall and the locations of landslide events in the Isle of Wight Undercliff (after Lee et al., 1998a; Halcrow, 2003)

example, it was recognised that the increasing average 4AER introduced a level of uncertainty about the future relationship between rainfall and landslide activity.

Figure 4.12 shows the 4AER totals that may be expected to be equalled or exceeded, on average, for particular recurrence intervals.

4 *Assessment of threshold conditions.* This involved relating the historical record for each landslide system to the 4AER data series to identify the minimum return period rainfall that is associated with landslide activity in a particular area. For example, in the westernmost system, Blackgang, significant movements are a frequent occurrence, and the minimum rainfall threshold needed to initiate significant movement appears to have been a 1 in 2 year event.

The 4AER associated with recorded ground movement events in particular areas is indicated in Fig. 4.12 to highlight the varying degrees of sensitivity of different parts of the Undercliff. Thus, 4AER totals that may be expected to be equalled or exceeded, on average, every year can lead to ground movement at a number of highly sensitive areas including The Landslip and near Mirables (see Fig. 4.12). Elsewhere, on low sensitivity areas such as Upper Ventnor and The Orchard, near Niton (see Fig. 4.12), ground movement has been associated with 4AER totals that have occurred, on average, every 10–30 years. During extreme winter rainfall periods many more landslide systems may become active as witnessed in 1960/61 and in January 1994. However, parts of Ventnor

Table 4.10. Example 4.4: An indication of the estimated annual probabilities of significant movement in four areas within the Isle of Wight Undercliff (after Lee et al., 1998b)

Location	Annual probability of threshold 4AER: P(4AER)	Annual probability of threshold 4AER triggering movement: P(O|4AER)	Estimated conditional probability of significant movement: Pm
Blackgang	0.9	0.1	0.09 (1 in 11)
Luccombe	0.25	0.2	0.05 (1 in 20)
Upper Ventnor	0.02	0.5	0.01 (1 in 100)
St Lawrence	0.005	0.5	0.0025 (1 in 400)

and St Lawrence have not displayed increased ground movement even when affected by events with return periods of 100 years or more.

5 *Assessment of the probability of landsliding.* That ground movement does not always occur when the 4AER thresholds shown on Fig. 4.12 are exceeded highlights the importance of other factors in controlling landslide activity, that is preparatory and triggering factors. An assessment was made, therefore, of the annual probability of a 4AER event of a particular magnitude actually triggering landslide activity. An estimate was made of the number of times that landsliding in a particular system was actually initiated when 4AER exceeded the threshold value, compared with the number of times that this threshold had been exceeded over the last 150 years, that is the proportional response.

The conditional probability of significant ground movement in a particular landslide system was calculated as follows:

$$Pm = P(4AER) \times P(O \mid 4AER)$$

where Pm = the annual probability of ground movement in a system, $P(4AER)$ = the annual probability of a threshold 4AER being equalled or exceeded in a particular year, $P(O \mid 4AER)$ = the annual probability of an event given the occurrence of the threshold 4AER being equalled or exceeded.

Table 4.10 provides an indication of the estimated probabilities of significant movement in a number of parts of the Undercliff.

Example 4.5
Landslides triggered by rainfall cause widespread damage throughout New Zealand, with average annual losses up to the mid-1980s estimated to be around US$33 million (Hawley, 1984). The New Zealand Earthquake Commission have paid US$14.8 million for landslide insurance claims since the mid-1970s, at an average of US$0.67 million per year (Glade and

Crozier, 1996). Costs of remedial and preventative measures (e.g. soil conservation, erosion control, sustainable land management programmes, education etc.) amounted to US$38.15 million for the period 1990–95 (Glade and Crozier, 1996).

The probability of landsliding has been established by compiling a database of landslide events and the daily rainfall totals associated with these events (Glade and Crozier, 1996; Glade, 1996, 1997, 1998). At a regional level, the probability of landsliding was estimated as the probability of landslide activity given a triggering event of a particular magnitude:

Prob. (Landsliding) = Prob. (Triggering Event)

× Prob. (Landslide | Triggering Event)

Triggering events were identified through an analysis of daily rainfall data and landslide records (Glade, 1998; Crozier and Glade, 1999). The 24 hour rainfall records from climate stations in Hawke's Bay, Wairarapa and Wellington were compiled and assigned to 20 mm-wide rainfall classes. The 24 hour rainfall totals associated with landslide occurrence were classified in the same way. Thresholds were defined that marked the limits to the relationship between rainfall and landsliding (Fig. 4.13). A minimum threshold corresponded with the daily rainfall class below which no landslide activity has been recorded and above which it may occur under certain conditions (i.e. the 24 hour rainfall *sometimes* causes landsliding). The maximum threshold was defined by the 24 hour rainfall class above which landsliding has always occurred (i.e. probability = 1). Table 4.11 summarises these thresholds for each of the three areas.

The probability of these thresholds occurring was defined through analysis of the daily rainfall records. Figure 4.14 presents the return period of the 24 hour rainfall totals for the Wellington area. This indicates that the maximum threshold associated with landsliding of 140 mm has a return period of 20.1 years (probability ≈ 0.05). The minimum threshold (20 mm) has a return period of less than 1 year.

The probability of landslide occurrence was calculated for each 20 mm rainfall class as:

Prob. (Landslides, Class r)

$$= \frac{\text{Number of Landslide Triggering Events, Class } r}{\text{Total Number of Rainfall Events, Class } r}$$

In each area the probability of landsliding increases with increased 24 hour rainfall (Fig. 4.13).

Table 4.12 summarises the *conditional probabilities* of landslide activity given the occurrence of a rainfall event of a particular magnitude.

202

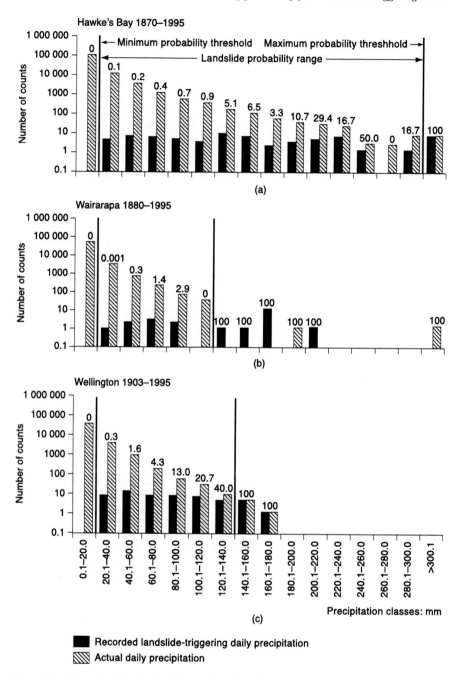

Fig. 4.13. Example 4.5: Probabilities (%) of landslide occurrence associated with rainfall of a given magnitude: (a) Hawke's Bay, (b) Wairarapa, (c) Wellington.
Note. A value of 50 means 50% of all measured daily rainfalls in a given category that produced landslides in the past

203

Table 4.11. Example 4.5: Daily rainfall probability thresholds associated with landsliding in different areas of New Zealand (based on Glade, 1998)

Area	Minimum threshold: mm	Maximum threshold: mm
Hawke's Bay	20	>300
Wairarapa	20	>120
Wellington	20	>140

The Isle of Wight Undercliff and New Zealand examples demonstrate how the analysis of triggering events can help provide a framework for estimating the probability of landsliding at the local and national level, respectively. However, both examples are the product of detailed research programmes that have drawn on lengthy rainfall records and an extensive database of historical landslide incidents. Unfortunately, in many areas the absence of reliable historical data will present a major constraint on the use of this type of approach.

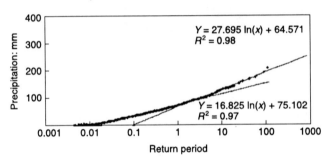

Fig. 4.14. Example 4.5: return periods of daily precipitation in the Wellington region, New Zealand (after Crozier and Glade, 1999). Note: the data set is presented in two parts (<100 mm and >100 mm)

Table 4.12. Example 4.5: Conditional probabilities of landslide activity triggered by different rainfall events in the Wellington area (based on Crozier and Glade, 1999)

Daily rainfall event: mm	Approximate probability of daily rainfall	Probability of landslide response	Conditional probability of landsliding
>140	0.05	1	0.05
120–140	0.1	0.4	0.04
100–120	0.15	0.2	0.03
80–100	0.33	0.13	0.043
60–80	0.66	0.04	0.026
40–60	1	0.016	0.016
20–40	1	0.003	0.003

Example 4.6

Earthquakes are a major cause of landslides in neotectonic regions such as California, USA. Kovach (1995) provides the hypothetical example of a geothermal power plant to be located in landslide prone terrain near San Francisco. The site is threatened by first-time failure triggered by seismic activity at any one of three nearby active strike-slip faults: the Maacama, Healdsburg–Rodgers Creek and San Andreas faults.

The probability of an earthquake triggered landslide at the proposed site was estimated as follows:

1 *Determining the maximum-magnitude earthquake* that could occur on each of the three faults. This assessment is based on studies of historical earthquake records and empirical investigations that correlate fault length with earthquake magnitude. The larger the fault rupture length, the greater the size of the potential earthquake. The maximum size events were:

Fault	Maximum size earthquake
Maacama Fault	M = 6.5
Healdsburg–Rodgers Creek Fault	M = 7.0
San Andreas Fault	M = 8.5

2 *Establishing the return period for the maximum-magnitude earthquake.* Earthquake recurrence data along individual fault segments can be expressed in the form: $\log N = a - bM$, where $N(M)$ is the number of earthquakes with magnitude larger than M and a and b are constants. From this relationship the return periods for the predicted maximum events on the faults were:

Fault	Maximum size earthquake	Return period (Annual Probability)
Maacama Fault	M = 6.5	1 in 4000 (0.00025)
Healdsburg–Rodgers Creek Fault	M = 7.0	1 in 380 (0.0026)
San Andreas Fault	M = 8.5	1 in 498 (0.0020)

3 *Calculating the probability of the earthquake occurrence over the project design life* (30 years), using the Poisson distribution:

Prob. (Maacama Fault) $= (1 - \exp(-30/4000)) = 0.007$

Prob. (H–R Creek Fault) $= (1 - \exp(-30/380)) = 0.08$

Prob. (San Andreas Fault) $= (1 - \exp(-30/498)) = 0.06$

4 *Estimating the probability that the maximum-size earthquake would actually trigger a landslide event.* This judgement was based on a geotechnical

205

appraisal of the site conditions and the level of ground acceleration and shaking that would occur at the site during the earthquake. The probability of landsliding in response to earthquake activity on each of the three faults was judged to be: Maacama Fault $= 0.2$; H–R Creek Fault $= 0.03$; San Andreas Fault $= 0.16$.

5 *Calculating the conditional probability of landsliding given a maximum-size earthquake.* The individual landslide probabilities during the 30-year design life of the power plant were calculated as follows:

$$\text{Prob. (Landslide; Fault } f) = \text{Prob. (Earthquake)} \times \text{Prob. (Landslide)}$$

$$\text{Prob. (Landslide; Maacama Fault)} = 0.007 \times 0.20 = 0.001$$

$$\text{Prob. (Landslide; H–R Creek Fault)} = 0.08 \times 0.03 = 0.002$$

$$\text{Prob. (Landslide; San Andreas Fault)} = 0.06 \times 0.16 = 0.01$$

A landslide could be caused by activity on any of the three faults. These scenarios are not mutually exclusive and, hence, the overall probability of the landsliding is not the sum of the probabilities, but a function of the following equation:

$$\text{Prob.} = 1 - ((1 - \text{Prob. Maacama}) \times (1 - \text{Prob. H–R Creek})$$

$$\times (1 - \text{Prob. San Andreas}))$$

$$= 1 - ((1 - 0.001)(1 - 0.002)(1 - 0.01)) = 0.01$$

Estimating probability through expert judgement

Expert judgement involves the use of experience, expertise and general principles to assign probabilities to landslide scenarios, preferably in an explicit and consistent manner. Such judgements are inevitably subjective (see Section 'Subjective probability' in this chapter) but, by proposing several possible scenarios followed by the systematic testing and elimination of options as a result of additional investigation and discussion, it is possible to develop reliable estimates.

Most landslide experts will willingly acknowledge the uncertainty that is inherent in predicting future landslide activity. Nonetheless, when asked to make judgements under conditions of uncertainty, experts tend to adopt heuristics and biases that can distort their judgements (e.g. Tversky and Kahneman, 1974). It is important, therefore, that a structured approach is adopted to estimate probabilities through expert judgement. This may involve, for example, breaking down the problem into *event sequences* consisting of initiating events, slope responses and outcomes (see Chapter 1), or more complex logic diagrams known as *event trees*. Under

Table 4.13. Landslide classes for subjective probability assessment (adapted from Hungr, 1997; Australian Geomechanics Society, 2000)

Description	Indicative slope condition	Estimated annual probability	Class name
Landsliding is imminent	Actively unstable	>0.1 (1 in 10)	Frequent
Landsliding should be expected within the design lifetime	Unstable	>0.01 (1 in 100)	Probable
Landsliding is possible within the design life, but not likely	Marginally stable	>0.001 (1 in 1000)	Occasional
Landsliding is highly unlikely, but not impossible within the design lifetime	Stable	>0.0001 (1 in 10 000)	Remote
Landsliding is extremely unlikely within the design lifetime	Stable	>0.000 01 (1 in 100 000)	Improbable
Landsliding will not occur within the design lifetime	Stable	>0.000 001 (1 in 1 000 000)	None/negligible

many circumstances, the resulting predictions will combine some quantitative evidence relating to the historic frequency of landsliding with site-specific analysis and expert judgement. The judgements should, as far as is possible, be made transparent by fully documenting the sources of evidence and detailing the process by which expert judgements of probability have been obtained. Peer review should be a routine aspect of assuring the quality of expert judgements. Experimental evidence suggests that group judgements appear to be more accurate than judgements of a typical (i.e. randomly chosen) group member (see the discussion in Section 'Subjective probability' in this chapter).

At the simplest level, probabilities may be rated in one of a number of indicative 'bands' (e.g. Table 4.13), drawing on the available knowledge of site conditions. These bands give a broad indication of the judged event probability and should not be viewed as implying a rigorous quantification of the likelihood of slope failure. As the available knowledge increases, so the judgements should become more reliable. Simple field observation methods can also be of value in supporting the judgement of landslide probabilities. For example, Crozier (1984, 1986) developed a ranking system which relates geomorphological descriptions of slopes with increasing probability of landslide occurrence, as shown in Table 4.14. The key factors include:

Table 4.14. Landslide probability classification (modified after Crozier, 1984)

Class	Description	Estimated annual probability
I	Slopes which show no evidence of previous instability and which by stress analysis, and analogy with other slopes or by analysis of stability factors, are considered to be highly unlikely to develop landslides in the foreseeable future.	>0.0001 (1 in 10 000)
II	Slopes which show no evidence of previous landslide activity but which are considered *likely* to develop landslides in the future. Landslide potential indicated by stress analysis, analogy with other slopes or analysis of stability factors; several sub-classes may be defined.	>0.001 (1 in 1000)
III	Slopes with evidence of previous landslide activity but which have not undergone movement in the previous 100 years.	>0.01 (1 in 100)
IV	Slopes infrequently subject to new or renewed landslide activity. Triggering of landslides results from events with recurrence intervals of greater than five years.	>0.1 (1 in 10)
V	Slopes frequently subject to new or renewed landslide activity. Triggering of landslides results from events with recurrence intervals of up to five years.	>0.2 (1 in 5)
VI	Slopes with active landslides. Material is continually moving and landslide forms are fresh and well defined. Movement may be continuous or seasonal.	≈ 1 (Certain)

- the presence or absence of instability features at a site, adjacent sites or similar sites (i.e. the same geological and geomorphological setting);
- the presence or absence of visible signs of active or recent movement (Table 4.15);
- surface evidence will also give an indication of the type of movement that could be anticipated at a site. At the simplest level – unfailed but unstable slopes may be prone to first-time failure (i.e. rapid, large displacements), whereas failed slopes (slid areas) may be prone to reactivation (i.e. usually slower, less dramatic movements).

A slope model developed from historical evidence, map sources and geomorphological assessment, together with site investigation and monitoring results, can provide a framework for estimating the probability of landsliding. Figure 4.15 presents a flow chart developed by Fell *et al.* (1996b) that uses observational factors to provide an indication of the probability of landsliding. The method involves a structured form of expert judgement, with the probability values based on the authors' experience of

Table 4.15. Features associated with active and inactive landslides (from Crozier, 1984)

Active	Inactive
Scarps, terraces and crevices with sharp edges	Scarps, terraces and crevices with rounded edges
Crevices and depressions without secondary infilling	Crevices and depressions infilled with secondary deposits
Secondary mass movement on scarp faces	No secondary mass movement on scarp faces
Surface-of-rupture and marginal shear planes show fresh slickensides and striations	Surface-of-rupture and marginal shear planes show old or no slickensides and striations
Fresh fractured surfaces on blocks	Weathering on fractured surfaces of blocks
Disarranged drainage systems, many ponds and undrained depressions	Integrated drainage system
Pressure ridges in contact with slide margin	Marginal fissures and abandoned levees
No soil development on exposed surface-of-rupture	Soil development on exposed surface-of-rupture
Presence of fast-growing vegetation species	Presence of slow-growing vegetation species
Distinct vegetation differences 'on' and 'off' slide	No distinction between vegetation 'on' and 'off' slide
Tilted trees with no new vertical growth	Tilted trees with new vertical growth above inclined trunk
No new supportive, secondary tissue on trunks	New supportive, secondary tissue on trunks

conditions on the interbedded sedimentary rocks of the Sydney Basin, Australia. It is likely that the approach has generic value and is applicable in areas of similar geology elsewhere, after calibration.

Event trees provide a structured and auditable approach for the use of expert judgement and subjective probability assessment. Event trees are a specific form of branching logic diagram which allows all likely sequences of events, or combinations of scenarios, arising from an initial postulate (event or trigger) to be mapped as a branching network, with estimated probabilities at each bifurcation (e.g. Cox and Tait, 1991). Once the structure of the tree has been developed by establishing all the likely sequences of events and resultant outcomes, then the individual probability of achieving a certain outcome is the product of the annual probability of the initial causal factor and the conditional probabilities of all intervening responses along a pathway leading to that specific outcome (i.e. an individual event sequence

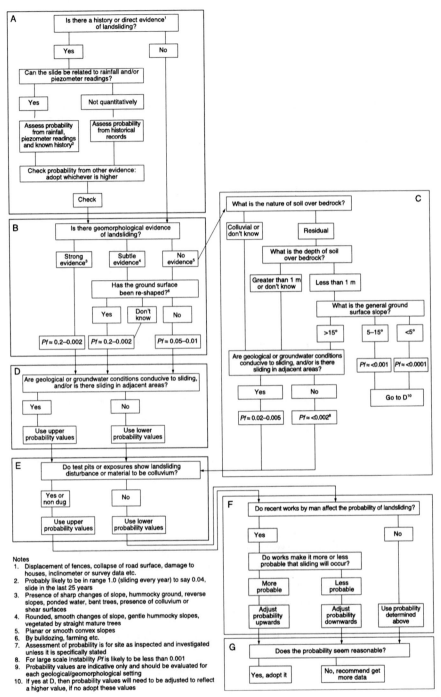

Fig. 4.15. A flow chart for estimating the probability of landsliding: soil slides (after Fell et al., 1996)

or accident sequence). For example, suppose an initiating event (E) has a probability $P(E)$. Given that this event occurs, the failure mechanism, M, has the probability $P(M|E)$. Likewise, the outcome O has a conditional probability $P(O|M)$. The probability of this scenario, or chain of events, occurring is

Scenario Probability $= P(E) \times P(M|E) \times P(O|M)$

It should be noted that each fork in the branching network represents a mutually exclusive alternative (i.e. they are *Boolean* parameters) with a cumulative probability of 1, and that the sum of the outcome probabilities for all the scenarios must also equal 1. In some cases more than one pathway, sequence or chain can result in the same outcome, in which case the resulting outcome probabilities or scenario probabilities for a specific outcome are combined (as in Example 4.6).

The following examples rely on the judgement of an experienced panel of experts. However, all draw on an understanding of the site conditions, based on geomorphological and sub-surface investigations, an awareness of the main controls on landsliding and the historical evidence of landslide behaviour or activity in the area of interest (e.g. past landslide events or the occurrence of triggering events).

Example 4.7
In 1891 a debris flow of about $30\,000\,m^3$ of material occurred on the slopes of Mount Dandenong in Montrose, Australia, and travelled up to 2 km across a colluvial fan at the base of the slopes. A house was destroyed and two people were lucky to survive. It is believed that the debris flow source area had been cleared of large trees prior to the event. The site now forms part of the Melbourne suburbs and a similar event could cause significant loss of life and widespread damage. In response to this threat the local authority, Shire of Lillydale, commissioned a study to quantify the risk to people and property (Moon *et al.*, 1992).

The estimated probability of landsliding was based on geomorphological mapping of catchments along the mountain front and inspection of the area by an 'expert panel' of four landslide specialists. As the 1891 event was initiated by a landslide, it was assumed that the potential for debris flows was related to the possibility of further landsliding in steep catchment heads. Potential source areas were identified and estimates were made of the expected volumes. On the basis of this inspection, the 1891 debris flow was considered to be the maximum credible event, in terms of both volume and run-out. The relative probability of debris flows within 26 catchments was then determined by the expert panel, based on a combination of topographic, geological and geomorphological data. Among the key factors guiding the

Catchment risk ranking Factor (1 and 2)	A1	A2	B1	B2	B3	C	D1	D2	D3	E	F1	F2	G1	G2	H1	H2	H3	H4	I	J1	J2	J3	J4	K1	K2	K3
Topography (spur or gully)																										
Relative amount of outcrop																										
Height of steep slope (3)																										
Proportion of steep slope																										
Size of colluvial fan																										
Number of modern landslips																										
Volume of modern landslips																										
Catchment risk ranking assessment (4)			L	L	M	L	L	L	M	M	M	X	X	H	L	L	L	L	M	L	L	L	L	L	M	L
Very large debris flow															H	H	H					H	H	H	H	
Small or large debris flow						H				H																
Debris torrent in gully								H	H																	

Notes.
1. Factors 1 to 4 apply to that part of the catchment underlain by the Ferry Creek Rhyodacite.
2. The shading in each box indicates the relative influence of the risk factor (dark grey – 'high'; light grey – 'medium'; no shading – low.
3. Steep refers to those slopes greater than 50%.
4. The symbols X, H, M and L refer to the relative risk, as indicated below.

Symbol	Term	Assumed recurrence interval (years)	Assumed prob. of occurrence in 50-year period %
X	Extremely high	1 in 100 to 1 in 300	15 to 39
H	High	1 in 100 to 1 in 1000	5 to 39
M	Medium	1 in 1000 to 1 in 10 000	0.5 to 5
L	Low	>1 in 10 000	<0.5

Fig. 4.16. Example 4.7: a summary of the approach to assessing the probability of debris flows in Montrose, Australia (after Moon et al., 1992)

judgements were:

- that the 1891 event occurred on natural slopes and the slopes had remained in a natural state since the event;
- a lack of evidence from mapping and inspection of the upper soil profiles, that any other major debris flows had occurred in the last 10 000 years;
- field assessment of the likelihood of slope failure in each of the potential debris flow source areas;
- the environmental conditions in the Montrose area, including rainfall records. Of note, large trees had probably been cleared from the landslide source area prior to the 1891 event. The potential effect of bushfires on the probability of landsliding was also taken into account.

Figure 4.16 illustrates the judgemental process which involved a 3-point ranking scale for a range of catchment hazard factors (high, medium, low). The factors were:

1 topography – spur or gully (gully being ranked as high hazard);
2 relative amount of outcrop (the higher the outcrop, the lower the hazard);
3 height of steep slope (the higher the slope, the greater the hazard);
4 number of recent landslides (the more landslides in the catchment source areas, the higher the hazard);
5 proportion of steep slopes (the larger the proportion of steep slopes, the higher the hazard);
6 size of colluvial fan (large fans were associated with high hazard);
7 volume of recent landslides (large events were ranked as high hazard).

Factors 1–4 provide a measure of the potential for a landslide, whereas factors 1, 5 and 6 reflect the likelihood of a debris flow resulting from the landslide event.

The combined scoring against each of these factors was used as the basis for assigning each catchment into 1 of 5 subjective probability classes (Table 4.16). Figure 4.17 presents part of the debris flow zoning map developed for the area. This distinguishes between different geomorphological units within each catchment on the basis of the judged probability of debris flow activity (detachment, throughput and run-out; Table 4.17).

Example 4.8
An existing chemical products pipeline passes through a number of landslide prone terrains in the Caucasus. Although the majority of landslides in these areas are of some antiquity, they remain prone to reactivation during periods of heavy rainfall or by earthquakes. It was recognised that landslide activity could present a threat to pipeline integrity, as ground displacements could

213

Table 4.16. *Example 4.7: Debris flow probability categories, Montrose, Australia (from Moon et al., 1992)*

Probability category	Range
Extremely high	0.01–0.0033 (1 in 100–1 in 300)
High	0.0033–0.001 (1 in 300–1 in 1000)
Medium	0.001–0.0001 (1 in 1000–1 in 10 000)
Low	>0.0001 (>1 in 10 000)
Very low	>0.000 01 (>1 in 100 000)

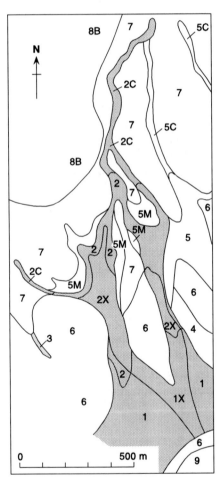

Fig. 4.17. *Example 4.7: part of the Montrose, Australia, debris flow hazard map (after Moon et al., 1992)*

Table 4.17. Example 4.7: Descriptions of the debris flow probability zones, Montrose, Australia – see Fig. 4.17 for a sample of the zone map (from Moon et al., 1992)

Zone number	Probability category	Description of hazard zone
1	High	Steep slopes where landslides may occur, some of which may become debris flows. Initiation and transportation zone for a high probability debris flow.
1X	Extremely high	Subdivision of Zone 1 to include Modern (post European) disturbed ground of $10\,000\,m^3$ or greater. Initiation and transportation zone for an extremely high probability debris flow originating from disturbed ground.
2	High	Likely extent of deposition area of a high probability event originating in Zone 1.
2X	Extremely high	Sub-division of Zone 2 to include likely extent of deposition for an extremely high probability event originating in Zone 1X.
2C	High	Gullies downstream of Zone 2 where debris may be deposited by the extremely high probability event.
3	High	Gullies where parts of the gully floor are steep (greater than 40% slope) and parts of the immediate catchment are very steep (greater than 50% slope). Debris torrents may affect the sections of gully covered by Zone 3.
4	Medium	Medium probability equivalent of Zone 1.
5	Medium	Medium probability equivalent of Zone 2.
5C	Medium	Medium probability equivalent of Zone 2C.
5M	Medium	Marginal area to Zones 2 and 2C. Medium probability because of the difficulty of predicting extent of deposits resulting from high probability events. The difficulty includes the uncertainty associated with assessing the proportion of debris flowing down particular gullies, and with assessing the extent of the fringe area. Zone 5M also takes into account the medium probability of larger than design debris flows occurring in the Zone 1 or Zone 1X areas.
6	Low	Low probability equivalent of Zone 1.
7	Low	Low probability equivalent of Zone 1. Extended to include all areas of flatter slopes in which deposits of colluvium or alluvium derived from the steep slopes could occur.
8A	Low	Foothills not included in Zone 7, where steeper slopes occur.
8B	Very low	Foothills or alluvial flats not included in Zones 7 or 8A.
9	Very low	Crestal ridge of the Dandenong Ranges.

result in pipe rupture. As part of a broader assessment of the risks to the pipeline, a preliminary review of the likelihood of pipeline rupture due to landsliding was undertaken by an 'expert panel' comprising local geologists and two international landslide specialists.

Prior to the expert panel visit, the local geologists had concentrated on the identification of all recognisable landslides along the route. An aerial photograph interpretation and subsequent fieldwork programme were directed towards the identification of 'pipe rupture' scenarios that provided a framework for assessing the probability of landsliding. In the absence of historical records of movement and a detailed understanding of landslide behaviour, the review had to rely on expert judgments to estimate the probability of landslide-related pipe ruptures. In doing so, the 'expert panel' considered:

- the potential for a particular type of landslide to generate a pipeline failure scenario;
- the annual probability of landslide triggering events (as described earlier for earthquakes and heavy rainfall);
- the expected frequency of pipeline failure events at a particular site, given the occurrence of a landslide-triggering event, (i.e. not all landslide movements will lead to pipeline failure).

The assessment involved:

1 Developing a broad understanding of the causes and mechanisms of the local landsliding.
2 Developing an appreciation of the stability (i.e. activity state) of the landslide systems that could threaten the pipeline. Landslide activity was assessed from field evidence, based on the following categories: dormant/inactive, relatively inactive (no signs of contemporary movement), relatively active (localised signs of contemporary movement), periodically active (likely to be active in most wet years), seasonally active (likely to be active in most wet seasons).

   Judgements about the expected *landslide behaviour* were made from the field assessment of landslide form and the materials likely to be involved (Table 4.18).
3 Identification, through group discussion involving the field party, of plausible sequences of events (scenarios) which could lead to landslide activity and subsequent pipeline failure. The field judgements on landslide activity and behaviour were used to consider the following range of pipeline rupture scenarios at each site:
   - *Displacement.* Pipe rupturing as a result of differential horizontal and/or vertical movement of the landslide mass.

216

Table 4.18. Example 4.8: Classification of landslide behaviour types

Mechanism (flow)	Type of behaviour (fluid-type movement)
Sliding	Plastic deformation
	Plastic/block deformation
	Slab/block deformation
	Brittle failure
Heave	Plastic deformation
	Plastic/block deformation
	Slab/block deformation
	Brittle failure
Fall	Brittle failure
	Creep and slow settlement

- *Spannng.* Pipe failure as a result of removal of support along a signifcant length following landslide movement.

4 Estimating the likelihood of these scenarios occurring using an event tree approach. This approach considers the probability (i.e. chance) of a sequence of events progressing, for example, from an initial trigger (an *initiating event* such as an earthquake) to the progressive reactivation of successive parts of the pre-existing landslide systems. The chance of each stage of a scenario occurring was determined through group discussion based on the broad appreciation of landslide behaviour and stability, that is the probabilities assigned to each scenario were the agreed 'judgements' of a team of experts.

Estimates of the likelihood of a triggering event were based on the historical frequency of earthquake-triggered landslide activity (around 2.5 events/10 years) and rainfall-triggered landslide activity (around 2 events/10 years; Kuloshvili and Maisuradze, 2000; Tatashidze *et al.*, 2000). These two figures can be combined to provide an estimate of the annual probability of a landslide triggering event in the order of 0.45 (4.5 events every 10 years). However, this relates to a country-wide likelihood of such an event occurring. It was felt that, on average, only 1 in 4 of these events were likely to act as potential landslide-triggers along the pipeline route as it only passes through part of the country and avoids the more seismically active, high rainfall areas. Thus, the expected annual probability of a triggering event was judged to be 0.1 (1 in 10).

Two generic models were developed specifically for the study (Figs 4.18 and 4.19) to provide a framework for the systematic estimation of the annual probability of a landslide response to a triggering event (not all potential triggers will actually initiate movement of a specific landslide)

217

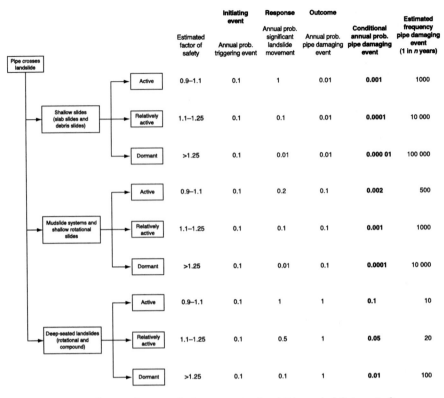

Fig. 4.18. *Example 4.8: framework for estimating landslide probabilities: pipeline crosses a recorded landslide*

and subsequent damaging outcomes (not all landslide movements will cause pipeline failure). These models address two key situations:

1 *Where the pipe crosses a recorded landslide*, the potential for pipeline failure is a function of the likelihood of landslide reactivation resulting in significant lateral displacement within pipe depth. It was assumed that this varies with landslide type (i.e. behaviour mechanism) and activity state – parameters that were estimated in the field. An active landslide (i.e. one with a low factor of safety) will be more sensitive to a potential triggering event (either heavy rainfall or an earthquake), and will have a higher annual probability of significant movement, than a dormant one with a higher factor of safety. For example, an active mudslide may respond to every triggering event (i.e. a probability of response of 1.0), whereas a dormant mudslide might only be reactivated by less frequent, larger triggers (say 1 in 100 year events, that is a probability of response of 0.01).

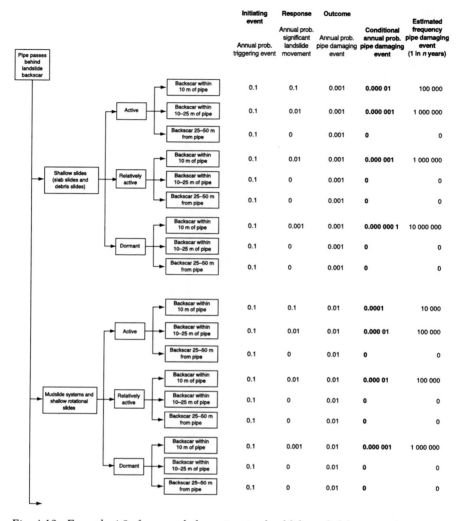

Fig. 4.19. Example 4.8: framework for estimating landslide probabilities: pipeline passes upslope of a recorded landslide

Landslide type and behaviour mechanism are important in determining whether landslide movement results in pipeline damage. For example, shallow slab slides involving plastic deformation of the displaced materials are less likely to cause pipeline failure than large deep-seated landslides characterised by brittle/slab styles of movement (estimates of the probability of pipeline damage, given the occurrence of significant movement, were 0.01 and 1 for these two slide types, respectively).

2 *Where the pipeline passes behind (i.e. upslope of) a recorded landslide*, the chance of pipeline failure is a function of the potential for a first-time

Table 4.19. Example 4.8: Indicative probability bands used in the review of the existing chemical products pipeline

Estimated annual probability of a pipe fracturing event	Likelihood of pipeline damage and annual chance of fracturing event
0	None
>0.000001	Extremely Improbable (1 in 1 000 000)
>0.00001	Improbable (1 in 100 000)
>0.0001	Remote (1 in 10 000)
>0.001	Possible (1 in 1000)
>0.002	Occasional (1 in 500)
>0.01	Probable (1 in 100)
>0.1	Frequent (1 in 10)
1	Certain (1 in 1)

failure of the ground upslope of the landslide backscar undermining the pipe. This potential reflects a combination of landslide type and activity state, together with the distance between the pipe and the current position of the landslide backscar. As a general guide it was considered extremely unlikely that any of the landslide situations had the potential to generate more than 25 m of cliff top retreat in a single event, or experience an average annual cliff top recession rate in excess of 0.5 m/year over an extended period.

The *conditional probability* for each scenario was calculated as follows:

Scenario Prob. = Prob. (Initiating Event) × Prob. (Response)

× Prob. (Outcome)

The structured use of event trees enabled the conditional annual probability of a landslide-related pipeline failure to be rated in one of a number of 'bands' based on an 9-point scale (Table 4.19).

Example 4.9
The coastal slopes at Lyme Regis on the Dorset coast, UK, are covered by active landslide systems which form the seaward part of a larger area of dormant landsliding that extends inland for around 0.5 km (see Example 2.8). Since the 1960s the landslides have displayed progressive reactivation and future movements present a significant hazard to the local community (e.g. Hutchinson, 1962; Lee, 1992). As a consequence, an 'expert panel' was convened in 1998 with the task of establishing possible landslide reactivation scenarios in order to provide a framework for assessing landslide risk, and for testing the economic viability of different coast protection and landslide management options (Lee *et al.*, 2000; Lee *et al.*, 2001b).

The expert panel consisted of representatives from the local authority (West Dorset District Council), the consultant and two landslide specialists. The range of landslide reactivation scenarios it identified were based on the understanding of the causes and mechanisms of landslide behaviour developed during an extensive, on-going programme of ground investigations undertaken as part of the Lyme Regis environmental improvements scheme (Clark *et al.*, 2000; Sellwood *et al.*, 2000; Fort *et al.*, 2000). Of particular importance were the likely reactivation sequences and an in-depth appreciation of the stability of the landslide systems and the interrelationships between adjacent landslide units.

In general, each scenario involved the progressive expansion inland of the zone of active instability, as pre-existing landslide units are unloaded, in turn, by the movement of the downslope (seaward) landslide units (which provide passive support to the upslope units); each phase of reactivation is promoted by the occurrence of high groundwater levels (see Fig. 2.25). While these sequences of events might, or might not, be expected to occur at some time within the timescale under consideration (i.e. the next 50 years), the precise timing of the initiating events and subsequent responses will be controlled by the almost-random occurrence of potential initiating events and the antecedent conditions at that time. In addition, the coastal slope conditions are progressively deteriorating, with the chance of failure expected to increase over time, due to a combination of the decline in structural integrity of the seawalls and the future increases in both storminess and winter rainfall totals predicted as a result of climate change.

For each landslide system along the coastal frontage (see Fig. 2.24), a series of event trees (reflecting different initiating events) and associated estimates of scenario probabilities were established as follows:

1 *Identification and characterisation of landslide systems.* Detailed geomorphological mapping of the coastal slopes delimited a series of discrete landslide units within broader landslide systems. Within each system there is a complex arrangement of individual landslide units which reflect the wide variety of landslide types and processes. The recognition of these systems and units formed the framework for understanding the contemporary ground behaviour of the Lyme Regis area, which was later refined as a result of a detailed ground investigation (Sellwood *et al.*, 2000).

2 *Identification of landslide reactivation scenarios.* Combining an understanding of the form and character of the landslide systems (i.e. aerial photograph interpretation, surface mapping and sub-surface geotechnical data) with monitoring data (ground movements and piezometric levels; Fort *et al.*, 2000) and analysis of past events and building damage, led to the development of a range of credible reactivation scenarios at each

221

site. Each scenario was developed from an initiating event (i.e. seawall failure, wet years sequences), with subsequent responses and outcomes identified as the effects of the initiating event were transmitted inland and upslope through the adjacent landslide units (Fig. 2.25). It should be noted that all of these scenarios are 'do nothing' scenarios, in the sense that nothing is considered to have been done to prevent an initiating event or to control the subsequent responses; that is landslide problems would be allowed to develop unchecked.

3 *Development of event trees.* Each scenario comprised an initiating event followed by a response (i.e. movement of parts of the landslide system; Response 1). In turn, this response may act as an initiating event for a second response (Response 2) and so on. Ultimately the combination of initiating event and the responses will lead to a particular outcome (i.e. impacts of coastal slope assets; Scenario outcome elements S1, S2 etc.). Each sequence of initiating event–response–outcome was simplified to a series of simple event trees (Fig. 4.20), with responses to a previous event either occurring or not occurring (i.e. yes/no options).

4 *Estimation of the annual probability of initiating events.* This involved the estimation of the likelihood of seawall failure and wet years sequences in each year from years 1–50.

The annual probability of seawall failure was assessed for individual sections of seawall by the local authority, West Dorset District Council, who also estimated an expected annual rate of increase in the chance of failure to reflect the gradual deterioration of the structures (under a 'do nothing scenario').

Analysis of rainfall records for the period 1868–1998 indicated that there have been eight 'wet year' sequences of 3–6 years duration in 130 years, suggesting an annual probability of around 1 in 16 (0.06). The frequency of these sequences (and possibly the duration) appears to have increased over the last three decades, suggesting a current annual probability of around 0.1 (1 in 10); this latter figure was used in the analysis.

5 *Estimation of the probability of responses.* Probabilities were assigned to each of the event tree 'branches/forks' at each site, mindful that the sum of probabilities at each fork must equal 1.0. This was achieved by first obtaining the 'expert judgement' of the individual project team members and then holding discussions to reach consensus on the 'best-guess' figures.

The approaches used to develop convergence or agreed consensus in this study were *open forum* and a *Delphi panel* (Roberds, 1990; see Section 'Subjective probability' in this chapter). Each individual in the project team was provided with the same set of background information and

S1 Localised damage to property, seawalls, services etc.
S2 Extensive loss of amenity gardens, sea front property, seawalls, services etc.
S3 Loss of up to 20 m of cliff top land, including gardens, tennis courts, access lane to Gardens, property

A. Initiating event = seawall failure (leading to toe failure, landslide reactivation and rear cliff failure)

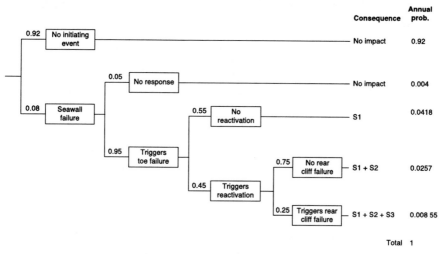

B. Initiating event = high groundwater levels

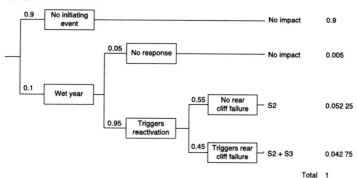

Fig. 4.20. Example 4.9: Langmoor Gardens, Lyme Regis, UK: event trees (from Lee et al., 2000)

was asked to conduct and document (in writing) a self-assessment. These assessments were then consolidated to identify areas of disagreement where further discussion was required in an open workshop meeting. Typically, the individual assessments tended to converge after discussion. Such iterations were continued until consensus was achieved.

It was found through discussions that the most acceptable approach to identifying probabilities at each branching of the event tree, or 'fork', was to identify a time period over which the team believed there was a 95% chance of the 'failure' route being realised. This was used to identify a corresponding annual probability that would deliver this cumulative probability over the agreed time period (this assumes a *binomial* distribution of events). The estimated annual probability, cumulative probability and the time by which an event is almost certain to have occurred are related as follows:

Probability of occurrence in x years $= 1 - (1 - \text{annual probability})^x$

6 *Calculation of conditional probabilities for each scenario.* For Year 1 (the initiating event and response occur in Year 1) the conditional probability associated with each 'branch' of an event tree (i.e. a unique sequential combination of scenario outcome elements, e.g. S1 + S2 + S3 etc.) was calculated as follows:

Scenario Prob. $= P(\text{Initiating Event}) \times P(\text{Response 1})$

$\times P(\text{Response 2}) \times P \ (\text{Response } n)$

For subsequent years (the initiating event and response occur in the same year) the calculation is essentially the same as the above, with the exception that the annual probability of the initiating event is changing over time (e.g. the estimated probability of seawall failure increases at 5% per year). In addition, the probability of a combination of scenario elements occurring in Year 2 needs to take into account the possibility that the scenario actually occurred in Year 1 and, hence, could not occur in Year 2. Thus, the annual probability for Year 2 (and subsequent years) was modified as follows:

Probability of failure in Year i

$= \text{Annual Probability of failure Year } i$

$\times (\text{Prob. failure not occurred in Year } (i - 2)$

$- \text{Prob. failure occurred in Year } (i - 1))$

However, a response might be delayed or lagged, so as to occur in any year after an initiating event; that is if the initiating event occurred in Year 1 the response need not be in Year 1, but could be in Year 2 or any year up to Year 50. Thus, calculating the combined probability of a response occurring in a particular year is more complex. For example, the probability of an initiating event leading to a response occurring in Year 4 involves the combination of four possibilities: $P(\text{seawall failure in Year 1 and the response three years later}) + P(\text{breach in Year 2 and response two years later}) + P(\text{breach in Year 3 and response one year later}) + P(\text{breach in Year 4 and response nil years later})$. For the probability of the response in Year 50 there would be 50 combinations of probabilities.

Site: East Cliff: Failure Scenario A
Model: Probability of Response 1 (Lower zone failure) following seawall failure

1 minus Probability S1 not occurred Year 2

Combined probability: initiating event (seawall failure) and response 1 (lower zone reactivation)

| Year of response | Annual prob. initiating event | Prob. response 1 lower zone event | Year of initiating event | | | ... | 48 | 49 | 50 | S1 Loss of property total probability | Prob. S1 Year i | Prob. S1 not occurred | Cumulative prob. S1 |
			1	2	3								
1	0.020	0.450	0.009							0.01	0.01	0.99	0.01
2	0.021	0.248	0.005	0.009						0.01	0.01	0.98	0.02
3	0.022	0.136	0.003	0.005	0.010					0.02	0.02	0.96	0.04
4	0.023	0.075	0.001	0.003	0.005					0.02	0.02	0.94	0.06
5	0.024	0.041	0.001	0.002	0.003					0.02	0.02	0.92	0.08
45	0.171	0.000	0.000	0.000	0.000					0.16	0.01	0.04	0.96
46	0.180	0.000	0.000	0.000	0.000					0.17	0.01	0.04	0.96
47	0.189	0.000	0.000	0.000	0.000					0.18	0.01	0.03	0.97
48	0.198	0.000	0.000	0.000	0.000		0.089			0.19	0.01	0.02	0.98
49	0.208	0.000	0.000	0.000	0.000		0.049	0.094		0.20	0.00	0.02	0.98
50	0.218	0.000	0.000	0.000	0.000		0.027	0.051	0.098	0.21	0.00	0.02	0.98
			0.020	0.021	0.022		0.165	0.145	0.098				

Prob(Initiating event Year 1) × Prob(Response 1 Year 3)

Prob(Initiating event Year 2) × Prob(Response 1 Year 4)

The cumulative probability of response 1 occurring over next 47 years if the initiating event occurs in Year 3 (Sum of column)

Probability of S1 occurring in Year 47 (Sum of row)

Probability of Scenario occurring in Year 48 × Probability that Scenario has not already occurred by previous year (Year 47)

Probability Scenario 1 not occurred in Year 46 – Probability Scenario 1 occurs in Year 47

Fig. 4.21. Example 4.9: An annotated example of the worksheet used to define the conditional probability of an outcome (Scenario outcome element S1) following an initiating event (seawall failure) and subsequent response (lower zone landslide reactivation) (after Lee et al, 2000)

The analysis necessitated the development of a sequence of related work-sheets for each landslide system. Each worksheet comprises a 50×50 matrix of probabilities derived from multiplying P(Initiating event) by P(Response) for all possible combinations of timings. Figure 4.21 presents an annotated worksheet, which illustrates how the analysis was built up. The example produces the probability of Response 1 following the occurrence of an initiating event. The results from this sheet (the total probability column) then form the input data (along with the probability distribution for Response 2) to the next sheet, and so on.

The basic principles of this method for assessing landslide risks in situations where there is no historical precedent are relatively straightforward. The aim was to assess the annual likelihood of a landslide event (here, a reactivation) and how this likelihood may change through time. In view of the long time scales, complex processes involved and sparse data, it is inevitable that expert judgement had an important role.

Estimating probability of cliff recession through simulation models

For eroding clifflines, the focus is generally concerned with determining how much recession will occur over a particular time period. Effort needs to be directed towards estimating the probability of sequences of events, rather than trying to estimate the likelihood of a single landslide event. Cliff recession in over-consolidated materials proceeds primarily via occasional landslide episodes followed by periods of relative inactivity, which may last for more than a hundred years on some coastlines (Lee, 1998; Lee and Clark, 2002). This is very different to the continuous process that has been implicit in many deterministic approaches to predicting cliff recession, such as the extrapolation of historical trends. The recession process is complex and far from random. Recession is not an inevitable consequence of the arrival of a storm that removes material from the cliff base or raises groundwater levels in the cliff. In order to fail, the cliff must already be in a state of deteriorating stability, which renders it prone to the effects of an initiating storm event (e.g. Fig. 4.1).

The pattern of past recession events is the result of a unique set of wave, tide, weather and environmental conditions. A different set of conditions would have generated a different recession scenario. The inherent random-ness in the main causal factors (e.g. wave height, rain storms etc.) dictates that the future sequence of recession events cannot be expected to be an accurate match with the historical records; there could, however, be a similar average recession rate with contrasting variability between measurements, trends and periodicity. Probabilistic methods offer an improvement on conventional deterministic predictions, because they aim to represent the variability and uncertainty inherent in the recession process (Lee *et al.*, 2001b).

The main elements in developing a probabilistic model are (Meadowcroft *et al.*, 1997; Lee and Clark, 2002) as follows:

1   Establishment of a cliff behaviour model, with particular emphasis on assessing potential events in terms of size (i.e. retrogression potential) and timing (i.e. recurrence intervals).

2   Assigning probability distributions to represent variability and uncertainty in the key parameters (e.g. event size, event timing, extreme wave heights etc.) Some parameters, such as extreme wave heights, have been extensively studied and probability distributions for these can be established using standard methods. Other aspects, such as future beach levels, may be established on the basis of historical data combined, if possible, with the results of modelling. Some factors, however, are more difficult to quantify and may call for a degree of subjective assessment, but this should be guided, wherever possible, by informed arguments about what ranges of values are likely and with what degrees of confidence. Distributions do not have to conform to the standard analytical forms; any probability distribution that can be envisaged can be simulated.

3   Developing a probabilistic prediction framework and selecting a simulation strategy. Simulations may be *static*, that is assessing responses at a given point in time, or *dynamic* to simulate a given time period using a time-stepping approach. The static approach is simply a Monte Carlo simulation of the model. There is no attempt to simulate any variation with time, though future prediction can be made by setting, for example, climate parameters to their predicted future values over particular time periods.

The dynamic approach, ideal for long-term prediction, involves repeating many simulations of the required time period in order to establish a histogram of probability distribution of the given response at a given point in time. The dynamic approach means that events that will occur in the future can be included. As well as random loadings this could include deterioration of a structure, or management intervention.

4   Running repeated simulations. The key requirements here are for a pseudo random number generator which produces a stream of values between 0.0 and 1.0, and the inverse probability distribution functions from which values for each of the variables are selected, based on the value of the random variable. Correlated variables require additional functions to ensure that sampled values are correctly correlated. After a large number of simulations, the frequency distributions and correlations of the sampled data should conform to the specified probability distributions, and the

227

result will be a stable frequency distribution, reflecting the variability of the input data and the form of the response function.

Simple response functions and models can be accommodated on a spreadsheet and can be set up and run quickly. More complex models can be built on the basis of existing numerical models, provided that these are not prohibitively slow to run. An advantage of the ever-increasing speed of computers is that multi-simulation techniques can now be used even with relatively complex process–response models, and can include long-term prediction.

Cliff recession data and predictions can be presented in a variety of ways, including probability density functions of the cliff position at a given time or the time required for cliff recession to reach a given point (Fig. 4.22). Alternatively, it may be useful to present the results as a hazard zoning based on the cumulative probability distribution of cliff recession over a given time (Fig. 4.23). For example:

Zone 1 It is certain that land within this zone will be lost through recession within a given time period.

Zone 2 There is a 50% chance that land within this zone will be lost through recession within a given time period.

Zone 3 There is a 10% chance that land within this zone will be lost through recession within a given time period.

Zone 4 There is a 1% chance that land within this zone will be lost through recession within a given time period.

Note that the probabilities that demarcate the zones are arbitrary and can be varied to suit the purpose. More detail (i.e. more zones) may be justified in areas with more assets at risk, although too many zones can be confusing and problematic. This form of presentation does not differentiate between different locations within the same zone, although in reality properties at the landward and seaward margins of a zone will have different probabilities of being affected by cliff recession.

Example 4.10
A *two distribution* model has been developed that assumes that the cliff foot can withstand a given number of storms before the cliff fails (Meadowcroft *et al.*, 1997; Lee *et al.*, 2001b; Hall *et al.*, 2002; Lee and Clark, 2002). In this model, a storm that causes undercutting of the cliff foot is defined as a wave height and water level with a certain return period. The return period, together with the number of storms required to initiate failure of the cliff, defines the average time interval between recession events. If a recession event does occur, then a second probability distribution can be

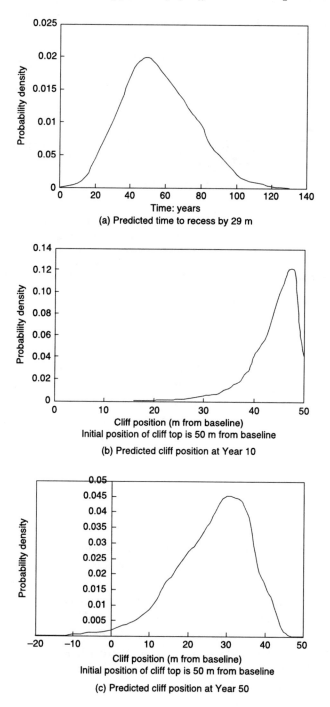

Fig. 4.22. Sample results from the two-distribution probabilistic model (from Lee and Clark, 2002)

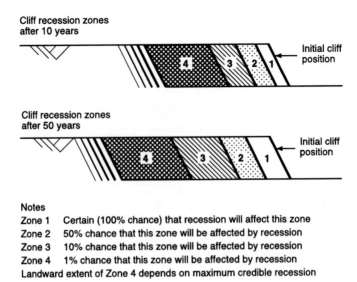

**Cliff recession zones
after 10 years**

Initial cliff
position

**Cliff recession zones
after 50 years**

Initial cliff
position

Notes
Zone 1   Certain (100% chance) that recession will affect this zone
Zone 2   50% chance that this zone will be affected by recession
Zone 3   10% chance that this zone will be affected by recession
Zone 4   1% chance that this zone will be affected by recession
Landward extent of Zone 4 depends on maximum credible recession

Fig. 4.23. Zoning of the cumulative probability of cliff recession over a given time (after Lee and Clark, 2002)

used to estimate the likely magnitude of the event, that is the amount of cliff top recession. This model, therefore, has the ability to differentiate between high and low sensitivity Cliff Behaviour Units (CBUs) by representing the number and magnitude of storm events needed to initiate recession events.

Cliff recession is assumed to proceed by means of a series of discrete landslide events, the size and frequency of which are modelled as random variables. A discrete model for the probabilistic cliff recession (X_t) during time period t, is

$$X_t = \sum_{i=1}^{N} C_i$$

where N is a random variable representing the number of cliff falls that occur in time period t, and C_i is the magnitude of the ith recession event. This model can be used to simulate synthetic time series of recession data, which conform statistically to the cliff recession measurements (e.g. Lee *et al.*, 2001b; Hall *et al.*, 2002). Three typical realisations of the model are shown in Fig. 4.24. The time series are stepped, reflecting the episodic nature of the cliff recession process. Multiple realisations of the simulation are used to build up a probability distribution of cliff recession.

The model is defined by two distributions:

1 An *event timing distribution* describes the timing of recession events. The model incorporates physical understanding of the cliff recession process

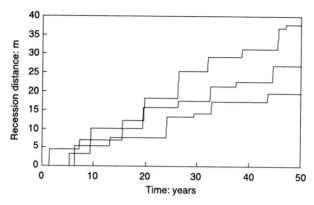

Fig. 4.24. Example 4.10: typical realisations of the two-distribution simulation model (see text for explanation of the model; from Lee et al., 2001b)

by representing the role which storms have in destabilising cliffs and initiating recession events. Note that in this example it is assumed that recession is driven by storm events; in other instances, groundwater and other factors will be important. The approach has links to renewal theory (Cox, 1962) inasmuch as the cliff is considered to be progressively weakened by succeeding storms. The arrival of damaging storms is assumed to conform to a Poisson process, that is successive storms are assumed to be independent incidents with a constant average rate of occurrence. After a number of storms of sufficient severity, a cliff recession event occurs. The time between successive recession events can therefore be described by a gamma distribution. The shape of this distribution is defined by a scaling parameter λ (the reciprocal of the return period of the significant storm event) and a shape parameter k (the number of storms above a certain threshold which cause damage to the foot of the cliff that is sufficient to trigger failure).

2 An *event size distribution* describes the magnitude of recession events in terms of the mean size and variability. The form and parameters of this distribution should reflect the frequency distribution of actual cliff failures and is likely to be site specific. The model uses a log-normal distribution, following the conclusions of the wave basin tests on a model cliff undertaken by Damgaard and Peet (1999). A log-normal distribution is non-negative which corresponds to the non-existence of negative cliff recession events. The probability density rises to a maximum value and then approaches zero as the recession distance becomes large, (i.e. very large cliff recession events are unlikely).

The cliff recession model is therefore characterised by four parameters, λ and k from the gamma distribution, and the mean, μ and variance, δ of the

231

Table 4.20. Example 4.10: Historical cliff recession rates (average annual recession rates expressed in metres/year; from Lee and Clark, 2002)

Location	Sea cliff 1907–29	Sea cliff 1929–36	Sea cliff 1936–62	Sea cliff 1962–91	Sea cliff 1907–91
1	0.09	1.5	1.31	0	0.57
2	2.09	2.0	0.31	0.06	0.83
3	1.63	0.57	0	0.34	0.57
4	0.91	0.57	0.23	0.31	0.48
5	0.64	0.28	0.31	0.28	0.5
6	0	0.28	0.15	0.34	0.19
7	0.09	0	0.54	0.18	0.24
8	1.27	0	0.54	0	0.26
Mean	0.84	0.65	0.42	0.19	0.45
SD	0.72	0.68	0.38	0.14	0.20

Note. SD = standard deviation

log-normal distribution. The model parameters can be estimated by maximum likelihood or Bayesian estimation methods (Hall *et al.*, 2002). The maximum likelihood method is based on optimally fitting the parameters to the available data, while the Bayesian method also makes use of expert knowledge about the size and frequency of landslide events. There is, therefore, scope to include geomorphological knowledge of event size and timing, which may not necessarily be revealed through examination of the historic data record.

This method has been tested using historical recession data for 20 m high cliffs, in Sussex, UK, developed in sandstones overlain by Wadhurst Clay, both dating from the lower Cretaceous. The position of the cliff top was obtained from 1:2500 scale topographic maps produced in 1907, 1929, 1936, 1962 and 1991. Cliff top locations were extracted at eight positions along the 400 m length of coastline. For each period (*epoch*) between map dates, the mean recession rate (m/year) was calculated for the eight locations. In addition, overall recession rates from 1907 to 1991 were calculated. For each of the five measurement periods, the standard deviation of recession rate between the different locations was calculated as well as the mean rate (Table 4.20).

The event timing distribution was chosen using a maximum likelihood parameter estimation model (Hall *et al.*, 2002), with parameters $k = 0.8$ and $\lambda = 0.046$. With more frequent events, the statistical model would not generate sufficient variability as compared with the data. Furthermore, the number of zero recession rates in the data record indicated that the characteristic time between recession events was quite long. For example, during the seven-year period from 1929 to 1936, two of the locations

Table 4.21. Example 4.10: Two distribution probabilistic model – simulation results (compare the mean and standard deviations with historical data in Table 4.20; from Lee and Clark, 2002)

Simulation 1

Location	Number of years and period				
	22 1907–1929	7 1929–1936	26 1936–1962	29 1962–1991	84 1907–1991
1	0.00	0.00	0.18	0.37	0.18
2	0.26	0.36	0.75	0.88	0.63
3	0.85	0.60	0.68	0.36	0.61
4	0.23	0.90	0.20	0.77	0.46
5	0.40	0.43	0.51	0.39	0.44
6	0.05	0.07	0.30	0.40	0.25
7	0.76	0.00	0.70	0.45	0.57
8	1.34	1.64	0.24	0.54	0.75
Mean	0.49	0.50	0.45	0.52	0.49
SD	0.43	0.52	0.23	0.19	0.18

Simulation 2

Location	Number of years and period				
	22 1907–1929	7 1929–1936	26 1936–1962	29 1962–1991	84 1907–1991
1	0.34	0.60	0.42	0.42	0.42
2	0.58	0.74	0.90	0.69	0.73
3	0.53	0.94	0.38	0.53	0.52
4	0.97	0.00	0.46	0.67	0.63
5	0.50	0.51	0.08	2.53	1.07
6	0.13	0.00	0.28	0.13	0.17
7	0.20	0.39	0.31	0.24	0.27
8	0.54	0.19	0.22	0.97	0.56
Mean	0.47	0.42	0.38	0.78	0.55
SD	0.24	0.32	0.23	0.71	0.26

showed no recession at all, indicating a significant probability (about 0.25) that the interval between recession rates could be greater than seven years. This type of reasoning was used to constrain the simulation model parameters.

Monte Carlo simulation was used to generate multiple realisations (simulations) of the recession process. Table 4.21 shows results of two simulations from the calibrated model. These were obtained by simulating the time period 1907 to 1991 and extracting results at the relevant years so that these could be compared directly with the measured values. As this is a sampling approach, different simulations give different results, so the two example simulations shown in Table 4.21 give different individual values. Nevertheless, the general characteristics of the model results are similar to the measured values in Table 4.20.

The statistical model was then used to make probabilistic predictions of:

- the time for the cliff to undergo recession of a certain distance, to assess when in the future a hypothetical fixed asset currently 29 m from the cliff top will be lost (Fig. 4.22a);
- the cliff position after 10 and 50 years (Fig. 4.22b and c). Cliff position is measured relative to a fixed baseline. The baseline is 50 m landward of the initial cliff position, so greater than 50 m recession appears as a negative value (i.e. it is landward of the baseline).

Since these are numerical simulation results the final distribution is not completely smooth.

Examination of Table 4.20 reveals that the average annual recession rate appears to have been progressively declining during the time period covered by the topographic maps, that is there is a trend in the data set which cannot be viewed as a stationary series. This is a common problem for coastal sites, where future conditions are not expected to resemble past conditions, due to significant natural or man-induced change in the physical processes. In such circumstances, historic records alone cannot be used for recession prediction. However, under changing conditions, a statistical prediction based on an assumption of stationary long-term average recession rate can be used to provide upper or lower bounds on future recession, past recession rates and rates of change (Hall *et al.*, 2002). Under changing conditions (especially where coast protection schemes have been recently implemented or are planned), deterministic methods combined with engineering judgement have tended to dominate in the past, although process simulation models are becoming available (e.g. Lee *et al.*, 2002).

Estimating probability through use of stability analysis

Stability analysis provides a quantitative measure of the stability of a slope or pre-existing landslide (e.g. Graham, 1984; Bromhead, 1986; Nash, 1987; Duncan, 1996). The slope form and materials are modelled theoretically, together with the loadings on the slope. A failure criterion is then introduced. The analysis indicates whether the failure criterion is reached and allows a comparison between the modelled conditions and those under which the

234

slope would just fail. The results are generally presented as a Factor of Safety (F):

$$F = \frac{\text{Shear strength}}{\text{Shear stress}}$$

The Factor of Safety is best viewed as the numerical answer to the question 'by what factor would the strength have to be reduced to bring the slope to failure by sliding along a particular potential slip surface' (Duncan, 1996).

The approach is deterministic, as single values of shear strength, pore water pressure and loading are the input parameters to the model. However, there can be significant uncertainties associated with the modelling, not least because of the variability of the soil and the possible presence of geotechnical 'anomalies', such as thin weak layers and discontinuities that may have a major influence on stability.

Stability analysis can be used to generate estimates of the probability of slope failure, that is the probability that the Factor of Safety will be less than 1, based on many simulations using variable parameter values.

In *probabilistic stability analysis*, the probability of failure is defined with respect to a *performance function* (Gx) that represents slope stability as a function of a combination of slope parameters, each with a continuous probability distribution (i.e. random variables; note that x denotes a particular combination of values).

The performance function of a slope is usually defined as follows:

Performance Function (Gx) = Factor of Safety -1

As a result, it is possible to use the performance function (Gx) to define the stability state:

$Gx > 0$ 'Safe' conditions

$Gx \leqslant 0$ 'Unsafe' or 'failure' conditions

One approach to obtaining an estimate of the probability of failure from stability analysis is to run repeated iterations of a stability model, using different combinations of input parameters selected by Monte Carlo sampling (see Section 'Continuous probability distributions and sampling' in this chapter; e.g. Priest and Brown, 1983). This approach can produce a probability distribution for the Factor of Safety (F) and, hence, the performance function (Gx).

An alternative approach is to estimate the probability of failure through the use of a *reliability index* (β) given as

$$\text{Reliability index } (\beta) = \frac{\text{Mean } Gx}{\text{Standard deviation } Gx}$$

235

The reliability index is a measure of the reliability of an engineering system (e.g. a man-made slope) that reflects both the mechanics of the problem and the uncertainty in the input variables. The index is defined in terms of the expected value and standard deviation of the performance function. It enables comparisons to be made about the reliability of different slopes without having to calculate absolute probability values. It can be regarded as providing a measure of the confidence in the ability of a slope to perform in a satisfactory manner, that is not to fail.

The reliability index can also be expressed as:

$$\text{Reliability index } (\beta) = \frac{(\text{Mean Factor of Safety} - 1)}{\text{Standard deviation Factor of Safety}}$$

The *mean Factor of Safety minus 1* measures the difference between the mean Factor of Safety and a Factor of Safety (F) of 1.0, that is it indicates the margin of stability (Wu et al., 1996). By dividing this value by the standard deviation of F, the margin of stability becomes relative to the uncertainty in F.

If the reliability index is assumed to be the number of standard deviations by which the expected value of a normally distributed performance function (Gx) exceeds zero, then the probability of failure can be approximated by the cumulative standard normal distribution (Φ) evaluated at $-\beta$ (see Table 4.22):

$$\text{Probability of Failure} \approx \Phi(-\beta)$$

(Note that the standard normal distribution is widely tabulated and available as a built-in function on many spreadsheet programs).

The choice of distribution can, however, have a significant impact on the estimates of the probability of failure. Baecher (1987) demonstrated that for a slope with a mean Factor of Safety of 2 and a standard deviation of 0.4, the following probabilities of failure could be derived:

- 2×10^{-3}, using a normal distribution;
- 2×10^{-4}, using a log-normal distribution;
- 2×10^{-4}, using a gamma distribution.

Probabilistic stability analysis is not for the faint-hearted and involves a level of statistical competence that is beyond the scope of this book. Interested readers are advised, therefore, to access some of the detailed papers that address this subject in theoretical and practical terms, before moving further down this route (e.g. Rosenblueth, 1975, 1981; Vanmarcke, 1977, 1980; Ang and Tang, 1984; Nguyen and Chowdhury, 1984; Harr, 1987; Chowdhury, 1988; Li, 1991, 1992; Mostyn and Li, 1993; Alén, 1996; Yu and Mostyn, 1996; Mostyn and Fell, 1997).

Table 4.22. Reliability index and probability of failure for various slopes (expanded, after Chowdhury and Flentje, 2003)

Mean factor of safety: F	Standard deviation of F	Reliability index: β	Probability of failure: Pf^*
1	0.01	0.0	0.5
1.1	0.01	10.0	0
1.2	0.01	20.0	0
1.3	0.01	30.0	0
1.4	0.01	40.0	0
1.5	0.01	50.0	0
1	0.05	0.0	0.5
1.1	0.05	2.0	0.022 75
1.2	0.05	4.0	3.17E-05
1.3	0.05	6.0	9.9E-10
1.4	0.05	8.0	6.66E-16
1.5	0.05	10.0	0
1	0.1	0.0	0.5
1.1	0.1	1.0	0.158 655
1.2	0.1	2.0	0.022 75
1.3	0.1	3.0	0.001 35
1.4	0.1	4.0	3.17E-05
1.5	0.1	5.0	2.87E-07
1	0.2	0.00	0.5
1.1	0.2	0.50	0.308 538
1.2	0.2	1.00	0.158 655
1.3	0.2	1.50	0.066 807
1.4	0.2	2.00	0.022 75
1.5	0.2	2.50	0.006 21
1	0.3	0.00	0.5
1.1	0.3	0.33	0.369 441
1.2	0.3	0.67	0.252 492
1.3	0.3	1.00	0.158 655
1.4	0.3	1.33	0.091 211
1.5	0.3	1.67	0.047 79
1	0.4	0.00	0.5
1.1	0.4	0.25	0.401 294
1.2	0.4	0.50	0.308 538
1.3	0.4	0.75	0.226 627
1.4	0.4	1.00	0.158 655
1.5	0.4	1.25	0.105 65

Note. *Using Excel, for example, simply insert the function NORMSDIST($-\beta$) to obtain the probability value.

Example 4.11

Rockfalls from a steep-sided railway cutting at Bethungra, New South Wales, Australia, have damaged trains and caused at least one derailment since 1946. Concerns about rail safety led the National Railway Corporation to consider stabilisation works to reduce the risks. Moon *et al.* (1996) describe how a probabilistic stability model was developed to estimate the frequency of falls (predominantly topples) of different magnitude that could be expected over the design life of any works (75 years), and to demonstrate that the stabilisation works would reduce the risk by a factor of 100.

The assessment involved the following stages:

1 *Establishment of a 'do nothing' rockfall magnitude/frequency distribution.* This involved repeated analysis of the stability of the cut face with no stabilisation works, and was undertaken using a toppling failure model with random combinations of input parameters (three sets of joint orientations, water in tension cracks; see Moon *et al.*, 1996) selected by Monte Carlo sampling. The procedure involved running a deterministic model over 70 000 times to build up a frequency distribution for different rockfall size classes to represent the 75-year design lifetime period (see Table 4.23, column 3).

2 *Estimation of the proportion of rockfall events that might impact on the railway.* Experience at Bethungra and the judgement of railway personnel from elsewhere in New South Wales, indicated that the risk associated with rockfall activity was influenced by a number of constraints:
   - the minimum size of event likely to damage or derail a train was judged to be $0.2\,m^3$ (the *critical size boulder*),
   - only around 1 in 10 rockfall events reached the tracks.

   The estimated probability of a critical size boulder reaching the track was established through expert judgement and is presented in Table 4.23, column 2 for each different rockfall size class.

3 *Calculation of the number of rockfall events, by size class, that could be expected to impact on the railway over the 75 year design lifetime.* This *hazard factor* was established as the product of the modelled frequency of events (Table 4.23, column 3) and the proportion of critical size boulders that are expected to reach the track (Table 4.23, column 2):

   Hazard Factor (Rockfall class *r*)

   = Frequency of Events (Rockfall class *r*)

   × Proportion of Critical Boulders on Track (Rockfall class *r*)

   The overall hazard factor was defined as the sum of the hazard factors for each rockfall size class.

   The hazard factors are presented in Table 4.23, column 4.

4 *Calibration of the probabilistic model results by comparison with the historical pattern of events.* This involved the establishment of a magnitude/frequency distribution for *historical* rockfall events, based on railway file

238

Table 4.23. Example 4.11: Rockfall hazard factors, Bethungra, New South Wales, Australia (after Moon et al., 1996)

Rockfall size class: m³	Probability of critical boulder on track (P_{crit})	Modelled rockfalls (>70 000 iterations)	Hazard factor (modelled rockfalls)	Extrapolated historical rockfalls (75 years design lifetime)	Hazard factor (historical rockfalls)	Adjusted modelled rockfalls (calibration factor = 2.5)	Adjusted hazard factor (modelled rockfalls)
330–1000	1	0	0	0.3	0.3	0	0
100–300	1	0	0	1	1	0	0
30–100	1	4	4	3.2	3.2	2	1.6
10–30	0.9	35	31.5	11.3	10.2	14	12.6
3–10	0.6	130	78	38	22.8	52	31.2
1–3	0.2	381	76.2	135	27	152	30.5
0.3–1	0.04	799	32	450	18	320	12.8
0.1–0.3	0.005	211	1.1	1500	7.5	0	0.4
Total			222.8		90		89.1

Note. Hazard Factor = Rockfalls × P_{crit}.

239

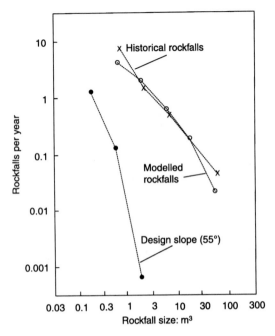

Fig. 4.25. Example 4.11: Bethungra railway cutting: rockfall magnitude/frequency distributions

records, discussions with railway inspectors and observation of recent rockfall events (Fig. 4.25).

The predicted number of rockfall events over a 75 year design life was established by simple extrapolation of the historical magnitude/frequency distribution (see Table 4.23, column 5).

The hazard factors derived from the historical magnitude/frequency distribution are presented in Table 4.23, column 6.

It was found that the toppling failure model overestimated the event frequency by a factor of 2.5, when compared with the historical rockfall records. This factor was adopted as a Calibration Factor and used to adjust the modelled event frequencies (Table 4.23, column 7).

5 *Establishment of a 'do something' rockfall magnitude/frequency distribution.* The toppling failure model was used to assess the stability of the cut face with the stabilisation works in place. Once again, this involved repeated iterations with randomly sampled input parameters. The predicted rockfall yield for a design slope of 55° was shown to be considerably lower than the 'do nothing' yield (the untreated cut slopes stood at 65–85°) and delivered the desired 100 times reduction in the total hazard factor and, hence, risk (see Fig. 4.25).

240

Example 4.12

The coastal cliffs at Whitby, UK, are developed in a highly variable sequence of overconsolidated sandy clay tills. Part of the cliffline, West Cliff, has been protected by a seawall and stabilised since the late 1920s, with the most recent works undertaken between 1988 and 1990 (see Clark and Guest, 1991). Concerns have been raised subsequently about the potential for a deep-seated failure of the stabilised slope that would threaten the integrity of the seawall. For example, a major breach in the seawall occurred in about 1962 as a result of deep-seated landsliding.

An assessment of the probability of deep-seated landsliding was undertaken, based on the use of stability analysis and the calculation of a reliability index for the slopes. This involved the following stages:

1 *Stability analysis.* A slope stability model was developed using the Bishop Simplified method for circular failures (Bishop, 1955). The range of input parameters (unit weight, effective cohesion and angle of friction) were derived from previous studies (e.g. Clark and Guest, 1991) and more recent site investigation results (High Point Rendel, 2003).

   A number of iterations (50) of the stability model were carried out, using pre-selected combinations of the input parameters. The model runs included:
   - a 'best estimate', which was assumed to correspond with the mean slope conditions,
   - a lower bound estimate, based on an assumed combination of input parameters that were believed to correspond to the 'worst case' scenario,
   - an upper-bound estimate, based on an assumed combination of input parameters that were believed to correspond to the 'best case' scenario.
   The deep-seated slip circle with the lowest factor of safety was considered to represent the stability of the slope for each model run.

   The results of the stability modelling indicated a mean Factor of Safety (F) of 1.21 (i.e. the mean value of all the lowest factors of safety), with a standard deviation of F of 0.08.

2 *Calculation of a reliability index (β).* The reliability index was calculated from

$$\text{Reliability index } (\beta) = \frac{(\text{Mean Factor of Safety} - 1)}{\text{Standard deviation Factor of Safety}}$$

$$= 0.21/0.08 = 2.62$$

3 *Estimation of the probability of failure.* The probability of failure was approximated by the cumulative standard normal distribution (Φ) evaluated at $-\beta$. This indicated a probability of 0.004 332 (i.e. 4.3×10^{-3} or 1 in 230.8).

241

Estimating probability: precision or pragmatism?

This chapter has presented the wide range of approaches that can be used to estimate the probability of landsliding, all of which rely on a combination of the historical record of landsliding and an understanding of slope conditions (i.e. geomorphology, geology or geotechnics). None of these methods can, however, be *guaranteed* to provide a reliable estimate. For example:

- Incomplete or inaccurate data sets can mean that the historical frequency of recorded landsliding is an imprecise guide to the future probability of failure. The record may be too short to contain evidence of high-magnitude low-frequency events or yield a frequency which does not conform to the long-term probability.
- Empirical relationships between landslide activity and the occurrence of triggering events assume that the critical landslide threshold conditions have remained constant over the period of the historical record (i.e. the data series is *stationary*). However, the fundamental nature of the relationship between the occurrence of triggering events and landsliding may have changed over the period in question because of land use change or progressive weathering of the slope; that is the data series is *non-stationary*. In addition, future environmental conditions and slope responses could be significantly different to those in the past because of the combined effects of human activity and predicted climate change.
- The use of expert judgement can lead to estimates that are heavily biased by the experience and personality of the experts. This may lead to a wide range of estimates, even when using exactly the same basic information.
- The estimates derived from probabilistic stability analysis and the use of reliability indices can be very sensitive to the amount of geotechnical information available about the slope conditions.

It is important to appreciate, therefore, that estimates of the probability of landsliding can only be *estimates*. The commonly expressed desire for increasing precision in the estimation process needs to be tempered by a degree of pragmatism that reflects the reality of the situation and the limitations of available information. Estimates will generally need to be *fit for purpose* (i.e. supporting the development of a risk assessment and guiding decision-making; see Section 'Risk assessment as a decision-making tool' in Chapter 1) rather than the product of a lengthy academic research programme.

Perhaps of greater concern to those seeking precision, when estimating the probability of landsliding, is the fact that slope behaviour does not completely conform to a probabilistic model. Indeed, natural systems can be classified in terms of their degree of complexity and randomness of *behaviour* (Weinberg, 1975):

- highly organised simple systems that can be modelled and analysed using deterministic mathematical functions (e.g. the prediction of tides),
- unorganised complex systems that exhibit a high degree of random behaviour and, hence, can be modelled by stochastic methods (e.g. the probability of waves of particular heights or flood discharges),
- organised complex systems, where the behaviour involves both deterministic and random components.

Slopes can be viewed as organised complex systems (e.g. Lee, 2003). The response of a slope to triggering events of a particular size is controlled by the antecedent conditions. Landsliding may not conform particularly well to either the deterministic or the random model. Landslide events are not independent but can be influenced by the size and location of previous events. In other words, landsliding is a process with a *memory* (insofar as the current and future behaviour is influenced by the effects of past events on the system). Indeed, the chaotic nature of the short-term preparatory and triggering factors (e.g. Essex *et al.*, 1987), and of geomorphological systems in general (e.g. Hallet, 1987), suggests that there may be a limit to the predictability of the landsliding process.

5

Estimating the consequences

Introduction

What would happen if a landslide occurred? This question is crucial to the determination of risk. However, to answer it requires that some form of *consequence assessment* be undertaken to estimate the likely damages and losses that could be expected. These losses or detriment can be extremely diverse in nature and involve people, property, structures, infrastructure, economic activity, welfare, amenity, ecology and environment.

Hewitt (1997) suggested four main groups of detriment:

- loss of life, injury and impairment of persons;
- destruction of property, resources and heritage;
- disruption of activities, and denial of supplies and services;
- cultural, spiritual and ethical violations.

Only a proportion of landslides cause detriment (see Fig. 2.1). Of those that do, most cause temporary, short-lived losses or detriment but some can have long-term affects. Usually the losses are confined to a limited and readily definable area and are, therefore, relatively easy to identify, but sometimes the adverse consequences spread across neighbouring areas like ripples on a pond becoming ever more difficult to distinguish with distance and the passage of time. In some instances the losses are easily quantified, such as repairing a building or replacing a cow, but often they include adverse consequences that are extremely difficult to value in monetary terms, such as the psychological effects experienced by survivors of a destructive landslide (see Lacey, 1972).

From the perspective of the scale and complexity of the adverse consequences that they generate, it is possible to distinguish a range of landslide events, including:

- *Simple events*, which cause detriment as a direct consequence of a single or repeated movement (e.g. a rotational failure or mudslide).
- *Compound events*, which cause detriment when a triggering event produces a cascade of different types of landslide (falls, flows and slides) that, if relief conditions and material availability allow, can achieve great size, speed and violence (e.g. the 1970 Huascaràn rock/ice avalanche which destroyed the town of Yungay, Peru, killing between 15 000 and 20 000 people; Plafker and Ericksen, 1978).
- *Multiple events*, involving widespread landslide activity and considerable detriment. This type of event is often associated with earthquakes, intense rainstorms, volcanic eruptions and forest clearance by humans. For example, when Hurricane Mitch came virtually to a halt over Central America for six days in October 1998 and dropped 2 m of rain over mountainous terrain, more than 9000 landslides were triggered in Guatemala alone (Bucknam, 2001); the Northridge (California) earthquake of 17 January 1994 triggered more than 11 000 failures (Harp and Jibson, 1995, 1996) and over 4000 landslides were generated in Umbria (Italy) by snowmelt following a sudden change in temperature on 1 January 1997 (Cardinali *et al.*, 2002).
- *Complex events*, in which a significant proportion of the overall damage is the product of the generation of *secondary geohazards*, such as floods, tsunamis or volcanic eruptions (e.g. Section 'Secondary hazards' in Chapter 2). For example, collapse of the northern flank of Mount St Helens, Washington State, USA, on 18 May 1980, was triggered by a large earthquake ($M = 5.1$). The huge landslide exposed the gas-rich magma that had been intruded into the volcano during the preceding months. Decompression caused the magma to explode and disintegrate into a cloud of gas and volcanic debris that raced downslope at 300–500 km/hour. This flow, together with the blast from the air pushed ahead of it, laid waste some 600 km^2 of land within 2 minutes (Lipman and Mullineaux, 1981).

Landslide magnitude does not, in itself, determine the potential for loss; some major failures do not produce adverse consequences. The nature of the adverse consequences is a reflection of the huge range of *hazard potential* (due to the character, magnitude, timing and location of landslide events) interacting with a distribution of humans and objects, activities and environments valued by humans that is spatially variable and displays differing degrees of susceptibility to impact by landsliding. This can be termed *exposure potential* (see Section 'Hazard and vulnerability' in Chapter 1) and includes both static items as well as the potential for loss through *accidents of timing*, or *temporal vulnerability* and relates to the assets *being in the wrong place at*

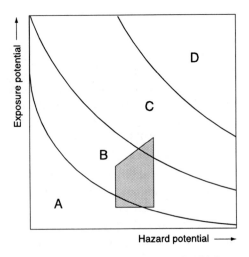

Fig. 5.1. The hypothetical division of contemporary landslides into four groups with reference to hazard significance, based on the relationship between hazard potential and exposure potential (see text for discussion). The shaded box represents the envelope of scenario values that might be produced as part of a risk assessment of a specific site

the wrong time. If these two measures are plotted as the axes of a diagram (Fig. 5.1) then four categories of landslide can be identified:

- *Category A* landslides which, due to small size or location, result in little or no impact from a human perspective and present low risk or even no risk.
- *Category B* landslides can be of any size and result in significant impacts and detriment.
- *Category C* landslides result in major impacts, some of which are termed disasters by the media.
- *Category D* landslides result in severe impacts, sometimes resulting in zones of total destruction; these are the events that tend to be referred to as major disasters, megadisasters, calamities or catastrophes by the media, as well as some sections of the academic literature.

Although *catastrophe potential* tends to be associated with high magnitude–low frequency (rare) events, it can also be the product of smaller landslides if they affect particularly vulnerable locations or sites, or through *accidents of timing.*

The approach to estimating the consequences has to be somewhat different in each of these cases. However, the basic principle of consequence assessment is the development of *consequence models* or *scenarios* which attempt to define and quantify the possible adverse outcomes. In each case, the key questions that will need to be considered include:

246

- What is there to be impacted?
- How will it be impacted?
- What is the likelihood that the landslide event will generate significant losses through the development of *follow-on hazards*, such as accidents and fires?
- How can the adverse consequences be valued?

Many consequence assessments will involve a series of scenarios covering a range of landslide magnitudes (*hazard potentials*) and possible impacts (*exposure potentials*). The shaded area in Fig. 5.1 illustrates the range of possible situations with respect to a hypothetical landslide risk assessment and shows that although most of the scenarios are likely to point to variations within Category B, the *best case scenario* will indicate an outcome within Category A, while the *worst case* will show a much higher level of detriment associated with Category C.

Using the historical record

The development of consequence models and scenarios must lean heavily on the historical record of landslide impacts in order to find a range of suitable *analogues*. Finding out how landslides occurred in similar situations elsewhere, how they impacted on the societies concerned and how the costs were calculated, are all important ingredients in the development of realistic consequence assessments. Clearly, the use of local examples is preferable. Should examples be from abroad, then cultural and societal differences have to be taken into account when using such analogues. It is also necessary to note that the character and behaviour of landsliding varies across the globe due to differences in tectonic setting, geological make-up, geomorphological history and prevailing climatic regime, which means that the applicability of landslide events must also be ascertained.

The historical record provides a vital source of data regarding the adverse consequences associated with past events in general, which provide guidance as to what *could* happen and the magnitude–frequency characteristics of losses (see Section 'Acquiring information: landslide investigation' in Chapter 2). It also contains descriptions and studies of what *actually* occurred in specific events, together with estimations of loss, which can be used as analogues for predicting impacts in the future. However, it has to be recognised that this record is of dubious quality in a number of ways and for a variety of reasons.

It is generally recognised that global data on geohazard impacts (disasters) is both incomplete and unreliable, especially prior to the mid-20th century when many significant events failed to be recorded because of remoteness

247

or for political reasons. The situation has improved considerably in recent years, for a number of reasons. First the United Nations sponsored International Decade for Natural Disaster Reduction (IDNDR) stimulated better systematic record keeping, including the establishment of a number of institutions dedicated to gathering information on contemporary disasters, most especially The Centre for Research on the Epidemiology of Disasters (CRED) at the University of Louvain, Belgium (Sapir and Misson, 1992). Second, the work of the International Federation of Red Cross and Red Crescent Societies (IFRCRCS) has contributed greatly to improving the level of information, not least through the production of annual reports on disasters. Third, much data collection and analysis has been undertaken by the insurance industry and most especially some of the major reinsurance companies such as Munich Re and Swiss Re. However, because landsliding is a phenomenon that becomes most acute in mountainous terrains subjected to seismic shaking and intense rainfall events, there seems little doubt that the actual significance of landsliding as a hazard is understated in the record, especially prior to 50 years ago.

The fact that widespread landsliding is often caused by other, more conspicuous, geohazard events such as tropical revolving storms, volcanic eruptions or earthquakes (*landslide generating events*) means that the losses caused by landsliding are often attributed to these other hazards (Jones, 1995a). This is especially the case with reference to seismic activity, where the huge losses of life and destruction associated with the Shansi (1556), Kansu (1920), Armenian (1988), Tajikistan (1988) and Iranian (1990) earthquakes were, to a large extent, the result of landsliding (see US Geological Survey, 1989). This further reduces the perceived significance of landsliding as the cause of detriment, despite the work that has been undertaken to show just how much landsliding can be generated by seismic activity (Keefer, 1984; see Chapter 2, Example 2.7). For example, the Assam earthquake of 1950 is thought to have caused the displacement of a total of $50\,000\,Mm^3$ of material over an area of $15\,000\,km^2$ (Kingdon-Ward, 1952, 1955), while the Guatemala City earthquake of 1976 ($M = 7.5$) caused $10\,000$ landslides over $15\,000\,m^3$ and 11 over $100\,000\,m^3$ (Harp *et al.*, 1981).

Similarly, the spatial (geographical) record is extremely variable and generally of medium to poor quality. While records kept in Developing Countries may be adequate, especially in the case of landslides that caused major impacts or were conspicuous/dramatic, no such records exist for extensive tracts of the Developing World, although attempts have been made to address this problem as part of IDNDR. Dependence on the media as a source of information has tended to introduce significant bias in terms of emphasis on events that were of interest to readers or viewers (i.e.

close to home), conspicuous, rapid on-set, had *human interest* and involved human deaths. These influences were clearly identified in a study of American TV coverage of natural disasters (Adams, 1986), which revealed a western prioritisation dominating over severity so that, in terms of newsworthiness, the death of one western European equalled three eastern Europeans, equalled nine Latin Americans, equalled 11 Middle Easterners, equalled 12 Asians. It is clear that only the undertaking of detailed, systematic studies of landsliding will reveal both the geographical patterns of landslide hazard and the relative significance of past events.

There undoubtedly exist additional accounts of landslides for most areas held in files, archives, books, reports, newspapers and journals, which can be searched to improve the record, as was attempted in the case of the UK Government sponsored 'Review of Research on Landsliding in Great Britain' (see Jones and Lee, 1994). The results can be disappointing, however, even if augmented by information held in confidential company files or by oral histories, because of chance factors that determined whether landslides were recorded or not, the nature of the information that was actually recorded and how the information was recorded (see Section 'Acquiring information: landslide investigation' in Chapter 2 and Example 4.2). Research has shown that the record tends to be biased towards larger, more conspicuous events (Brunsden *et al.*, 1995) and that so-called 'trivial events' tend to be largely unreported and unrecorded.

Thus the historical record in the public domain provides a biased and incomplete picture of how landslides cause detriment. The significance of landsliding undoubtedly remains underestimated (Jones, 1995a) and will continue to be so despite the publication of regional and global reviews (e.g. Brabb and Harrod, 1989). Nevertheless, there now exist numerous well-documented case histories of the impact of major geohazard events, such as catastrophic landslides, which reveal that an enormous variety of adverse consequences or detriment can be produced. People are killed and injured, buildings and property are damaged or destroyed, services are cut-off, 'normal life' is disrupted and the survivors may experience hunger, exposure, illness and despair. In every major event, innumerable individual accident sequences (see Section 'Risk assessment' in Chapter 1; Fig. 1.5) result in a cascade of detriment that may last for years and which can be almost impossible to fully unravel after an event, let alone predict in advance of some unspecified time in the future. Thus consequence assessments have, through necessity, to be increasingly pragmatic and broad-brush with the increasing extent of the envisaged impact. For small sites and well defined events, however, more specific detail can be modelled.

Finally, it has to be noted that the range of reported impacts does not cover all possibilities for the future, not simply because the record is incomplete but because changes in society and technology will undoubtedly result in new forms and combinations of impacts. Thus the historical record provides an important guide but not a straight jacket; consequence assessments need to consider new possibilities.

A framework for adverse consequences

As was shown in the introduction to this chapter, the adverse consequences of landsliding can be extremely varied in terms of nature and timing. In order to develop consequence models and scenarios, this complexity needs to be distilled into a simple and clearly structured framework in order to facilitate prediction. Perhaps the most fundamental issue that needs to be considered is the recognition and classification of the range of adverse consequences. The simplest division is into *direct effects* and *indirect effects*. These consist of both *gains* and *losses* and can be further divided into *tangible* and *intangible* categories, depending on whether or not it is possible to assign generally accepted monetary values to the losses/gains:

- Direct impact on people, that is loss of life or injury.
- Direct and indirect economic losses.
- Intangible losses.

Direct effects are the first-order consequences that are intimately associated with an event, or arise as a direct consequence of it (e.g. destruction), while *indirect effects* emerge later, such as mental illness, longer-term economic problems or relocation costs (Smith, 2001). These categories are discussed more fully in the following sections.

The distinction between 'direct' and 'indirect' can be somewhat arbitrary and a matter of subjective preference. The main focus of uncertainty concerns whether the impact of secondary hazards (e.g. fires and floods) and follow-on hazards (e.g. looting) count as direct or indirect losses.

There are strong grounds for including the physical impacts of secondary hazards, including fires and directly attributable accidents, within direct economic losses because they arise primarily as a consequence of ground movement. It is also often difficult to distinguish where one set of destructive forces ceases and another begins. For example, if a landslide descends into a lake or dams a valley, then it is illogical for the costs of inundation and downstream flooding to be excluded from the losses attributable to the landslide (e.g. the Vaiont dam disaster of 1963). The same logic is applicable in the case of tsunamis even though the costs may be borne by other societies. Similarly, if a landslide directly causes an accident, then the costs of the accident should be attributed to the landslide. Thus care should be taken not to confuse *secondary impacts* and *indirect effects* with indirect losses.

An alternative approach advocated by Hewitt (1997) is to subdivide losses into *primary, secondary* and *tertiary damages*, with secondary damages largely the product of secondary hazards but including fire, and tertiary damages arising from the impairment of general functions such as disease, delayed economic effects and forced out-migration. This is a useful framework for use in hindsight studies concerned with assessing the effects of past major impact events, where the passage of time allows the identification of longer-term political, economic, social and environmental consequences. However, it is of limited value for determining adverse consequences as part of a risk assessment, where a simpler and more pragmatic approach is required.

Timing is crucial and the actual losses incurred in a particular event can be dependent on a unique combination of circumstances. For example, it is estimated that 66 people were killed in 1988 by a landslide along the Trabzon–Erzerum highway at Çatak, Turkey (Jones et al., 1989a,b). A small landslide blocked the road at midnight on the wet night of 22/23 June and led to the delayed travellers congregating in a road-side coffee house. A grader was despatched to clear the debris, but operations were delayed because of the difficulty of night-time working in heavy rainfall. Eight hours later a much larger slide occurred from the same slope, burying the coffee house and killing all inside. Similar sized landslide events elsewhere along the road would not have caused so many fatalities.

Events are the result of particular and unique sets of physical and social conditions. At Çatak, a different set of conditions could have generated a different landslide damage scenario. The inherent randomness in the main factors involved in such disasters (e.g. the build-up of people in a particular location, night-time etc.) dictates that future landslides cannot be expected to cause the same levels of adverse consequences, even if they are of the same magnitude.

Although the damage information contained in the historic record provides an important guide as to the potential adverse consequences of future landslides, despite inevitable incompleteness and bias (see Section 'Using the historical record' in this chapter), it cannot properly indicate the true range of possibilities. The past, therefore, is not a complete guide to the future because circumstances change with respect to both the 'physical' and 'human' or 'socio-economic' environments. New hazards are identified, new vulnerabilities develop, new accident sequences emerge.

It follows that there is a need to develop a structured approach to assessing adverse consequences that is supported by the historical record, but based on an awareness of the key factors that determine landslide damage. These factors are:

- the nature of likely ground movements (*ground behaviour*),
- the nature and value of the *assets or elements at risk*,

- their *vulnerability* to the predicted ground movements,
- and the extent to which they are likely to be *exposed* to the hazard of the predicted ground movements.

These factors are considered in greater detail in Sections 'Ground behaviour' to 'Vulnerability' later in this chapter.

Loss of life and injury

One of the main priorities of any risk assessment is the identification of any possibility of human loss of life or physical injury. This is not merely a response to an increasingly litigious society but reflects a combination of other factors: the human-centred nature of risk assessment, the fact that humans generally tend to value human life above all other things; and because casualty figures often convey a sense of loss far more effectively than does any other form of impact statistic. The communicability of human loss figures, together with their availability relatively quickly after a hazard impact, have made them a favourite with the media. As a consequence, death tolls have become 'the currency of disaster', often determining the threshold at which serious impacts should be called 'disaster' (i.e. 10 deaths, 24, 50, 100 etc.) and even employed by some to indicate the relative and absolute significance of disasters. This focus does, however, ensure that records of numbers killed are likely to be the best kept statistics of landslide impacts.

The historical record reveals that *multiple events* can occasionally cause 100 000 deaths, with the figure possibly exceeding 500 000 under quite exceptional circumstances (e.g. Shansi earthquake, 1556). However, death tolls attributable to the landsliding caused by such events are more usually less than 15 000, with relatively frequent records in the range 100–1000. The maximum death toll in a single landslide event is around 20 000, in the cases of the catastrophic failures at Huascaràn (1970; Plafker and Ericksen, 1978) and Nevado del Ruiz (1985; Herd *et al.*, 1986). More usually the numbers killed by individual landslides, or the floods created by the failure of landslide dams, are less than 1000 and the frequency appears to increase with diminishing casualty figures.

Because of the poor quality of the historical record there remains uncertainty as to the actual numbers of people killed by landsliding per year, the magnitude–frequency distribution of landslide generated death-tolls and the regional patterns of casualty figures. Early disaster statistics produced by the Red Cross indicated 3006 deaths due to landslide disasters over the period 1900–76 (Crozier, 1986), or a trivial 40 deaths per annum out of a combined total of 60 000 deaths a year due to natural disasters. These data are now seen to be gross underestimations. Hewitt (1997) claims that there were 54 landslide disasters with death-tolls of over 100 during the period 1963–92, which works out at 1.8 disasters per annum or at least 200 people a year. More recent statistics produced by the Red Cross/Red Crescent reveal 14 landslide disasters per year for the period 1988–97, with an annual average death-toll of 790 (IFRCRCS, 1999). Information compiled by the Office of US Foreign Disaster Assistance (OFDA) and analysed by Parfitt (1992) indicates that landslides which kill more than ten people have occurred about three times per year on average. If landslide generating events are included, then there is an average of one event per year that causes more than 100 deaths.

Among the best-known high fatality events are:

- The Vaiont disaster of 9 October 1963 in Italy when a 250 Mm³ rockslide entered the reservoir, sending a wave around 100 m high over the crest of the concrete dam. The flood wave destroyed five downstream villages and killed around 1900 people (Kiersch, 1964; Hendron and Patton, 1985).
- The Aberfan disaster of October 1966 in south Wales. Parts of a 67 m-high colliery spoil tip collapsed and flowed downslope into the village below. Twenty houses and a primary school were overwhelmed and 144 persons died, including 107 out of 250 pupils in the school (Bishop *et al.*, 1969; Miller, 1974). The disaster led to stricter control over the disposal of mine waste, under the Mines and Quarries (Tips) Acts of 1969 and 1971.
- Over 250 people were killed by landslides in Hong Kong during a period of heavy rainfall in June, 1972. Most of the deaths were as a result of two major landslides (Government of Hong Kong, 1972a,b). At Sau Mau Ping Resettlement Estate, Kowloon, a 40 m-high embankment collapsed and the resulting flowslide left 71 persons dead and 60 injured. A 12-storey building collapsed under the impact of an extremely rapid flowslide from the slopes above Po Shan Road, in the Hong Kong mid-levels, killing 67 persons and injuring a further 20. The public outcry from these events, and the 18 deaths caused by flowslides on the Sau Mau Ping Resettlement Estate in 1976, led to the establishment of a Geotechnical Control Office (now named the Geotechnical Engineering Office, GEO) in 1977. This is a central policing body to regulate planning, investigation, design, construction, monitoring and maintenance of man-made slopes (Malone, 1998; Chan, 2000).
- In October 1985 the tropical storm Isabel dropped over 560 mm of rainfall within a 24-hour period on the island of Puerto Rico. The rainfall triggered the Mameyes rock slide which destroyed 120 homes and killed an estimated 129 people, though only 39 bodies were recovered (Jibson, 1992). This death toll is the largest from a single landslide in North America.

- On 30 July 1997 a landslide occurred in the Thredbo ski village, Australia. The landslide caused a section of the Alpine Way, the major road in the area, to collapse. The Carinya Lodge, directly under the Alpine Way, moved down the slope about 100 m before hitting the Bimbadeen Lodge. Eighteen people were killed in the landslide, with only one survivor emerging from the rubble (Hand, 2000).

In addition to these high-fatality events, there are numerous incidents which cause fewer deaths. The following examples can only begin to show something of the diversity of the unfortunate circumstances:

- Five soldiers drowned and three workers were injured in 1999, when a boat dredging for gold in the River Nechi, Colombia, was hit by a landslide and capsized.
- In July 1989 a large rockfall ($1400 \, m^3$) collapsed onto a rock shed protecting traffic on the National Highway Route 35 in Echizen-cho, Japan. The crest of the shed was displaced, crushing a micro-bus that had been passing at that moment; all 15 passengers were killed (Yoshimatsu, 1999).
- In April 1988 a 17 m-high railway embankment at Coledale, a suburb of Wollongong, Australia, failed during heavy and prolonged rainfall. The mudslide enveloped a house situated about 20 m from the base of the embankment, trapping four occupants. Two were rescued from the debris, but a mother and her two-year-old son were killed (Davies and Christie, 1996).
- A rockfall from a cutting on the British Columbia Highway 99 in 1982 fell on a vehicle that was stopped in traffic, killing a woman and disabling her father (Cory and Sopinka, 1989; Bunce *et al.*, 1997).
- A school party from Surrey were studying the geology of Lulworth Cove on the English coast in 1977, when they were buried beneath a sudden rock slide (Jones and Lee, 1994). One of the pupils was killed and two more pupils seriously injured, one of whom died later in hospital.

Most events like these will only be recorded locally and, therefore, will only be counted if researched either locally or at a national level. When this is done, the numbers of landslide-induced deaths rises dramatically. For example, Ikeya (1976) found that in Japan over the five years 1967–72 no fewer than 662 people had been killed by debris flows and a further 682 by landslides, which gives a death rate of 269 per year. Similarly, research in Peru (Taype, 1979) identified 34 975 deaths in 50 years due to 168 slides, 37 mudflows and 3734 *huaycos* (debris torrents), that is 700 deaths per year of which an unspecified proportion should be attributed to flooding. Taken together, these data indicate that death-producing landslide events probably number in the low hundreds per annum and result in over 2000 deaths per year. These figures will rise dramatically if the effects of individual impacts within landslide generating events are included, but these data have yet to be properly disaggregated. Irrespective of the actual figures, the conclusion has to be that deaths from landsliding are a distinct possibility and need to be considered carefully in the majority of risk assessments.

The uncertainty regarding death-tolls is compounded further in the case of physical injuries. Red Cross/Red Crescent statistics indicate that for the period 1973–97, 267 people a year were injured in landslide disasters (IFRCRCS, 1999). However, the actual number injured by landslides is undoubtedly several times this number. It also has to be noted that the term 'injury' covers a wide spectrum of differing levels of adverse consequence in terms of scale, costs involved in recovery and the extent to which full recovery can be achieved. The simple division of injuries into *light*, *moderate*, *severe* and *critical* is a useful first approximation. Alternatively, the *triage* injury categories can be used:

- dead or unsaveable;
- life threatening injuries requiring immediate medical attention and hospitalisation;
- injury requiring hospital treatment;
- light injury not requiring hospitalisation.

A scheme based on degree of incapacitation might also be adopted:

- injury causing major incapacitation;
- injury causing minor incapacitation;
- injury not affecting performance.

It is unfortunate, therefore, that despite much research on measures of injury severity (see Olser, 1993), there is still no internationally agreed standard injury classification and no agreed basis for data collection. Thus existing statistics must be viewed as of dubious quality, especially as more trivial injuries tend to be under-recorded. Also some schemes fail properly to address the differing types of physical injury, or the extent to which pre-impact normal life can be resumed due to individuals being crippled, blinded, deafened or experiencing other forms of long-term physical disability.

It follows from the above that casualty estimation is extremely problematic. Although there has been much work in the context of earthquake impacts, there is relatively little information available for other geohazards, such as landslides. Estimation must, therefore, be based on the development of scenarios, using, for example, the following categories of impact on humans (Alexander, 2002):

255

- *no injury;*
- *slight injury.* Minor medical attention will solve the problem and transport to hospital is not required; medical assistance is not urgently required;
- *serious injury that does not require immediate priority treatment.* The patient will be taken to hospital but will not be among the first to be treated. Injuries are not life threatening and will not lead to a worsening of the patient's condition if he or she must wait to receive treatment;
- *serious injury that requires priority treatment* in order to produce some significant improvement in the patient's long-term prognosis, or simply to avoid deterioration in the patient's condition;
- *instantaneous death;*
- *subsequent death* due to stress-induced heart attacks, catastrophic deterioration of pre-existing medical conditions or complications arising from initially non-fatal injuries.

As for the costs that can be attributed to injuries, Alexander (2002) quotes figures for the mid-1990s from the USA based on *earnings foregone* calculations, which indicate that if death results in an average value of $2.2 million, then equivalent values for 'moderate' and 'slight' injuries are $5000 and $200, respectively.

While it is clear that landslides can have significant direct impacts on humans, estimating what these impacts are likely to be is problematic. Unlike physical structures and infrastructure which are *fixed* or *static assets*, humans are *mobile assets* whose presence and concentration varies in time and over space. Thus it is necessary to estimate the likelihood of differing numbers of people being at specified locations with reference to temporal factors such as times of the day, weather conditions or seasons. The notion of *occupancy ratios* is often used for major hazards such as earthquakes, and is the anticipated number of people likely to be in a building or at a location at a particular time, as a proportion of the maximum number that could be expected. Thus *occupancy levels or rates* tend to be highest for houses at night, cinemas in the evenings but with variations over the week, schools or offices during the day but only during the working week, recreational beaches during the middle of the day in hot weather and especially during the holiday season etc. The same principles need to be applied to modes of transport, where there are similar fluctuations in activity and numbers of people.

The physical juxtaposition of people and a landslide does not necessarily imply that there will be human losses. The behaviour of the landslide is important in terms of its unexpectedness, suddenness, speed, ground disturbance characteristics and violence. In some circumstances the landslide may be the main cause of human losses, but often it is the result of secondary events such as the collapse of buildings, transport accidents, fires and local flooding. How people are likely to be affected is also important in determining deaths and injuries. Are people going to be hit by missiles, engulfed by torrents of debris, buried alive or trapped within collapsed or partially collapsed structures? Will they be crushed, battered, suffocated, burned, drowned or suffer hypothermia, and will the effects of shock and resulting depression also take their toll? Death-tolls should properly include post-event deaths, especially as mortality rates tend to rise after a disaster because of suicides and depression among the elderly. Indeed there is evidence that in major disasters, the proportion of deaths within different age-groups tends to increase with age. Thus estimating deaths and injuries is a complex problem which requires scenario development based on suitable analogues. For larger-scale events, simulation models are of considerable value and there is much to be gained from drawing on the approaches that have been developed in the study of earthquakes (e.g. Coburn and Spence, 1992; Coburn, 1994; Noji, 1994).

Although it is an integral part of risk assessment, the valuation of deaths and injuries is very contentious. In the case of injuries it is possible to estimate the costs of rescue operations, of emergency treatment and of hospitalisation for differing categories of injury (i.e. light, moderate, severe etc.). This will produce average costs per type of injury. Rather more difficult is the problem of identifying long-term costs in terms of loss of function (i.e. blindness) and loss of earnings. Research can produce figures for these, based on certain assumptions, but it is a time-consuming and expensive exercise, so for most landslide risk assessments these items will be considered to be *intangibles* (see Section 'Intangible losses' in this chapter). The same is generally true of *death*, although there have been efforts to attempt to place a value on human life in an abstract sense. The basis of this work is well described in Mooney (1977), Jones-Lee (1989) and Marin (1992) and focuses on establishing *the value of a statistical life* (VSOL), a notion which a significant proportion of the population find morally repugnant and ethically unacceptable, largely because of the confusion between VSOL and an actual person.

Pearce *et al.* (1995) describe the approach as follows:

> Attributing a monetary value to a 'statistical life' is controversial and raises a number of difficult theoretical and ethical issues. It is important to understand that what is valued is a change in the risk of death, not human life itself. In other words, the issue is how a person's welfare is affected by an increased mortality risk, not what his or her life is worth. If 100 000 people were exposed to an annual mortality risk of 1:100 000, there will, statistically, be one death incidence per year.

Removing the risk would thus save one 'statistical life'. It is this statistical life that has an economic value. It would make no sense to ask an individual how much he or she is willing to pay to avoid certain death. Nor is that the context of social decision making. But it can make sense to ask what individuals are willing to pay to reduce the risk of death or what they are willing to accept to tolerate an increased risk of death.

The reality is that safety is not 'beyond price'. If it were, most of the world's wealth would be spent trying to save lives by reducing accidents and preventing disease. Risks are taken every day, both by individuals and by governments in choosing their social and economic expenditures, some of which are specifically directed at protecting and extending human life. For example, if a government introduces a programme of inoculation for childhood diseases that costs $10 000 000 per year and saves an average of 80 lives per year, a statistical life is implicitly valued at $125 000 at a minimum.

Several methods have been applied in attempts to calculate VOSL, all of which suffer from problems. The *prescriptive* or *normative* approach attempts to set VOSL in terms of what life *ought* to be worth and tends to be dismissed as lacking a proper economic basis. By contrast, the *'descriptive'* approach essentially involves establishing how much people are actually willing to spend to avoid the risk of death. Two approaches have mainly been used (Pearce *et al.*, 1995):

1 *The human capital approach.* This involves treating an individual as an economic agent capable of producing an output that can be valued in monetary terms. The value of this output, less any consumption that the individual would have made, is assumed to occur when he/she is killed by a landslide. The approach tends to produce extremely low values for those with low earnings, discriminating against the poor.
2 *The willingness-to-pay/willingness to accept method.* This involves valuing a statistical life on the basis of what individuals are willing to pay or accept for risk changes. Such values can be derived from *contingent valuation*, where individuals are asked directly how much they would be willing to pay to reduce risks. Other measures include finding out how much people are spending on safety and disease-preventing measures, or by how much wages differ between safe and risky jobs (the *hedonic approach*). For example, suppose 100 000 workers are paid an additional $15 each to tolerate an increased risk of mortality of 1/100 000. The increased risk will result in one statistical life lost, valued at $15 × 100 000 = $1 500 000.

The *willingness-to-pay* (WTP) approach is the more favoured, involving as it does both asking people about their views of risk (*expressed preference*) and studying responses to risk (*revealed preference*). The resulting calculations have revealed quite variable values for VOSL within Developed Countries, with some support for the view that VOSL for the UK should be set at £2–3 million at 1990 prices (Marin, 1992). Other work has produced VOSLs for Developed Countries in the range $1.8–$9 million, with a best guess average estimate of $3.5 million. However, estimates for other countries were significantly less at $300 000 for Russia and only $150 000 for China, India and Africa, which provoked outrage in certain quarters. A global average of $1 million was proposed by Pearce *et al.* (1995).

Direct and indirect economic losses

Direct and indirect economic losses are those losses capable of being given monetary values because of the existence of a market, with all other losses classified as *intangibles*. *Direct economic losses* arise principally from the physical impact of a landslide on property, buildings, structures, services and infrastructure.

Estimating and predicting economic losses is both complex and problematic. Quite naturally there is a tendency to restrict direct losses so as to make them easier to compute. This can be achieved by restricting direct losses to physical damage and destruction, thereby excluding costs of transport disruption, pollution etc. Alternatively, temporal limits can be placed on direct costs so that only those caused during the period of ground motion are considered, thereby neatly side-stepping the problems posed by fires, floods and tsunamis. Lastly, spatial limits can be placed on direct losses so that only those costs caused by the landslide within its limits are considered, as is exemplified by Schuster and Fleming's (1986) definition of direct losses as 'costs of replacement, repair, or maintenance due to damage to installations or property within the boundaries of the responsible landslide' (Schuster and Fleming, 1986). According to this definition, direct losses are only those losses that are caused by the ground disruption that occurs during landsliding and the physical contact of displaced materials with properties and their contents. The definition could be expanded to include the contents of buildings, equipment, machinery, items of personal property, works of art, ornamental vegetation, crops and domesticated animals.

The potential for landslides to cause direct economic damage is enormous because human constructions and wealth are concentrated at, or very near to, the Earth's surface. Thus movements of the ground itself, or of material over the surface of the ground, has the potential to inflict serious impact on fixed assets that may either be shallowly buried or located on the surface, or raised above the surface but founded in the ground. Adverse impacts may range in scale from minor cracks in walls, buildings and roads

259

due to limited displacements, via the destruction of individual buildings or even groups of buildings, to scenes of total devastation which, in the case of the Huascaràn catastrophic failure of 1970, resulted in the total destruction of the town of Yungay and the death of most of its inhabitants (Plafker and Ericksen, 1978).

Data on the global effects of landsliding in terms of direct economic costs is recognised to be of very poor quality, for the reasons discussed in Section 'Using the historical record' in this chapter. However, the widespread occurrence of landsliding suggests that it must be significant and this can be supported by the work of Alfors *et al.* (1973) in California, USA, which concluded that landsliding was ranked second to earthquakes in terms of loss potential, accounting for 25.7% of projected losses from geohazards over the period 1970–2000.

Among the more dramatic impacts in terms of direct losses have been:

- Mudflows that accompanied the 1980 eruption of Mount St Helens, Washington State, USA, damaged or destroyed over 200 buildings and 44 bridges, buried 27 km of railway and more than 200 km of roads, badly damaged three logging camps, and disabled several community water supply and sewage disposal systems (Schuster, 1983). Landslide damages associated with the eruption are estimated to have been $500 million.
- The Ancona landslide, Italy (Marche Region), occurred on 13 December 1982. It involved the movement of 342 hectares of urban and suburban land, damage to two hospitals and the Faculty of Medicine at Ancona University, damage to or complete destruction of 280 buildings with a total of 865 apartments, displacement of the main railway and coastal road for more than 2.5 km, one (indirect) death, and the evacuation of 3661 people (Crescenti, 1986; Catenacci, 1992). The economic loss was estimated at US$ 700 million (Alexander, 1989).
- Heavy rainfall associated with a strong El Niño in the winter and spring of 1998 caused over $158 million in landslide damage in the San Francisco Bay region, USA. This led to ten counties being declared eligible for disaster assistance from the Federal Emergency Management Agency (Godt and Savage, 1999). Similar events have occurred on three other occasions since 1969 (Table 5.1).
- The widespread landslide damage in the Basilicata region of southern Italy where it is estimated that 18.5% of the land area is affected by landsliding, including 1800 deep-seated landslides (Regione Basilicata, 1987). Damaged buildings, unsafe bridges and heavily disturbed roads are ubiquitous. As the majority of the 131 towns (communes) in the region are built on hill-tops, it is not surprising to learn that no less than 115 are

Table 5.1. San Francisco Bay landslide damage costs (from Godt and Savage, 1999)

Event	CPI: 1998	Reported costs: $ million	Percentage change	Total: 1998 prices $ million
1969	34.4	33	467	204.9
1973	38.1	10	412	51.2
1982	97.5	66	100	132
1998	195	158	0	158

Note. CPI – Consumer Price Index; historical landslide costs were adjusted to 1998 prices by: Percentage adjustment of reported costs = (CPI Current Period − CPI Previous Period)/(CPI Previous Period) × 100.

sufficiently endangered by the progressive upslope development of land-sliding as to require either extensive engineered structural measures (consolidation), abandonment in favour of relocation to a new, safer location (transferral), or a combination of the two (Fulton *et al.*, 1987; Jones, 1992). The abandonment of the village of Craco is a particularly well documented example (Del Prete and Petley, 1982; Jones, 1992) but is not particularly unusual, for there are numerous examples of abandoned villages in the area due to a combination of earthquake shaking and slope failure. Indeed, in the adjacent region of Calabria the cost of damage to roads, railways, aqueducts and houses was estimated at US$200 million for 1972–1973 alone (Carrara and Merenda, 1976), and over the centuries nearly 100 villages have had to be abandoned due to landsliding, involving the displacement of nearly 200 000 people.

• The town of Ventnor, England, has been built on an ancient landslide complex (Lee and Moore, 1991). Contemporary movements within the town have been slight; however, because movement occurs in an urban area with a permanent population of over 6000, the cumulative damage to roads, buildings and services has been substantial. Over the last 100 years about 50 houses and hotels have had to be demolished because of ground movement. The average annual loss in the Ventnor area has been estimated by the local authority as exceeding £2 million.

These events are merely extreme examples of what can be a common problem in urban areas within mountainous or hilly terrain. Typical problems include the costs of clearing-up landslide debris from roads and pavements, rebuilding or repairing damaged property, replacing or repairing lengths of road, sewers, water pipes and fractured gas mains. For example, Murray (2000) reports that in New Zealand there were 800 insurance claims for damage caused by landslides during 1997–1998, at a cost of $4.7 million.

Over time significant direct losses can arise as a result of cliff recession, especially on the coast. It is known that the Holderness coast of England has retreated by around 2 km over the last 1000 years, and in the process destroyed at least 26 villages listed in the Domesday survey of 1086; 75 Mm^{-3} of land has been eroded in the last 100 years (Valentin, 1954; Pethick, 1996). Perhaps the most famous example of the effects of rapid recession on the English coast is at Dunwich, Suffolk, where much of the town was lost over the last millennium (e.g. Bacon and Bacon, 1988). Gardner (1754) records that by 1328 the port was virtually useless and that 400 houses, together with windmills, churches, shops and many other buildings, were lost in one night in 1347.

Indirect economic losses or *indirect costs* are those that subsequently arise as a consequence of the destruction and damage caused by a primary hazard (i.e. the landslide itself), secondary hazards or follow-on hazards. They include the costs due to the long-term disruption to transport, loss of production of agricultural products, manufactured goods and minerals, loss of businesses and retail income, costs of cleaning up, resulting water pollution or contaminated land, the various costs incurred during the recovery process such as unemployment and the extra costs resulting from increased illness etc.

Examples of this type of loss are:

- An earthquake in 1987 triggered landslides in the Andes of north-eastern Ecuador, which resulted in the destruction or local severance of nearly 70 km of the Trans Ecuadorian oil pipeline and the only highway from Quito to the eastern rain forests and oil fields (Nieto and Schuster, 1999). Economic losses were estimated at $1 billion. Oil exports were disrupted for almost six months, reducing the Government's income by 35% (Stalin Benitez, 1989).
- Landslides near Lake Tahoe, USA, led to the closure of the US-50 highway. The direct loss associated with highway repairs was $3.6 million (Walkinshaw, 1992). However, the road was closed for 2.5 months, causing access disruption and loss of tourist revenues of $70 million (San Francisco Chronicle, 1983).
- The Thistle landslide, Utah, in 1983 severed the Denver and Rio Grande Western Railroad main line. The temporary closure resulted in $81 million of lost revenues to the railroad company (University of Utah, 1984).
- The Fraser River salmon fisheries, Canada, were severely affected by a small rockfall that occurred during construction of the Canadian National Railway in 1914. The fall prevented migrating salmon reaching their spawning grounds. Between 1914 and 1978, the losses to both the sockeye and pink salmon fishery were in the order of $2600 million (International Pacific Salmon Fisheries Commission, 1980).

Indirect costs should also include the costs incurred due to the threat of hazard. These include the costs of planning for and implementing emergency action and evacuation, including the provision of temporary shelter, food and medical services to those who may have had to be moved; relocation costs arising from the need to permanently re-house people in safer areas; the extra costs of increased geotechnical investigation, monitoring and mitigation measures required as a consequence of the recognition of increased risk following a hazard impact; additional costs to future construction arising from the development and enforcement of new building regulations and codes; the costs of increased insurance premiums arising from the recognition of landsliding as a problem in an area; any fall in property prices as a consequence of an event or the implementation of a new zoning scheme etc.

A complete assessment of indirect losses can be difficult to achieve because many elements may still be incurred a considerable time after the 'hazard event' and are not specifically restricted to the immediate area physically impacted by the event. They are, therefore, extremely difficult to predict in advance for the purposes of risk assessment.

Intangible losses

Intangible losses are the vague and diffuse consequences that arise from an event and which cannot easily be valued in economic terms. They include effects on the environment, nature conservation, amenity, local culture, heritage, aspects of the local economy, recreation and peoples' health, as well as their attitudes, behaviour and sense of well-being. Possible examples are landslide scars disfiguring a famous view or landsliding causing the destruction of an important historical monument, both of which could result in 'costs' in terms of distress and reduced pleasure, as well as on the local economy in terms of reduced tourism. The effects of post-traumatic stress on the populations affected can also be significant.

Landsliding can cause significant ecological damage. Changes in terrain, vertical displacement of ground, alterations to slope aspect and drainage, burial of existing ground, changes in river flow, inundation behind landslide dams, catastrophic floods and increased sediment loads in rivers represent some of the more obvious sources of ecological stress. Alteration or destruction of habitat can have profound consequences.

The following three examples illustrate the problematic nature of intangible losses and indicate how extremely difficult it is to incorporate them within risk assessments:

- A major landslide occurred during the construction of a tailings dam to store waste from the Ok Tedi copper mine, Papua New Guinea, in

1984. The landslide led to the abandonment of the dam site and the dumping of 80 000 tonnes of mine waste containing lead, cadmium, zinc and copper, directly into the river. According to the Australian Conservation Foundation, nearly 70 km of the Ok Tedi river became 'almost biologically dead', with 130 km 'severely degraded', while fish stocks declined by 90%. The chief executive of Broken Hill Properties, the leader of the Ok Tedi Mining Ltd consortium, described the problems that followed the landslide as an 'environmental abyss'. Broken Hill Properties agreed to a $430 million write-down of the value of the asset and to hand its stake to the Papua New Guinea Government.

- In the aftermath of the Aberfan disaster of October 1966, which killed 107 out of 250 pupils at the village school, Gaynor Lacey, a consultant psychiatrist at Merthyr Tydfil Child Guidance Centre, saw 56 children who had developed behavioural problems since the landslide (Lacey, 1972). The most common problems were sleeping difficulties, nervousness, insecurity, enuresis and unwillingness to go to school. Many children had lost all their friends, so there was little point in going out, and once having stayed in for a lengthy period, it was difficult to start going out again. The educational development of many of the children was delayed, until they began to overcome the trauma.

- In recent years concerns have been raised that the ancient Incan stronghold of Machu Picchu in Peru is threatened by an active rockslide (e.g. Carreño and Bonnard, 1999; Hadfield, 2001). The citadel was used by the Incas as a refuge from the Spanish conquistadores in the 16th century and only re-discovered by Hiram Bingham in 1991. The ruins were declared a UNESCO World Heritage Site and are widely regarded as a priceless monument.

Intangible costs or losses may be produced as both direct and indirect consequences of landsliding. Because they cannot easily be valued in economic terms, their existence results in the production of impact studies and risk assessments that consist of an unfortunate mixture of statistics (e.g. so many dead and injured), monetary costs and value judgements (e.g. reduced amenity value), which make comparisons of risks an extremely difficult and subjective process. As a result, there has been much effort in developing approaches and techniques designed to allow monetary values to be assigned to different types of loss, as has been the case in the value of human life (see Section 'Loss of life and injury' in this chapter). Many of these are described in Bateman *et al.* (2002) and DTLR (2002).

For example, ecosystems can be considered to have value because people derive utility either from their use or because of their existence. They provide goods and services and, as a consequence, any reduction due to an impact (i.e. landslide) can be measured in terms of *change in productivity* and expressed in monetary values. The *resource restoration cost method* can be used to calculate or predict the expenditure required to restore productivity to its level prior to disturbance. The most generally applicable method of valuing *loss of amenity* is *contingent valuation* (see Bateman *et al.*, 2002), which involves questioning people in order to ascertain their willingness to pay for environmental improvements or demand compensation for environmental losses. The most difficult aspects of ecosystem losses to value are those associated with the general benefits that are, or could be, derived from the ecosystems. These *indirect ecosystem values* consist of two groups: *option values*, which measure the willingness of individuals to pay in order to retain the option of having future access to a species or resource (an important element in the concept of *sustainability*); and *existence value*, which is the value people attach merely to knowing of the existence of a species or habitat, even though they may never encounter it first hand. Contingent valuation is the main technique used to produce values.

Ground behaviour

Landslides cause 'damage' in two ways. The first is by removing material from its pre-existing position so that usable land is lost through the retreat of a cliff or the formation of a scar. The second results from movements of the landslide mass and is a function of the static or dynamic loads exerted on those assets that are affected (Leone *et al.*, 1996). These loads may include:

- horizontal displacement and deformation;
- vertical displacement and deformation;
- lateral pressure;
- pressure from impact of moving debris or boulders;
- accumulation or loading from run-out, either instantaneous or the progressive build up of debris over several events.

Different parts of a landslide mass come under tension, compression, tilt, contra-tilt, heave etc. according to their location and the stresses involved. A particular asset may suffer damage from a number of different loads, depending on its location relative to the landslide. For example, in the case of a buried pipeline, full-bore rupture and a spill could occur as a result of the following processes:

1 *Lateral displacement.* Pipe rupturing as a result of differential horizontal and/or vertical movement of the landslide mass. The potential for pipeline displacement is a function of landslide depth, the behaviour of the materials (e.g. plastic or block type deformation), the speed of movement and the cumulative displacement that could occur over time.

265

Table 5.2. *Examples of landslide velocity and damage (from Cruden and Varnes, 1996)*

Landslide velocity class	Landslide name/location	Source	Estimated landslide velocity	Damage
7	Elm, Switzerland	Heim (1932)	70 m/sec	115 deaths
	Goldau, Switzerland	Heim (1932)	70 m/sec	457 deaths
	Jupile	Bishop (1973)	31 m/sec	11 deaths, houses destroyed
	Frank, Canada	McConnell and Brock (1904)	28 m/sec	70 deaths
	Vaiont, Italy	Mueller (1964)	25 m/sec	1900 deaths by indirect damage
	Ikuta, Japan	Engineering News Record (1971)	18 m/sec	15 deaths, equipment destroyed
	St Jean Vianney, Canada	Tavenas et al. (1971)	7 m/sec	14 deaths, structures destroyed
6	Aberfan, Wales	Bishop (1973)	4.5 m/sec	144 deaths, some buildings destroyed
5	Panama Canal	Cross (1924)	1 m/sec	Equipment trapped, people rescued
4	Handlova	Zaruba and Mencl (1969)	6 m/day	150 houses destroyed, complete evacuation
3	Schuders	Huder (1976)	10 m/year	Road maintained with difficulty
	Wind Mountain, USA	Palmer (1977)	10 m/year	Road and railway require frequent maintenance, buildings adjusted periodically
2	Lugnez, Switzerland	Huder (1976)	0.37 m/year	Six villages on slope undisturbed
	Little Smokey	Thomson and Hayley (1975)	0.25 m/year	Bridge protected by slip joint
	Klosters	Haefeli (1965)	0.02 m/year	Tunnel maintained, bridge protected by slip joint
	Fort Peck Spillway	Wilson (1970)	0.02 m/year	Movements unacceptable, slope flattened

2 *Spanning.* Pipe rupture as a result of removal of support along a significant length, due to landslide movement evacuating material from beneath the pipeline. The potential for spanning is a function of the vertical displacement of the landslide mass or retreat of an eroding scar across the pipeline alignment.
3 *Loading.* Pipe rupture as a result of an imposed load (e.g. burial by landslide debris or the impact of falling or rolling rocks). This is a function of the depth of run-out of landslide debris, the size/height of fall of material onto the pipeline or the velocity of missiles.

These failure mechanisms illustrate how destructive intensity of a landslide is related to kinetic parameters, such as velocity and acceleration, along with its dimensions and the material characteristics (Leone *et al.*, 1996). The maximum movement velocity is a key factor in landslide destructiveness (Table 2.4; Table 5.2). An important limit appears to lie around 5 m/ second, approximately the speed of a person running away from a slide.

There is no accepted measure of landslide intensity (Hungr, 1997). Cardinali *et al.* (2002) considered landslide intensity (*I*) as a measure of the destructiveness of the landslide and defined it as a function of landslide volume (*v*) and maximum landslide velocity (*s*):

$$I = f(v, s)$$

Table 5.3 presents the estimated intensity for a range of landslides based on these criteria. For a given landslide volume, fast-moving rockfalls have the highest landslide intensity, while rapidly moving debris flows exhibit intermediate intensity and slow-moving landslides have the lowest intensity.

In first-time slides the available potential energy (a function of slope height and geometry) is progressively dissipated into several components

Table 5.3. Landslide intensity scale for different landslide types (after Cardinali et al., 2002)

Estimated volume: m³	Fast moving landslide (rockfall)	Rapid moving landslide (debris flow)	Slow moving landslide (reactivated slide)
<0.001	Slight		
<0.5	Medium		
>0.5	High		
<500	High	Slight	
500–10 000	High	Medium	Slight
10 000–50 000	Very High	High	Medium
>500 000		Very High	High
≫500 000			Very High

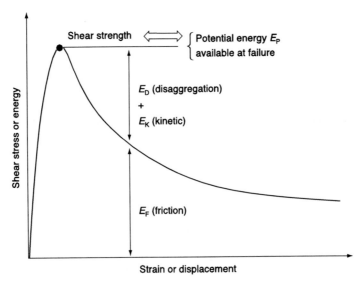

Fig. 5.2. Schematic diagram showing the redistribution of potential energy after failure (after Leroueil et al., 1996)

(Leroueil *et al.*, 1996). *Frictional energy* moves the detached material over the shear surface. The amount of frictional energy required depends on the stress–displacement behaviour of the material (Fig. 5.2). In elasto-plastic or ductile materials all the potential energy is dissipated as frictional energy, resulting in low movement rates and small overall displacements. Any remaining potential energy is dissipated in the break-up and remoulding of the moving material (*energy of disaggregation*) and in accelerating it to a particular velocity (*kinetic energy*). In brittle materials, where there is a large difference between the peak and residual strengths, the kinetic energy can be very large, giving rise to long run-out landslides. Figure 5.3 compares landslide volume and the normalised run-out distance (run-out length/slope height) for rock avalanches, quick clay failures and submarine slides. Run-out length generally increases with the volume of the failed mass, as the energy per unit volume increases with the slide height and, hence, its size (see Section 'Main phase hazards' in Chapter 2).

Landslide reactivation generally involves the sliding of a rigid block over a rigid base. The rate of displacement depends on a variety of factors, including the local slope geometry, the total stresses and pore pressures induced by the thrust exerted by the soil mass, the rate of pore pressure dissipation and fluctuations in the groundwater table (Leroueil *et al.*, 1996). Highly variable rates of displacement are common, ranging from 10 m/month to less than 1 mm/month. Often movements are progressive, starting in some

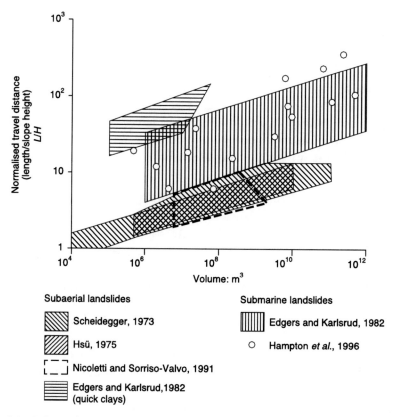

Fig. 5.3. *Relationship between landslide volume and the normalised travel distance (after Leroueil et al., 1996)*

sections and spreading downslope because of the thrust exerted by the moving mass. As a result, reactivated slides can behave like glaciers, with zones of tension and compression occurring at the same time in different parts of the slide.

The way the displaced materials behave as they move can have a significant impact on the degree of damage inflicted on assets. Taking a buried pipeline as an example, the potential for a rupture will be influenced by whether the movements involve the plastic deformation of clays around the pipe, as in some mudslides, or brittle-style block movements that build-up significant lateral pressures. Structures that happen to be located on a moving mass are damaged in proportion to their internal distortion (Cruden and Varnes, 1996). The extensive Lugnez landslide in Switzerland, for example, has moved down a 15° slope at up to 0.37 m/year since 1887. However, as the whole 25 km² landslide is moving as a single block

269

without internal distortion, properties within the six villages on the landslide have been unaffected.

Elements at risk

The term *assets* is used to describe all objects or qualities valued by humans. All assets occurring in an area that could be adversely affected by a hazard, such as a landslide, are known as *elements at risk*. These elements are extremely diverse in nature and are usually divided into the following major groupings:

1 *Populations*. At the simplest level this is the number of people present in the area likely to suffer impact. More detailed analysis could use age–sex distributions and an indication of the state-of-health (e.g. good health, poor health, infirm etc.), as these aspects do influence death rates and the nature and severity of injuries within a population (see Section 'Vulnerability' in this chapter). Detriment is usually expressed in terms of loss of life or injury, although the longer-term consequences of physical impairment, psychological effects and ill-health should also be considered (see Section 'Loss of life and injury' in this chapter).
2 *Buildings, structures, services and infrastructure*. The value of these physical assets can usually be determined from real estate agents or local authority tax bands for housing stock and from the owners/operators for services and infrastructure. Damage can be total or repairable (i.e. partial loss).
3 *Property*. This includes the contents of houses, businesses and retailers, machinery, vehicles, domesticated animals and personal property. Information on values may be obtained from trade organisations and the insurance industry.
4 *Activities*. All activities whether for financial gain or pleasure. The main components are business, commerce, retailing, entertainment, transportation, agriculture, manufacturing and industry, minerals and recreation. The losses incurred through landsliding are equivalent to the temporary or permanent disruption of these activities and are, in the main, expressed in terms of loss of revenue.
5 *Environment*, including flora, fauna, environmental quality and amenity.

The actual number of categories of elements at risk will depend upon the nature of the study area and the scope and the proposed risk assessment.

Exposure

In Section 'Hazard and vulnerability' in Chapter 1 the term exposure was defined as the proportion of the total value of *humans and the things humans value* that would suffer detriment in a hazard impact or, to put it another way, *total value × vulnerability*. But total value does not stay constant in time, for factors such as inflation, developmental growth, decline and depreciation have to be taken into consideration. Vulnerability also changes with time, due to improvements in health and safety, increased standards of living, improvements in housing quality and developments in technology, although it has to be recognised that such *advances* do not necessarily lead to reduced vulnerability with reference to geohazards such as landsliding (e.g. piped water supply and swimming pools have often increased the potential for slope instability). Exposure, therefore, displays trends and cycles with references to short, medium and long timescales.

Exposure also displays marked variations at short to very-short timescales due to the dynamic characteristics of a population and the mobile nature of certain assets. This can be termed *temporal probability* (Morgan et al., 1992; Australian Geomechanics Society, 2000) or *temporal vulnerability*. Thus estimations of exposure must, for the purposes of risk assessment, involve consideration of two distinct components:

1 *permanent*, where fixed assets, such as buildings or pipelines, could be damaged irrespective of the timing of the landslide event. As the assets are always within the zone of impact, the adverse consequences for a particular magnitude of event can be assumed to remain constant;
2 *temporary*, where the degree of risk can vary with the timing of the event, be it night or day, week-day or weekend, tourist season or off-season. The consequences will reflect the chance of the event occurring at a time when mobile assets are either within the zone of impact (e.g. a train passing through a rock cutting at the time of a rockfall) or at a relatively high level of concentration (i.e. occupancy rates for housing are higher at night). Temporary exposure can be represented by a factor on the scale from 0 (never present) to 1 (permanently present fixed asset).

As there is uncertainty regarding exactly when a landslide will occur, it is difficult to reliably predict how many people or what value of temporary assets will be present to be impacted by the landslide (i.e. in the wrong place at the wrong time). Thus for the purposes of risk assessment, two approaches can be adopted:

• gather data on the variation of mobile assets within the area under threat and then use that data to calculate the proportion of time that different values occur;
• use the data to estimate an average value.

The first approach results in a range of values and, therefore, a number of calculations. As a consequence, most pragmatic risk assessments to date have tended to adopt the second approach. This can involve calculating either:

- the average value of the mobile asset;
- the proportion of time spent by an individual or asset within the 'danger zone';
- the average number of people or vehicles within the 'danger zone' at any one time.

In essence, exposure involves being in the *wrong place* (i.e. the 'danger zone' where a landslide impacts – the spatial probability) at the *wrong time* (i.e. when the landslide occurs – the temporal probability). For example, a 500 m long seaside promenade is located directly beneath a high cliff that is regularly affected by small rock falls (5 m wide) that have occurred at a frequency of one per year. The proportion of the promenade that is affected by a fall (i.e. the 'danger zone') is

$$\text{Danger Zone} = \frac{\text{Size of Rockfall}}{\text{Length of Promenade}} = 5/500 = 0.01$$

The proportion of time that an individual walking at 2.5 km per hour spends on the promenade is

$$\text{Occupancy} = \frac{\text{Length of Promenade}}{\text{Speed (m/hour)}}$$

$$= 500/2500 = 0.2 \, \text{hours} = 0.2/(365 \times 24) \, \text{years}$$

$$= 0.000\,022\,8 \quad (2.28 \times 10^{-5})$$

The probability of being in the 'wrong place' at the 'wrong time' is

$$\text{Prob. (Wrong Place and Time)} = \text{Prob. (Wrong Place)}$$

$$\times \text{Prob. (Wrong Time)}$$

$$= \text{Danger Zone} \times \text{Occupancy}$$

$$= 0.01 \times 0.000\,022\,8$$

$$= 0.000\,000\,228 \quad or \quad (2.28 \times 10^{-7})$$

This represents the proportion of time spent by the individual within the 'danger zone'. The annual probability of the individual being hit by a rockfall is:

$$\text{Prob. (Individual Hit)} = \text{Prob. (Rockfall)}$$

$$\times \text{Prob. (Wrong Place and Time)}$$

$$= 1 \times 0.000\,000\,228$$

$$= 0.000\,000\,228 \quad or \quad (2.28 \times 10^{-7})$$

If a group of promenade users sit on a 3 m-wide bench admiring the view out to sea, the exposure (i.e. the spatial probability of being in the 'wrong place') is:

$$\text{Wrong Place} = \frac{\text{Length of Bench}}{\text{Length of Promenade}} = 3/500 = 0.006$$

If the group sat on the bench for an hour, the chance (annual probability) of being on the promenade at the 'wrong time', that is when a rock fall occurs, would be:

$$\text{Wrong Time} = \text{Occupancy Time}/(24 \times 365)$$

$$= 1/(24 \times 365) = 1/8760 = 0.000\,114 \ or \ (1.14 \times 10^{-4})$$

If the annual probability of a rockfall occurring along the cliffline was 0.5, the chance of the seated group being hit by a rockfall would be:

$$\text{Prob. (Group Hit)} = \text{Prob. (Event)} \times \text{Prob. (Wrong Place)}$$

$$\times \text{Prob. (Wrong Time)}$$

$$= 0.5 \times 0.006 \times (1.14 \times 10^{-4})$$

$$= 0.5 \times (6.85 \times 10^{-7}) = 0.000\,003\,42 \ or \ (3.42 \times 10^{-7})$$

A similar approach is used to estimate the exposure of people within moving traffic passing through a danger area. Thus the exposure of a single vehicle travelling through a 1 km-wide landslide area at 30 km/hour would be:

$$\text{Vehicle Exposure} = \text{Journey time through exposed area}/24$$

$$= \frac{\text{Distance/Speed}}{24} = 0.033/24 = 0.001\,389 \ \text{(days)}$$

$$= 0.001\,389/365 \ \text{(years)} = 0.000\,003\,8 \ or \ (3.8 \times 10^{-6})$$

If the vehicle occupancy was three people, then the annual exposure would be:

$$\text{Annual Exposure} = 0.000\,003\,8 \times 3 = 0.000\,011\,4 \quad (1.14 \times 10^{-5})$$

If the number of vehicles crossing the landslide is 1000 per day (all with the same journey time) with an average vehicle occupancy of 1.5 people, the overall exposure would be:

$$\text{Exposure} = \text{Number of Vehicles/Year} \times \text{Vehicle Exposure}$$

$$\times \text{Average Occupancy}$$

$$= (1000 \times 365) \times 0.000\,003\,8 \times 1.5 = 2.08$$

Thus, the passage of 1000 vehicles and 1500 occupants a day through the landslide area is equivalent to 2.08 people being within the area for 100% of the time.

The above example shows how an average value of exposure can be calculated. In those cases where the objective is to calculate how many people are likely to be at risk within an area, it is necessary to develop a *population model* in order to estimate the average number expected to be within the danger zone during a given time period, such as an hour or a day. To estimate the average daily exposure for a residential population it will be necessary to calculate variations in the actual number of people within each different type of building at different times of the day using an *occupancy model*, and then sum the values for the area as a whole. For example, an apartment block has a total population of 1000, but the actual population varies between 95% at night (some people are always away from home or on night shift) to around 25% in the middle of the afternoon.

Time of day	Residential population (% of total population)
00–06 hours	0.95
06–09 hours	0.75
09–12 hours	0.5
12–15 hours	0.5
15–18 hours	0.25
18–21 hours	0.75
21–24 hours	0.90

The average daily population at risk can be estimated as follows:

$$\text{Average Population} = \text{Total Population} \times (6/24 \times 0.95) + (3/24 \times 0.75)$$
$$+ (3/24 \times 0.5) + (3/24 \times 0.5) + (3/24 \times 0.25)$$
$$+ (3/24 \times 0.75) + (3/24 \times 0.9)$$
$$= \text{Total Population} \times 0.69 = 690$$

The daily variation in the number of people within the block is equivalent to a fixed population of 690 exposed for 100% of the time, that is 690 people with an exposure of 1. Alternatively, it can be expressed as 1000 people (maximum possible) with an exposure of 0.69. If this estimation is considered too crude because of the scale of the daily variation, then grouping the data results in four separate values:

1000 people × 0.9375 for 37.5% of the time

1000 people × 0.75 for 25% of the time

1000 people × 0.5 for 25% of the time

1000 people × 0.25 for 12.5% of the time

The statistics can be of value in terms of developing *least impact* and *worst case* scenarios.

Vulnerability

The impact of an event on the population or assets of an area can vary between *no detectable loss* to *total loss* (e.g. write-off of all properties or death). *Partial loss* is the usual outcome, where only a proportion of the population or assets are killed or destroyed. This variation in impact is generally expressed in terms of the *vulnerability* of the assets, including population and activity (e.g. Leone *et al.*, 1996; see the discussion in Section 'Hazard and vulnerability' in Chapter 1). According to Timmerman (1981), 'vulnerability is the degree to which a system, or a part of a system, may react adversely to the occurrence of a hazardous event. The degree and quality of that adverse reaction are partly conditioned by the system's resilience, the measure of a system's, or part of a system's, capacity to absorb and recover from the occurrence of a hazardous event'.

Vulnerability can also be envisaged as the level of potential damage, or degree of loss, of a particular asset (expressed on a scale of 0 to 1) subjected to a landslide of a given intensity (Fell, 1994). It can be difficult to quantify, being dependent on the nature and intensity of the mechanical stresses generated by the actual failure (e.g. differential ground movement, subsidence, heave, loading etc.) and the vulnerability characteristics of the 'elements at risk' that are involved.

Most of the work on vulnerability has focused on buildings, structures and infrastructure. Damage to structures and infrastructures can be classified as:

- *Superficial* (cosmetic or minor damage), where the functionality of buildings and infrastructure is not compromised, and the damage can be repaired, quickly and at low cost;
- *Functional* or medium damage, where the functionality of structures or infrastructure is compromised, and the repairs take time and significant expenditure;
- *Structural* or severe to total damage, where buildings or transportation routes are severely damaged or destroyed, necessitating extensive and extremely costly demolition and reconstruction.

Examples of structural damage classification schemes for buildings are presented as Tables 5.4 and 5.5.

Estimates of vulnerability can be based on the inferred relationship between the intensity type of the expected landslide (Table 5.3), and the likely damage the landslide would cause (Table 5.5). Table 5.6 presents a simple classification, developed by Cardinali *et al.* (2002), of the expected

275

Table 5.4. Structural damage classification (from Alexander, 2002)

Grade	Description
None	The building has sustained no significant damage.
Slight	There is non-structural damage, but not to the extent that the cost of repairing it will represent a significant proportion of the building's value.
Moderate	There is no significant non-structural damage, and light to medium structural damage. The stability and functionality of the building are not compromised, although it may need to be evacuated to facilitate repairs. Buttressing and ties can be used for short-term stability.
Serious	The building has sustained major non-structural damage and very significant structural damage. Evacuation is warranted in the interests of personal safety, but repair is possible, although it may be costly and complex.
Very serious	The building has sustained major structural damage and is unsafe for all forms of use. It must be evacuated immediately and either demolished or substantially buttressed to prevent it from collapsing.
Partial collapse	Portions of the building have fallen down. Usually these will be cornices, angles, parts of the roof or suspended structures such as staircases. Reconstruction will be expensive and technically demanding; demolition of the remaining parts will be the preferred option.
Total collapse	The site will have to be cleared of rubble. A few very important buildings may need to be reconstructed (usually for cultural reasons) even though they have collapsed totally, but most will not be rebuilt.

damage to buildings and roads by landslides of different type (rock fall, debris flow or slide) and intensity (slight, medium, high or very high). However, it has to be recognised that such intensity scales are of limited use unless properly related to local conditions in terms of the range of building types present, local materials, indigenous building practices, quality of workmanship, building styles and economic conditions in the area under consideration.

Typical building typologies are usually developed based on the following information (Alexander, 2002):

- Construction type: load-bearing walls (mudbrick, random rubble, dressed stone, bonded brick etc.).
- Nature of vertical and horizontal load-bearing members (e.g. brick wall, steel beam).
- Size of building, number of floors, number of wings, square metres of space occupied or cubic metres of capacity.
- Regularity of plan and elevation.

Table 5.5. *Landslide damage intensity scale (after Alexander, 1986)*

Grade	Description of damage
0	*None*; building is intact
1	*Negligible*; hairline cracks in walls or structural members; no distortion of structure or detachment of external architectural details
2	*Light*; building continues to be habitable; repair not urgent. Settlement of foundations, distortion of structure and inclination of walls are not sufficient to compromise overall stability
3	*Moderate*; walls out of perpendicular by 1–2°, or substantial cracking has occurred to structural members, or foundations have settled during differential subsidence of at least 15 cm; building requires evacuation and rapid attention to ensure its continued life
4	*Serious*; walls out of perpendicular by several degrees; open cracks in walls; fracture of structural members; fragmentation of masonry; differential settlement of at least 25 cm compromises foundations; floors may be inclined by 1–2°, or ruined by soil heave; internal partition walls will need to be replaced; door and window frames too distorted to use; occupants must be evacuated and major repairs carried out
5	*Very serious*; walls out of plumb by 5–6°, structure grossly distorted and differential settlement will have seriously cracked floors and walls or caused major rotation or slewing of the building (wooden buildings may have detached completely from their foundations). Partition walls and brick infill will have at least partly collapsed; outhouses, porches and patios may have been damaged more seriously than the principal structure itself. Occupants will need to be rehoused on a long-term basis, and rehabilitation of the building itself will not be feasible
6	*Partial collapse*; requires immediate evacuation of the occupants and cordoning off the site to prevent accidents with falling masonry
7	*Total collapse*; requires clearance of the site

- Degree to which construction materials and methods are mixed in the building, or alternatively whether there is a single technique and set of materials.
- Age category of the building; this can be generalised according to the dominant building material of the period (usually this involves categories such as pre-1900, 1900–1940s, 1950s–1965, 1966 onwards).
- The building's state of maintenance (excellent, good, mediocre, bad).

Using the three grade division described earlier, it is possible to develop a simple vulnerability index (on a scale of 0 to 1) by estimating the relative proportion of the overall value of assets that would be lost in an event that causes each of these three grades of damage. For example, Table 5.7 presents

Table 5.6. Vulnerability of a range of assets to landslide events of different intensity (based on Cardinali et al., 2002)

Landslide intensity		Buildings	Main roads	Secondary roads	Minor roads	Buried pipelines	Railway lines
Light	Rockfall	C	C	C	C	C	C
	Debris flow	C	C	F	F	C	C
	Slide	C	C	F	S	F–S	C
Medium	Rockfall	F	F	F	F	C	F
	Debris flow	F	F	F	F	C	F
	Slide	F	F	S	S	S	F
High	Rockfall	S	S	S	S	C	S
	Debris flow	S	S	S	S	C	S
	Slide	S	S	S	S	S	S
Very high	Rockfall	S	S	S	S	C	S
	Debris flow	S	S	S	S	C	S
	Slide	S	S	S	S	S	S

Notes. C = superficial/cosmetic damage; F = functional damage; S = structural damage

the results obtained when these principles are applied in estimating the vulnerability of buildings within a slowly moving (high to very high intensity) deep-seated ancient landslide complex.

Table 5.8 presents the estimated vulnerability indices for buildings and people within an area susceptible to debris flow activity at Montrose, Victoria, in Australia (Moon *et al.*, 1992; Fell and Hartford, 1997; see Example 4.7).

The Australian Geomechanics Society (2000) has identified a number of factors that determine *human vulnerability* in the context of landsliding:

- Volume of slide;
- Type of slide, mechanism of slide initiation and velocity of sliding;

Table 5.7. An example of the estimated vulnerability of a building to ground movements within an ancient landslide complex

Damage	Grade of damage (Table 5.5)	Vulnerability index	Estimated damage for a £100 000 property	Typical repairs required
Structural	4–7	0.55–1.0	£55 000–£100 000	Complete re-build
Functional	2–3	0.15–0.5	£15 000–£50 000	Underpinning, part re-build
Superficial	0–1	0.01–0.1	£1000–£10 000	Filling of cracks, redecoration

Table 5.8. Building vulnerabilities to debris flow activity, Montrose, Australia (from Moon et al., 1992; Fell and Hartford, 1997)

Debris flow risk zone	Factors	Intensity*	Vulnerability index (buildings)	Vulnerability index (people)
Extremely high	High velocity High depth	Very High	1.0	0.8
High	High–medium velocity Medium depth	High	0.7	0.5
Medium	High–low velocity Low depth	Medium	0.4	0.1
Low	Medium–low velocity Low depth	Slight	0.1	0.01

Note. *Intensities have been estimated from the classes presented in Table 5.6

- Depth of slide;
- Whether the landslide debris buries the person(s);
- Whether the person(s) is in the open or enclosed in a vehicle or building;
- Whether the vehicle or building collapses when impacted by debris;
- The type of collapse if the vehicle or building collapses.

Table 5.9 presents the vulnerability factors developed as part of a research programme into the application of risk assessment methods in areas exposed to natural terrain hazards in Hong Kong (Halcrow, 1999). The factors reflect the probability of the death of a person located in an area affected by landslide events of different magnitude and take account of whether the person is inside or outside an affected building. The vulnerability factors used by Michael-Leiba *et al.* (2000) in an assessment of landslide risk in the Cairns area of Australia are presented in Table 5.10.

Table 5.9. Vulnerability factors for open hillside landslides in Hong Kong (from DNV Technica, 1996; Halcrow, 1999)

Landslide volume: m^3	Vulnerability (indoor population)	Vulnerability (outdoor population)
50	0.0002	0.03
100	0.006	0.054
500	0.011	0.078
1000	0.026	0.11
2500	0.04	0.15
5000	0.17	0.48

Table 5.10. Vulnerability factors for residents in the Cairns area, Australia (from Michael-Leiba et al., 2000)

Landslide event	Vulnerability of resident people
Hillslope failure	0.05
Debris flow: proximal end	0.9
Debris flow: distal end	0.05

Even small slides and single boulders can kill. The ability to escape from a landslide is related to the velocity of movement. In Hong Kong, Finlay *et al.* (1999) found that a person is very vulnerable in the event of complete or substantial burial by debris, or the collapse of a building (Table 5.11). If the person is buried by debris, death is most likely to result from asphyxia rather than crushing or impact. If the person is not buried, injuries are much more likely than death. However, research on earthquake casualties (Coburn and Spence, 1992) has shown that building design and building materials have an important influence on survival rates and this should be taken into consideration in the case of landsliding.

Table 5.11. Summary of vulnerability factors for people affected by landsliding in Hong Kong (from Finlay et al., 1999; Australian Geomechanics Society, 2000)

Case	Range of data	Recommended value	Comments
Person in open space, if struck by a rockfall	0.1–0.7	0.5	May be injured but unlikely to cause death
Person in open space, if buried by debris	0.8–1.0	1.0	Death by asphyxia almost certain
Person in open space, if not buried	0.1–0.5	0.1	High chance of survival
Person in vehicle, if vehicle is buried/crushed	0.9–1.0	1.0	Death is almost certain
Person in vehicle, if vehicle is damaged only	0–0.3	0.3	High chance of survival
Person in a building, if the building collapses	0.9–1.0	1.0	Death is almost certain
Person in a building, if the building is inundated with debris and the person buried	0.8–1.0	1.0	Death is highly likely
Person in a building, if the debris strikes the building only	0–0.1	0.05	Very high chance of survival

Cruden (1997) has suggested that a ratio of three injured for every death might be a useful starting point for estimating the fatality rate associated with landslides. This may be satisfactory for sluggish and intermediate slides, but, for major, extremely rapid landslides the rate of fatalities to survival injuries is likely to be much higher, as in the case of volcanic eruption phenomena which 'leave few walking wounded; a sharp dividing line separates the quick and the dead' (Baxter, 1990). For example, the 1985 Armero mudflows in Colombia killed 21 000, but only 65 were rescued from the debris (e.g. Voight, 1990).

It should be stressed, however, that applying the concept of vulnerability to people and societies is more problematic than in the case of physical structures and infrastructure. Impacts on humans are not simply confined to death and injury but include long-term and more subtle effects (see Section 'Framework for adverse consequences' in this chapter), including loss of well-being. Thus it is not merely a case of measuring the robustness of humans in the context of the physical threats they may face. The frailties of people, together with any limitations in terms of lack of protection, survival capabilities or ability to recover, owe much to prevailing socio-political conditions or what Hewitt (1997) terms 'the making of vulnerability by human activity'. This notion applies equally to the creation of conditions prior to a hazard impact as to what happens during and immediately after an event. The potential for adverse consequences associated with many follow-on hazards (e.g. looting, disease), as well as some secondary hazards (e.g. fire), is largely a function of management failure. Similarly, incompetent search and rescue operations after an event, including failure to mobilise adequate medical services, can result in increased losses due to what Alexander (2000) calls *secondary vulnerability*. Human vulnerability is, therefore, a complex concept that is still being refined in an active and often confusing literature.

As Hewitt (1997) observed, 'vulnerability assessment is essentially about the human ecology of endangerment'. The potential for persons to be impacted by a landslide is a function of a large number of factors, some involving chance (i.e. temporal exposure, see Section 'Exposure' in this chapter), some relating to the robustness and resilience of the people who are being threatened, some relating to the social conditions prevailing, and some relating to the nature and violence of the ground movements and subsequent hazards. Clearly age, health, fitness and awareness are all important. Similarly, the stage in the evolution of a landslide at which a person or people become involved can also determine the outcome, as does the speed and efficiency of search and rescue operations. Finally, chance plays a role, not merely in involvement but also in survival or the limiting of injury. There are numerous tales of miraculous survivals including that of a man

281

who survived the Sale or Salashan, China, loess flowslide of March 1983 (D. K. C. Jones, 1992; Sassa, 1992) by clinging to a lone tree which floated upright on a small raft of ground for 700 m surrounded by 3 Mm^3 of flowing loess which obliterated three villages, killing over 200 people.

Taking a broader view, vulnerability also determines the variable impact of landslides on different sections of a society (i.e. poorer neighbourhoods and cheap or shanty housing tend to be more vulnerable) and the ability of a community as a whole to withstand a landslide impact and recover. *Societal vulnerability* includes three interrelated factors (Cannon, 1993; Blaikie *et al.*, 1994):

- *Resilience*, or the ability to maintain a system and to recover after impact, is used in the context of communities as well as of the economic or wealth generating systems within an area. An economy dependent on a limited number of transportation routes or a single industry may be highly vulnerable to landsliding.
- *Robustness*, or the ability to respond to a spectrum of uncertainties or actual physical threats, is applied to organisations, communities and individuals within a group, reflecting social, health and economic strength. Insurance, for example, can make individuals more able to recover from the losses incurred in a landslide.
- *Preparedness*, which can reflect both the protection provided to a group and their willingness to act on their own behalf.

To many social scientists and planners the vulnerability of groups within a society may be of greater importance in determining the impact of a landslide than its intensity. A highly vulnerable group, such as a squatter camp on the unstable slopes at the margins of a city, may often be badly affected by large, fast-moving landslides, whereas wealthy, low vulnerability groups might experience minimal impacts from similar events. Thus numbers of people displaying particular levels of vulnerability, with respect to landslides of a particular intensity, will provide a measure of the potential impact.

Thus to estimate societal vulnerability it is necessary to attempt to obtain some or all of the following information:

- scenarios of landslide development in terms of physical characteristics and intensity;
- estimations of patterns and intensities of damage/destruction to buildings and infrastructure;
- estimations of the likelihood, nature and severity of secondary hazards;
- estimations of the number of people expected to be within the area affected by primary and secondary hazards, after consideration of timing and the possibility of warning/evacuation;

- estimation of relevant characteristics of people within danger area (age/sex etc.);
- estimation of survival rates based on anticipated nature of destruction and likely efficiency of emergency response.

There are, of course, many other aspects of vulnerability. *Economic vulnerability*, or the *vulnerability of economic systems*, is the extent to which economic activity is reduced as a consequence of hazard impacts of differing severity. Damage or destruction of factories, businesses and sites of production, loss of cash crops, the dislocation of transport routes, disruption to power and water supplies and the reduction in available workforce are some of the factors that lead to economic consequences in short to long timescales. In the case of major disasters, such consequences are increasingly measured in terms of changes in Gross Domestic Product (GDP). Where the costs of destruction and damage also represent a significant proportion of GDP, then the financial burden of reconstruction will also have an adverse effect on economic performance. Landslides rarely achieve such impacts although they are important contributors in the case of landslide generating events. More usually, the adverse effects of landsliding exacerbate the effects of economic marginalisation. Economic marginality describes the situation where the returns of an economic activity barely exceed the costs. Destruction, damage and disruption due to landsliding can lead to both costs and loss of output, which can combine to render an activity uncompetitive or even determine that the activity should not recommence (i.e. no repair, reconstruction or restoration), both of which will have knock-on effects for the local economy. Thus economic consequences may increase with time after an event. Risk assessments may need to identify whether such economic vulnerability exists and the scale and duration of the consequences.

The term *environmental vulnerability* is used to describe the extent to which environmental quality and amenity can be adversely affected as a consequence of a hazard event. In the case of landsliding, a number of aspects can be identified. First, landslide generating events can initiate further erosion, resulting in widespread land degradation. Such developments occur where the terrain is particularly prone to change following disturbance (*high sensitivity*) and has difficulty in reproducing its former capability (*low resilience*). The consequences are loss of soil resources, unsightliness and restricted land use options, together with possible sedimentation and increased flood potential elsewhere. For individual landslides, the impacts are mainly in terms of the development of what can be perceived to be unsightly scars and lobes of debris, together with effects on streams and rivers, in terms of sediment load and discharge. Clearly, visual despoliation will gradually diminish as geomorphological processes and vegetation regeneration progressively blend the landslide features with their surroundings, but the effect may last for years to decades. Measuring such impacts in terms of loss of amenity is extremely difficult and is usually done qualitatively, although economic valuation techniques can be used (see Bateman *et al.*, 2002; DTLR, 2002).

Finally, there is *ecological vulnerability*, which is the extent to which fauna and flora can be adversely affected by a change in environmental conditions brought on slowly, as in the case of global warming, or more suddenly, as in the case of a pollution episode or landslide. When dealing with non-human organisms in the environment, it is usual to refer to disturbing influences (*stressors*) acting on spatially identifiable units known as *ecosystems*. The concept of ecosystem is of fundamental importance. It refers to areas of the Earth's surface characterised by recognisable and distinctive 'functional relationships among organisms and between organisms and their physical surroundings' (e.g. a pond, stretch of river, patch of marsh, a small wood or belt of scree slope). Alternatively, it can be described as a given community (i.e. a defined group of populations) within its habitat. This focus on ecosystems is the reason why the term vulnerability is frequently replaced by *ecosystem fragility*.

Consequence models

As discussed earlier, the impact of a landslide is controlled by the ground behaviour (e.g. the landslide intensity), the exposure of the elements at risk and their vulnerability to damage. Combining these factors enables landslide *consequence models* to be developed that reflect the damage signature of particular landslide scenarios. In each of the examples presented in Table 5.12, both the exposure (permanent and/or temporary) and vulnerability factors are used to establish the predicted damages compared with a *total loss event* in which the assets at risk would be completely lost. Thus, for a landslide of a particular intensity:

$$\text{Risk} = \text{Prob. (Event)} \times \text{Adverse Consequence}$$

$$= \text{Prob. (Event)} \times (\text{Total Loss} \times \text{Exposure} \times \text{Vulnerability})$$

In most cases it will be necessary to carry out this exercise for each of the individual elements at risk, or even each individual property:

Risk (Element 1)

$$= \text{Prob. (Event)} \times \text{Adverse Consequence (Element 1)}$$

$$= \text{Prob. (Event)} \times (\text{Total Loss} \times \text{Exposure} \times \text{Vulnerability})$$

Table 5.12. Examples of simple landslide consequence models

Landslide scenario	Consequence factor	Consequence model	Comment
Cliff recession, on-going cliff retreat causing progressive loss of cliff top property	Ground behaviour Assets at risk	Loss of cliff top land Properties set back from the cliff edge, at varying distances	Damage is *total loss* of property in year of land loss
	Exposure	Fixed assets, permanent exposure. No loss of life as population is evacuated before land loss	
	Vulnerability	Total loss of assets as they are declared unsafe before they drop off the cliff edge	
First-time landslide, buried pipeline crossing slope with potential for landsliding	Ground behaviour	Horizontal and vertical displacements and deformation	Uncertain outcome, damage is dependent on the vulnerability of the pipe to different intensity events $Damage = Total\ Loss \times Vulnerability\ Factor$
	Assets at risk	Buried pipeline runs through the potential landslide area	
	Exposure	Fixed asset, permanent exposure	
	Vulnerability	Potential for pipe rupture depends on the intensity of movement	
Landslide reactivation, periodic slow ground movement threatens properties built on the unstable slope	Ground behaviour	Horizontal and vertical displacements and deformation	Uncertain outcome, damage is dependent on the location of the property and its vulnerability to different intensity events $Damage = \sum Total\ Loss \times Vulnerability\ Factor$ for each property Note vulnerability factor varies with location
	Assets at risk	Properties located on landslide blocks, at varying distances from block boundaries No loss of life expected because of very slow speed of movement	

Table 5.12. Continued

Landslide scenario	Consequence factor	Consequence model	Comment
	Exposure	Fixed assets, exposure variable dependent on location relative to landslide block boundaries	
	Vulnerability	Potential for property damage depends on the intensity of movement	
Debris slide or rockfall, failure of a cliff threatens road users at the cliff foot	Ground behaviour	Detachment and boulder fall with instantaneous impact. Debris slide run-out and instantaneous impact	Uncertain outcome, damage is dependent on the exposure of the population and its vulnerability to different intensity events. $Damage = \sum Total\ Loss \times Exposure \times Vulnerability\ Factor$ for each event size
	Assets at risk Exposure	Vehicles using the mountain road. Temporary exposure, dependent on number of users per hour and vehicle speed	
	Vulnerability	Potential for loss of life depends on boulder and/or slide size and/or intensity	
Rockfall blocks road and causes temporary road closure and traffic diversion	Ground behaviour	Detachment and boulder fall with instantaneous impact	Damage (additional transportation costs and opportunity costs) dependent on period of road closure
	Assets at risk Exposure	Vehicles using road. Dependent on number of vehicles and occupants per hour and vehicle speed	
	Vulnerability	Vehicles and occupants travelling will be delayed	

Table 5.12. Continued

Landslide scenario	Consequence factor	Consequence model	Comment
Debris slide or rockfall, failure of a road cutting threatens road users and road-side property	Ground behaviour	Detachment and boulder fall with instantaneous impact Debris slide run-out and instantaneous impact	Uncertain outcome Loss of life is dependent on the exposure of the population and its vulnerability to different intensity events
	Assets at risk	Vehicles driving along the road, properties at varying distances from the base of the cutting	Damage $= \sum$ Total Loss $\times$ Exposure $\times$ Vulnerability Factor for each event size Property damage is dependent on the location of the property and its vulnerability to different intensity events
	Exposure	Road users: temporary exposure, dependent on number of vehicles per hour and speed Property: fixed assets, but exposure varies with location and event intensity	Damage $= \sum$ Total Loss $\times$ Vulnerability Factor for each event size $\times$ Exposure to that event size for every property Note exposure varies with location because of the variations in landslide footprint with differing intensities of event
	Vulnerability	Potential for loss of life depends on boulder and/or slide size and/or intensity Potential for property damage depends on the intensity of movement	

Risk (Element 2)

$$= \text{Prob. (Event)} \times \text{Adverse Consequence (Element 2)}$$

$$= \text{Prob. (Event)} \times (\text{Total Loss} \times \text{Exposure} \times \text{Vulnerability})$$

$$\text{Risk} = \text{Prob. (Event)} \times \sum \text{Adverse Consequences (Elements 1 to } n)$$

A slightly different approach has been adopted by Wong et al. (1997) in which the adverse consequence scenarios are considered in relation to the adverse consequences expected from a *benchmark* or *reference landslide* of known dimensions and impact. The approach was developed for estimating potential loss of life in Hong Kong, where the reference slide was taken to be a 10 m-wide, 50 m^3 volume, shallow hillside failure. The number of fatalities expected when such a reference slide affects various land use categories in differing locations is shown in Table 5.13. The loss of life associated with a potential landslide is then scaled up or down from the reference slide. The scaling factor used was based on the width of the landslide relative to the reference slide:

$$\text{Scale Factor} = \frac{\text{Potential Slide Width}}{\text{Reference Slide Width}}$$

A vulnerability factor is used to relate the loss of life associated with the reference landslide to that expected with the potential slide. The vulnerability is influenced by a number of factors, including the nature, proximity and spatial distribution of the facilities (assets) and the debris mobility (see Example 5.6, below).

Using the reference landslide approach, the risk is calculated as:

$$\text{Risk} = \text{Prob. (Event)} \times \text{Adverse Consequence}$$

$$= \text{Prob. (Event)} \times (\text{Reference Slide Expected Fatalities}$$

$$\times \text{Scale Factor} \times \text{Vulnerability})$$

Note that variable exposure is not specifically addressed in this reference landslide approach. However, the probability of a person being caught in the path of the debris (i.e. being in the 'wrong place at the wrong time') can be incorporated into the model, as follows:

$$\text{Risk} = \text{Prob. (Event)} \times (\text{Reference Slide Expected Fatalities}$$

$$\times \text{Scale Factor} \times \textit{Exposure} \times \text{Vulnerability})$$

This would involve the development of a *population model* in order to estimate the average number expected to be within the danger zone during a given time period, as described in Section 'Exposure' of this chapter.

288

Table 5.13. Fatalities associated with a reference landslide (10 m-wide, 50 m³ volume) in Hong Kong (from Wong et al., 1997)

Group number	Facilities at risk	Expected number of fatalities
1	A. Buildings with a high density of occupation or heavily used residential building, commercial office, store and shop, hotel, factory, school, power station, ambulance depot, market, hospital/polyclinic/clinic, welfare centre	3
	B. Others • bus shelter, railway platform and other sheltered public waiting areas • cottage, licensed and squatter area	3
2	A. Buildings with a low density of occupation or lightly used built-up area (e.g. indoor car park, building within barracks, abattoir, incinerator, indoor games sports hall, sewage treatment plant, refuse transfer station, church, temple, monastery, civic centre, manned substation)	2
	B. Others • road with heavy vehicular or pedestrian traffic density • major infrastructure facility (e.g. railway, tramway, flyover, subway, tunnel portal, service reservoir)	1
3	Roads and open space • densely-used open space and public waiting area (e.g. densely used playground, open car park, densely-used sitting out area, horticultural garden) • quarry • road with moderate vehicular or pedestrian traffic density	0.25
4	Roads and open space • lightly-used open-air recreation area (e.g. district open space, lightly used playground, cemetery, columbarium) • non-dangerous goods storage site • road with low vehicular or pedestrian traffic density	0.03
5	Roads and open space • remote area (e.g. country park, undeveloped green belt, abandoned quarry) • road with very low vehicular or pedestrian traffic density	0.001

Notes. (1) To account for different types of building structure with different detailing of window and other perforations etc. a multiple fatality factor 1–5 is appropriate for Group 1A, to account for the possibility that some incidents may result in a disproportionately larger number of fatalities than that envisaged. (2) For incidents that involve the collapse of a building, it is assumed that the expected number of fatalities is 100.

Table 5.14. Vulnerability factors for roads at the toes of cut slopes (from ERM-Hong Kong, 1999)

Failure volume: m³	Run-out of debris: degrees								
	20–25	25–30	30–35	35–40	40–45	45–50	50–55	55–60	>60
<20			0.0015	0.0065	0.019	0.042	0.072	0.095	0.1
20–50			0.03	0.1	0.23	0.37	0.47	0.5	0.5
50–500		0.0015	0.078	0.26	0.48	0.63	0.69	0.7	0.7
500–2000		0.01	0.15	0.48	0.83	0.95	0.95	0.95	0.95
>2000	0.01	0.15	0.48	0.83	0.95	0.95	0.95	0.95	0.95

The reference landslide approach has been used by ERM-Hong Kong (1999) in order to assess expected fatalities along selected roads in Hong Kong, including Castle Peak Road (a two-lane highway; see Example 6.14). The highway falls into category 2B 'road with heavy vehicular or pedestrian traffic density' in Table 5.13, for which the reference landslide would lead to one fatality. Table 5.14 presents the vulnerability factors for roads at the toe of cut slopes (slope angle = 60°). These factors are expressed in terms of the run-out angle (shadow angle) of debris for a range of landslide volumes and slope heights.

The shadow angle was calculated as follows:

$$\text{Shadow angle} = \tan^{-1} (\text{Slope Height}/\{\text{Distance from Slope Toe}$$
$$+ [\text{Slope Height}/\tan(\text{Slope Angle})]\})$$

For a given run-out angle, vulnerability factors have been estimated considering the different lane sections separately and then averaged. For example, considering a failure volume of <20 m³ and slope height of <10 m:

- *lane 1* (closest to the cut slope). The lane lies between 1.5 m and 4.5 m from the base of the cut slope and would be covered by debris with a shadow angle of over 50° (Table 5.15), indicating a vulnerability factor of 0.072 (Table 5.14);
- *lane 2* (furthest from the cut slope). The lane extends for between 4.5 m and 8 m from the base of the slope and would be covered by debris with a shadow angle of around 40°, indicating a vulnerability factor of 0.019.

The average vulnerability factor is 0.0455 i.e. ((0.072 + 0.019)/2). Table 5.16 presents the scale factors that relate the expected fatalities in the reference landslide (see Table 5.13) to the actual failures. For a failure involving less than 20 m³ of material a scale factor of 0.4 would be used. Thus, for this

Table 5.15. Landslide shadow angles that would cover different road lanes, for different cut slope heights (from ERM-Hong Kong, 1999)

Slope height (metres)	Lane 1 (1.5 m from slope base)	Lane 2 (4.5 m from slope base)	Lane 3 (8 m from slope base)	Lane 4 (12.5 m from slope base)
<10	52.1°	40.3°	31.3°	24.0°
10–20	55.9°	48.7°	42.0°	35.3°
>20	57.9°	54.0°	49.8°	45.2°

example the adverse consequences, expressed as the probability of a fatality, are:

$$\text{Adverse Consequence} = (\text{Reference Slide Expected Fatalities}$$

$$\times \text{Scale Factor} \times \text{Vulnerability Factor})$$

$$= 1 \times 0.4 \times 0.0455 = 0.0182$$

Table 5.16. Scale factors used to modify number of fatalities expected from a reference landslide (see Table 5.13; from ERM-Hong Kong, 1999)

Failure volume: m^3	Average width	Scale factor
<20	4	0.4
20–50	7	0.7
50–500	15	1.5
500–2000	20	2
>2000	25	2.5

Example 5.1
A small coastal community is located above an eroding cliffline that has been retreating at an average annual rate of 1.25 m/year. Continued recession over the next 25 years will threaten a number of properties and important services, including a gas main and sewer (i.e. exposure = 1.0). These losses will be *irreversible*, that is once the land has been eroded, or the cliff top is considered too close for the property to be used safely, then the land or property cannot be regained (i.e. vulnerability = 1.0). The potential for loss of life is considered negligible because the local authority will require occupants to move out of potentially unsafe properties before they become too close to the cliff edge.

The average erosion rate was used to project erosion contours for 25 years into the future in order to determine the expected year of loss of particular properties (Fig. 5.4). In addition, a narrow strip of agricultural land is

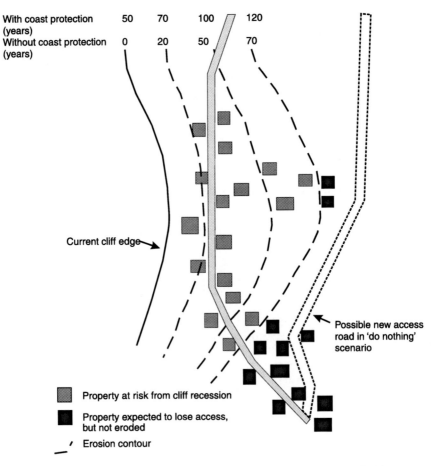

With coast protection (years)   50   70   100   120

Without coast protection (years)   0   20   50   70

Current cliff edge →

Possible new access road in 'do nothing' scenario

Property at risk from cliff recession

Property expected to lose access, but not eroded

Erosion contour

Fig. 5.4. Example 5.1: Erosion contours for modelling consequences of cliff recession (after Penning-Rowsell et al., 1992)

expected to be lost each year. Property prices and agricultural land prices were obtained from a local real estate agent, whereas the value of the services and infrastructure was provided by the operators.

The consequences of continued cliff recession were calculated for each year, from Year 0 to Year 25, as follows:

$$\text{Consequences (Year } t) = \text{Total Loss} \times \text{Exposure} \times \text{Vulnerability}$$

$$= \text{Market Value (Property and Land Lost)}$$

$$\times 1 \times 1$$

The overall losses arising from destruction within a 25-year period are the sum of the consequences for each year, and have a current market value of £6.242 25 million (Table 5.17).

Table 5.17. Example 5.1: Consequence model (after MAFF, 1993)

Year	Property lost due to recession	Market value: £ thousands	Agricultural land: hectares per year	Market value: £ thousands
0			0.125	0.625
1			0.125	0.625
2			0.125	0.625
3			0.125	0.625
4			0.125	0.625
5			0.125	0.625
6			0.125	0.625
7	Trunk gas main	450	0.125	0.625
8	1–5 Acacia Ave	350	0.125	0.625
9			0.125	0.625
10	Café	120	0.125	0.625
11	Sunnyview and Dunswimmin	180	0.125	0.625
12	Sewage Pump Station	2300	0.125	0.625
13	Hightrees House	1500	0.125	0.625
14			0.125	0.625
15			0.125	0.625
16			0.125	0.625
17	2–8 Acacia Ave	400	0.125	0.625
18	The Saltings	76	0.125	0.625
19			0.125	0.625
20			0.125	0.625
21	14–20 Rocco Blvrd	850	0.125	0.625
22			0.125	0.625
23			0.125	0.625
24			0.125	0.625
25			0.125	0.625
Total		6226	3.25	16.25

Note: Loss (Year t) = Property (Market value) + Land (Market value).

Example 5.2
An oil pipeline has been built across the dissected hills of Hong Kong, buried within the colluvium and residual soils that mantle most slopes. The hill-slopes are susceptible to shallow (<3 m deep) debris slides or debris avalanches. Concerns have been expressed about the risk within a 5 km section of the route, where it crosses a series of deeply dissected catchments that have a history of widespread slope movements.

The route was sub-divided into sections, based on bedrock geology and slope class. The Hong Kong Natural Terrain Landslide Inventory (NTLI; Evans and King, 1998; Evans *et al.*, 1997; King, 1997, 1999) statistics were

293

used to generate landslide densities for each section (see Example 2.1). It should be noted that the figures used were for *recent slides* (i.e. occurred within the last 50 years), as the distribution of these slides is believed to be a good indication of landslide density over a 50 year period (Evans and King, 1998). The frequency per year was calculated by dividing the landslide densities by 50.

The landslide densities generated from the NTLI statistics were expressed as landslides/km^2 and not per unit length of pipeline. Although the pipeline right-of-way is around 30 m wide, it can be impacted by landslides occurring on the adjacent upslope and downslope hillsides. It was assumed, therefore, that the route and that portion of the adjacent hillside that might generate impact on the pipeline could be considered to be a 250 m-wide corridor. Thus the NTLI landslide densities were adjusted by a factor of 0.25 to produce landslide frequencies per kilometre of route corridor, which could then be used to yield annual probabilities (Table 5.18).

The pipeline will only be vulnerable to a proportion of all the landslides that affect the right-of-way. It was assumed that only slides >20 m wide *could* damage the pipe. The NTLI statistics revealed that 15% of all recorded slides are greater than 20 m wide (Evans *et al.*, 1997). As these >20 m-wide slides could have a range of intensities and impacts on the pipeline, it was assumed that only 25% *would* lead to rupture of the pipe. Thus, a *pipeline vulnerability factor* of 0.0375 was used in the analysis:

Pipeline Vulnerability Factor $= 0.15 \times 0.25 = 0.0375$

No exposure factor was used in the analysis as the pipeline is a permanent asset (i.e. the exposure factor would be 1.0).

The consequences of a pipe rupture include the environmental damage and clean-up costs associated with an oil spill, estimated to be, on average, $5 million per event. In addition, supply interruptions lasting from days to weeks can be expected, depending on the scale and location of the pipe rupturing landslide event. The average length of disruption to the supply is 10 days. As the pipe carries 40 000 barrels of oil per day (estimated value of $1 million), a disruption of operation for 10 days would lead to a loss of $10 million in business.

For each particular section of route, the risk was calculated as follows:

Risk (Section 1) $=$ Prob. (Event) $\times$ Adverse Consequences

$\quad\quad\quad\quad\quad = $ Prob. (Event) $\times$ (Total Loss $\times$ Vulnerability Factor)

$\quad\quad\quad\quad\quad = $ Prob. (Event) $\times$ (15 million $\times$ 0.0375)

Table 5.18. Example 5.2: Consequence model

Section	Landslide prone length: km	Slope class	Geology	Recent landslide density/km² per 50 years	Recent landslide density: normalised for right of way	Potential natural terrain landslides/year	Vulnerability factor	Frequency of rupture/ year	Potential losses: $ million	Risk: $ million
1	1.5	35–40	GC	10.04	3.77	0.0753	0.0375	0.0028	15	0.042
2	0.4	30–35	GF	22.8	2.28	0.0456	0.0375	0.0017	15	0.026
3	0.7	30–35	FT	11.63	2.04	0.0407	0.0375	0.0015	15	0.023
4	0.1	35–40	GF	22.7	0.57	0.0114	0.0375	0.0004	15	0.006
5	1.3	25–30	CBS	17.46	5.67	0.1135	0.0375	0.0043	15	0.064
6	0.5	25–30	CS	19.81	2.48	0.0495	0.0375	0.0019	15	0.028
7	0.5	20–25	CS	14.04	1.76	0.0351	0.0375	0.0013	15	0.020
Total	5					0.37		0.014		0.209

Notes. Geology types are GC – coarse grained granite; GF – fine-grained granite; FT – fine ash tuff; CBS – conglomerate and breccia; CS – siltstone and sandstone. Landslide density statistics are derived from the Hong Kong NTLI (Evans and King, 1998).

Frequency of Rupture/Year = Potential Natural Terrain Landslides/Year × Vulnerability Factor.

Risk = Frequency of Rupture/Year × Potential Losses.

The overall risk along the 5 km of landslide susceptible route was the sum of the risks for each individual section (Table 5.18):

$$\text{Risk (Sections 1 to } n) = \sum \text{Prob. (Event)}$$
$$\times \text{Adverse Consequences (Sections 1 to } n)$$

This worked out at $0.209 million per annum, with around 50% of the total risk produced within Sections 1 and 5. However, these sections actually comprise 56% of the total route, which implies that other sections of the pipeline route have higher levels of risk. Dividing the risk in each section by the length of each section reveals that the highest levels of risk per unit length of pipeline appear to occur in the relatively short lengths of Sections 2 and 4.

Example 5.3
The small town of Runswick Bay, on the North Yorkshire coast, UK, has grown up on a pre-existing landslide complex developed in glacial till. In addition to almost continuous creep, high to very high intensity, very slow movements occur during periods of heavy winter rainfall. The majority of the building stock is brick-built and of Victorian age. Many buildings have been damaged by the cumulative effects of ground movement (e.g. Rozier and Reeves, 1979). The extent of the damage was determined by a systematic survey of contemporary damage to roads and structures. This survey involved the assessment of the type and magnitude of damage, using a classification based on guidelines provided by the Building Research Establishment (1981) and Alexander (2002; Table 5.4).

In order to provide an indication of the risk, the potential damages were estimated by an expert panel (see Section 'Estimating probability through expert judgement' in Chapter 4) for a range of landslide reactivation scenarios:

- Scenario 1. An episode of significant ground movement in response to the exceedance of a winter rainfall threshold level (annual probability estimated as 0.04).
- Scenario 2. An episode of major ground movement in response to the exceedance of an extreme winter rainfall threshold level (annual probability estimated as 0.01).

The effects of movements in terms of the damage to the properties (i.e. separate buildings) within the landslide was estimated on a landslide block by landslide block basis. For each block, it was assumed that the level of damage will vary across the unit depending on the precise location. The following method was used to accommodate the uncertainty in damage levels/locations:

- calculate the number of properties within each block and the average property value (total value divided by number of properties);
- establish, for both scenarios, the proportion of the block surface that would be affected by ground movements of differing magnitude (i.e. *the hazard factor*). Damage was found to be most commonly located at the boundaries between individual landslide blocks, where it is possible to recognise narrow bands of severe hazard. Indeed, the degree of hazard can vary dramatically within as little as a metre of the surface exposure of inter-block shear surfaces. While one property may be severely damaged by differential movement, an adjacent property may be largely unaffected;
- relating the pattern of hazard with the distribution of properties revealed the proportion of properties likely to be subjected to differing levels of ground movement (i.e. *the exposure factor*). These proportions were established through expert panel discussion, based on historical evidence;
- the proportion of each block affected by differing magnitudes of movement in each scenario was then converted into six damage levels (e.g. 'unaffected' to 'write-off'), each of which was assumed to be equivalent to specific percentages of the average property value. This *vulnerability factor* was derived through expert panel discussion.

The method can be illustrated with reference to the predicted ground movement damage for a single landslide block (termed Block 1). There are 25 properties located on this block, with an average property value of £100 000 (the values were obtained from a local real estate agent).

A range of *exposure factors* were developed to estimate the likely impact of ground movements associated with each scenario on the properties located on the block. The results are shown in Table 5.19. In Scenario 1, for example, it is assumed that 50% of all the properties would be unaffected, with 25% affected by negligible damage, 15% by moderate damage and 5% by both serious and severe damage. By contrast, in Scenario 2 some 80% of houses would be affected by serious or worse damage, with 25% considered to be 'write-offs'.

The economic implications of the variation in impact between landslide scenarios are reflected in different levels of loss, that is the *vulnerability* of the assets vary with the scale and intensity of the movement associated with each scenario. Vulnerability is defined as the level of potential damage, or degree of loss, of a particular asset (expressed on a scale of 0 to 1) subjected to a landslide of a given severity. Estimates of vulnerability are based on an inferred relationship between the severity of ground movement and the proportion of the property value that would have to be spent in undertaking repairs. For example, moderate damage is expected to result in losses equivalent to 10% of the property value, whereas write-off would result in 100% losses (Table 5.19).

Table 5.19. Example 5.3: Consequence model for Block 1 (note there are 25 properties with an average value of £100 000)

Scenario	Annual probability		Unaffected	Negligible–slight damage	Moderate damage	Serious damage	Severe damage	Write-off	Total: £ thousands	Risk: £ thousands
1 Episode of significant movement	0.04	Proportion of properties affected	0.50	0.25	0.15	0.05	0.05	0.00		
		Vulnerability factor	0	0.01	0.1	0.25	0.5	1		
		Number of properties	12.5	6.3	3.8	1.3	1.3	0.0		
		Total property value: £ thousands	1250	625	375	125	125	0		
		Damage: £ thousands	0	6.25	37.5	31.25	62.5	0	137.5	5.5
2 Episode of widespread movement	0.01	Proportion of properties affected	0.00	0.10	0.10	0.25	0.30	0.25		
		Vulnerability factor	0	0.01	0.1	0.25	0.5	1		
		Number of properties	0.0	2.5	2.5	6.3	7.5	6.3		
		Total property value: £ thousands	0	250	250	625	750	625		
		Damage: £ thousands	0	2.5	25	156.25	375	625	1183.75	11.8

Note. Damage = Number of Properties (25) × Proportion Affected × Average Value (£100 000) × Vulnerability Factor.

Total Damage (Scenario s) = $\sum$ Damage (Damage Class unaffected to write off).

Risk (Scenario s) = Scenario Probability × Total Damage.

Thus the risk in Block 1 being associated with Scenario 1 (an episode of significant ground movement) was calculated as follows (Table 5.19):

Risk (Block 1) = Prob. (Event) × Damage to Property within Block 1

The value of damage in Scenario 1 was estimated for each damage class, as follows:

Damage (Class *d*) (Block 1) = Number of Properties (Block 1)

$$\times \text{ Proportion Affected (Scenario 1)}$$

$$\times \text{ Average Value}$$

$$\times \text{ Vulnerability Factor (Damage Class } d)$$

Taking severe damage as an example:

Severe Damage (Block 1) = Number of Properties × Proportion Affected

$$\times \text{ Average Value} \times \text{Vulnerability Factor}$$

$$= 25 \times 0.05 \times 100 \times 0.5 = \pounds 62\,500$$

The overall damages for Scenario 1 were calculated as

$$\text{Damage (Block 1)} = \sum \text{Damage (Unaffected to Write-off)} = \pounds 137\,500$$

The exercise was repeated for Scenario 2 (Table 5.19), yielding an overall damage total of £1 183 750.

The risk determined for Block 1, based on these two Scenarios, is as follows:

Risk (Scenario 1) = Prob. (Event) × Damage within Block 1

$$= 0.04 \times 137.5$$

$$= \pounds 5500 \text{ per year}$$

Risk (Scenario 2) = Prob. (Event) × Damage within Block 1

$$= 0.01 \times 1183.75$$

$$= \pounds 11\,800 \text{ per year}$$

Example 5.4

A mountain road in British Columbia is below a high cliff that is regularly affected by rockfalls. In the past, rockfalls have occurred at a frequency of five per year, blocking the road and presenting a threat to people using the road.

The probability of a fatal accident to an occupant of a vehicle driving along the road is:

Risk = Prob. (Rockfall) × Exposure Factor × Vulnerability Factor

The *exposure factor* is the chance of the vehicle occupying the same space as the rockfall pathway.

Assuming that 100 vehicles per hour drive along the road at an average speed of 50 km per hour, then the proportion of the road that is *instantaneously occupied* by a vehicle is:

$$\text{Exposure} = \frac{\text{Number of Vehicles per hour} \times \text{Average Vehicle Length (m)}}{\text{Speed (m/hr)}}$$

$$= \frac{100 \times 5}{50\,000} = 0.01$$

The probability of a rockfall hitting a moving vehicle was modelled using the binomial distribution, and based on the number of falls per year (each is a separate trial in the binomial model; see Section 'Multiple events' in Chapter 4) and the chance of a particular outcome (hit or no hit, depending on the exposure):

$$\text{Prob. (Vehicle Impact)} = 1 - (1 - \text{Exposure})^{\text{Number of falls per year}}$$

$$= 1 - (1 - 0.01)^5 = 4.9 \times 10^{-2}$$

If it is assumed that the traffic is uniformly distributed in time and space throughout the year, (i.e. there is no temporal variability in exposure) then the probability of the vehicle occupying the same space as the rockfall pathway is 1.0.

The annual probability of a rockfall hitting a moving vehicle is:

Prob. (Accident) = Prob. (Vehicle Impact)

× Prob. (Vehicle Occupies Rock Fall Path)

$$= 4.9 \times 10^{-2} \times 1$$

Given that a rockfall has hit the vehicle, a vulnerability factor was used to estimate the probability of loss of life in the incident. The factor is based on the fact that only 25% of a vehicle length is occupied by passengers and that only 1 out of every 3 rockfall impacts are severe enough to cause death. Thus, the *vulnerability factor* is $0.25 \times 0.33 = 0.0825$.

The annual probability of a rockfall hitting a moving vehicle and causing loss of life is:

Prob. (Fatal Accident) = Prob. (Vehicle Impact)

$$\times \text{Prob. (Vehicle Occupies Rock Fall Path)}$$

$$\times \text{Vulnerability Factor}$$

$$= 4.9 \times 10^{-2} \times 1 \times 0.0825 = 4.04 \times 10^{-3}$$

Example 5.5

A mountain road between two small towns in the Scottish Highlands is regularly blocked by rockfalls and the traffic is forced to follow a lengthy diversion around the affected area. The consequences of temporary road closure can be measured in terms of the *additional transportation costs* that result from the diversion (i.e. additional fuel, oil and depreciation costs incurred in travelling further or at an additional speed) and the *opportunity costs* caused by the delay (i.e. time wasted because of the additional journey time). Note that road closure generates disruption not only on the blocked road but also on the diversion routes because the traffic already using these routes will be slowed down by the diverted traffic.

In the consequence model developed for this example, the *exposure factor* is represented by the average number of vehicles and their occupants using the road. The *vulnerability* of the traffic to the disruption is assumed to be 1.0, that is if a road user wants to travel between the two towns on the day of a rockfall, then he/she will incur the additional costs. Among the factors that influence the scale of traffic disruption costs are:

- the traffic flow;
- the traffic flow as a proportion of the route capacity;
- the rockfall frequency;
- the time period of road closure;
- the length of diversions;
- the road types of the diversion routes.

Three rockfall scenarios were identified by an *expert panel review* (see Section 'Estimating probability through expert judgement' in Chapter 4):

- Scenario 1. A small rockfall ($<100\,\text{m}^3$) which results in the road being closed for one day, during which the debris is removed by the road maintenance crews (estimated annual probability of 0.3).
- Scenario 2. A relatively small rockfall ($<1000\,\text{m}^3$) which results in the road being closed for five days because of the need for repairs to the road surface (estimated annual probability of 0.05).

Table 5.20. Example 5.5: Parameters used to calculate vehicle operating costs (from Highways Agency, 1997)

Vehicle type	Parameter (fuel and non-fuel combined)		
	a	b	c
Cars	3.87	32.47	0.000 043 0
LGV	4.76	62.83	0.000 057 7

- Scenario 3. A large rockfall ($>10\,000\,\mathrm{m}^3$) which results in the road being closed for 50 days, during which time emergency works are undertaken to stabilise the unstable rockfall backscar and repair the road surface (estimated annual probability of 0.01).

The effects of the traffic disruption associated with each of the three scenarios were calculated as the additional transportation costs (resource costs) and opportunity costs (delay costs) arising from the road closure, using the method presented in Parker *et al.* (1987). The losses include:

- *the marginal transportation costs*, which are a function of vehicle type and speed:

  Marginal Transportation Costs $= a + b/v + cv^2$

  where *a*, *b* and *c* are coefficients given in Table 5.20, and *v* is the speed in kilometres per hour.
- *opportunity costs* caused by the delay. A measure of the impact of the delay can be obtained from values assigned to a traveller's time (Table 5.21).

Traffic disruption was calculated in the following steps, for both a *without landslide* and a *with landslide* case:

1 Construction of a road network diagram that incorporates roads onto which traffic will be diverted when the route is blocked. On Fig. 5.5, the traffic using the main route A–B–C is diverted onto the secondary route D–E–F when the road is blocked by rockfalls along sections A or B. Each section was assigned a road class (Table 5.22).

Table 5.21. Example 5.5: Vehicle occupancy and resource values of time (from Highways Agency, 1997)

Vehicle type	Total occupancy	Value of time: pence/hour
Cars	1.65	673.6
Light goods vehicle (LGV)	1.47	1166.7
Average road vehicle	1.71	784.4

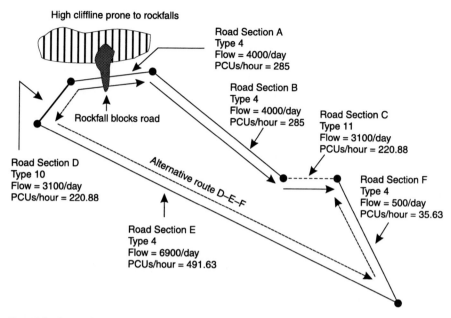

High cliffline prone to rockfalls

Road Section A
Type 4
Flow = 4000/day
PCUs/hour = 285

Road Section B
Type 4
Flow = 4000/day
PCUs/hour = 285

Road Section C
Type 11
Flow = 3100/day
PCUs/hour = 220.88

Rockfall blocks road

Road Section D
Type 10
Flow = 3100/day
PCUs/hour = 220.88

Alternative route D–E–F

Road Section F
Type 4
Flow = 500/day
PCUs/hour = 35.63

Road Section E
Type 4
Flow = 6900/day
PCUs/hour = 491.63

Fig. 5.5. Example 5.5: road network diagram

2 Establishing the average traffic flow (vehicles per hour) along each road section from local authority statistics, based on traffic count data.
3 The traffic flow along each road section was converted to Passenger Car Units (PCUs) to provide an indication of the number of users (i.e. drivers and passengers) and allow for the mix in the flow (i.e. cars, light vehicles, buses, HGVs etc.). In this example, the traffic flow was multiplied by 1.71, the occupancy of an average road vehicle (Table 5.21).
4 The average speed along each section was calculated from Table 5.22 which relates the free-flow speed to road type.

For a Class 4 road (single two lane carriageway) the traffic speed is 63 km/hour up to a flow of 400 PCUs. At the limiting capacity (LC) the

Table 5.22. Example 5.5: Speed flow relationships for roads (from Parker et al., 1987)

Road type	Class	Free flow speed: km/hour	Free flow limit: PCUs/hour/lane	Limiting capacity: PCUs/hour/lane	Speed at free flow limit: km/hour
4	Single two lane carriageway	63	400	1400	55
10	Urban two lane carriageway	35	350	600	25

303

speed of flow is 55 km/hour. Between the free flow limit (FFL) and the limiting capacity (1400 PCUs), the speed falls as the flow increases. At 600 PCUs per hour:

$$\text{Speed} = \text{Speed (FFL)} - \left((\text{Speed FF} - \text{LC}) \times \frac{(\text{PCUs} - \text{PCUs FFL})}{1000} \right)$$

$$= 63 - \left(8 \times \frac{(600 - 400)}{1000} \right) = 61.4 \, \text{km/hour}$$

5 The resource cost per hour was calculated from the marginal transportation cost ($a + b/v + cv^2$; see above), multiplied by the number of vehicles and the section length:

Resource Cost = Marginal Cost × Number of Vehicles

× Section Length

Values for the coefficients a, b and c were obtained from Table 5.20; for simplicity it has been assumed that the traffic flow is limited to cars ($a = 3.87$; $b = 32.47$; $c = 0.000\,043$).

6 The delay cost was calculated from the value of time figures presented in Table 5.21, adjusted for the road conditions:

Delay Cost = Value of Time × PCUs/hour

× Section Length/ Travel Speed

An average figure of 784.4 pence/hour was used for the value of time.

7 The costs of disruption (expressed as a cost per hour) is the difference between the combined resource cost and delay cost for the *with landslide* and *without landslide* cases:

Disruption Cost = With Landslide (Resource Cost + Delay Cost)

− Without Landslide (Resource Cost + Delay Cost)

The results of the analysis are presented in Table 5.23, which indicates a potential loss of £11 900 for each day the road is blocked.

The estimated total disruption costs associated with each of the three scenarios were calculated as

Total Disruption Cost = Daily Disruption Cost × Length of Delay

The disruption cost associated with each scenario is

Disruption Cost (Scenario 1) = 11.9 × 1 = £11 900

Disruption Cost (Scenario 2) = 11.9 × 5 = £59 500

Disruption Cost (Scenario 3) = 11.9 × 50 = £595 000

Table 5.23. Example 5.5: Traffic disruption calculations

	Length: km	Type	Flow/day	PCUs/hr	Traffic speed: km/hr	Resource costs: p/hr	Delay costs: p/hr	Total costs: p/hr
Without landslide (road section)								
A	3.2	4	4000.00	285.00	63.00	4156.04	11 355.12	15 511.17
B	5.5	4	4000.00	285.00	63.00	7143.20	19 516.62	26 659.82
C	1.8	11	3100.00	220.88	25.00	2066.07	12 474.31	14 540.38
Total	10.5						43 346.05	56 711.37
D	2.5	10	3100.00	220.88	35.00	2678.88	12 375.31	15 054.19
E	13.1	4	6900.00	491.63	63.00	29 348.80	80 186.69	109 535.49
F	4.4	4	500.00	35.63	63.00	714.32	1951.66	2665.98
Total	20					32 742.00	94 513.66	127 255.66
Overall total	30.5					32 742.00	137 859.71	183 967.03
With landslide (road section)								
D	2.5	10	6200.00	441.75	35.00	5357.75	24 750.62	30 108.37
E	13.1	4	10 900.00	776.63	59.99	46 463.86	133 034.14	179 498.00
F	4.4	4	4500.00	320.63	63.00	6428.88	17 564.96	23 993.84
Total	20					58 250.49	175 349.72	233 600.21

Note. Disruption Costs = With Landslide Total Costs − Without Landslide Total Costs = 233 600.21 − 183 967.02 = 49 633.18 pence/hour = £11 911.96 per day. Resource Costs = $(a + b/v + cv^2)$ × Section Length × PCUs/hour. Delay Cost = Value of Time (784.4 pence) × PCUs/hour × Section Length/Traffic Speed. PCUs/hr = Flow/hr × 1.71 (i.e. the average vehicle occupancy – see Table 5.21).

Example 5.6

A road and building located at the base of a cut slope in Hong Kong are threatened by a potential failure (500–2000 m³ volume) and run-out of the debris (Wong et al., 1997). The risk is, in part, a function of the travel distance of the debris. However, for a given slope type, landslides of similar mechanism can produce a range of run-out distances because of variations in slope conditions (Wong and Ho, 1996). In order to evaluate the risk it was necessary to develop a frequency distribution for the potential run-out distances.

In this example, the run-out distance was modelled in terms of the travel angle (α), defined as the inclination of the line joining the far end of the debris to the slope crest (Fig. 5.6). Based on historical events, the travel angle was assumed to range between 25° and 40°, with the following probability distribution:

- Scenario 1. $\alpha = 27.5° \pm 2.5°$; Probability = 0.05
- Scenario 2. $\alpha = 32.5° \pm 2.5°$; Probability = 0.60
- Scenario 3. $\alpha = 37.5° \pm 2.5°$; Probability = 0.35

The location of the assets at risk are represented by a 'shadow angle' (β), defined as the angle of a line that joins the asset to the slope crest (Fig. 5.6).

Wong et al. (1997) used expert judgement to develop a vulnerability matrix that defines the probability of loss of life for a range of scenarios

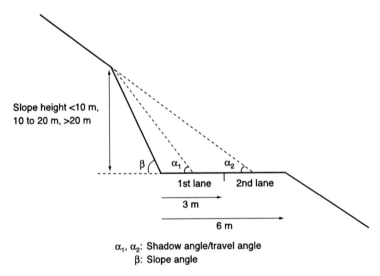

Fig. 5.6. Example 5.6: influence zone for cut slope failures (after ERM-Hong Kong, 1999)

Table 5.24. Example 5.6: Vulnerability matrix factors (from Wong et al., 1997)

Frequency of occurrence of landslides with different travel angles	Likely probability of death for different shadow angle ranges							
	>60°	55–60°	50–55°	45–50°	40–45°	35–40°	30–35°	25–30°
Scenario 1. $\alpha = 27.5° \pm 2.5°$ – Probability = 0.05	0.95 (0.95)	0.95 (0.95)	0.95 (0.95)	0.95 (0.95)	0.95 (0.95)	0.60 (0.95)	0.20 (0.60)	0.05 (0.20)
Scenario 2. $\alpha = 32.5° \pm 2.5°$ – Probability = 0.60	0.95 (0.95)	0.95 (0.95)	0.95 (0.95)	0.95 (0.95)	0.60 (0.95)	0.20 (0.60)	0.05 (0.20)	
Scenario 3. $\alpha = 37.5° \pm 2.5°$ – Probability = 0.35	0.95 (0.95)	0.95 (0.95)	0.95 (0.95)	0.60 (0.95)	0.20 (0.60)	0.05 (0.20)		
Vulnerability factor calculated	0.95 (0.95)	0.95 (0.95)	0.95 (0.95)	0.83 (0.95)	0.48 (0.83)	0.17 (0.48)	0.04 (0.15)	0.0025 (0.01)

Note. The upper figure in each cell is the vulnerability for a person within a building. The figure in brackets is the vulnerability for a person on a road

involving failures with different travel distances and facilities or assets located at different distances from the slope (i.e. shadow angles; Table 5.24).

The risk, expressed as the probability of loss of life, is calculated for each scenario, as follows:

$$\text{Risk (Scenario 1)} = \text{Prob. (Event)} \times \text{Vulnerability Factor}$$

$$\text{Overall Risk} = \sum \text{Prob. (Event)}$$

$$\times \text{Vulnerability Factor (Scenarios 1 to 3)}$$

For a person travelling along a particular lane of the road (shadow angle of 35–40°), the corresponding probability of death would be (from Table 5.24):

$$\text{Risk (Scenario 1)} = \text{Prob. (Event)} \times \text{Vulnerability Factor}$$

$$= 0.05 \times 0.95 = 0.0475$$

$$\text{Risk (Scenario 2)} = \text{Prob. (Event)} \times \text{Vulnerability Factor}$$

$$= 0.6 \times 0.6 = 0.36$$

$$\text{Risk (Scenario 3)} = \text{Prob. (Event)} \times \text{Vulnerability Factor}$$

$$= 0.35 \times 0.2 = 0.07$$

Should a landslide occur, the overall probability of death for a person at that location would be 0.48.

For a person within a building, again with a shadow angle of 35–40°, the risk of death would be (from Table 5.24):

$$\text{Risk (Overall)} = (0.05 \times 0.6) + (0.6 \times 0.2) + (0.35 \times 0.05) = 0.17$$

The difference in level of risk reflects the different degrees of protection afforded to the people in the potential run-out area. In this example, the vulnerability of a person within a building with reference to loss of life is assumed to be over 50% less than a person on a road at the same location.

Note that *variable exposure* is not addressed in this example. However, the probability of a road user being caught in the path of the debris (i.e. in the 'wrong place at the wrong time') can be estimated using the approach described in Example 5.4.

Multiple outcome consequence models

The previous examples have been based on a simple, deterministic view of the adverse consequences resulting from a landslide event. If a landslide of a particular magnitude or travel distance occurs, then a particular set of adverse consequences *will* result. In reality, the precise consequences can often reflect an almost unique and, perhaps, unexpected combination of circumstances that arise at the moment the event occurs and during its aftermath.

Take, for example, a small rockfall from a cutting which lands on the tracks of a mainline railway:

- *The event occurs at night (i.e. 5.00 a.m.) when the line is not in operation.* It is spotted by the track maintenance crew who alert the train operators and arrange for a clean-up team to remove the debris. The impact of the event is restricted to the costs of the clean-up operations, any minor repairs and the temporary delay to the commencement of services.
- *The event occurs at 8.00 a.m. immediately prior to a packed commuter train entering the cutting.* The train hits the debris and is derailed, forcing it into the path of an on-coming goods train including wagons containing hazardous chemicals. The goods train is also derailed and runs down an embankment spilling chemicals into the river which is important for producing salmon. The impacts of the event include the loss of life in the commuter train, the driver of the goods train, the trauma affecting many of the survivors and rescue teams, the destruction and damage to railway property, lengthy transport delays, long-term contamination of the river, loss of salmon stocks leading to a reduction in the tourist income to the local economy, followed by the bankruptcy of a number of local hotels and businesses.

Chance, therefore, is a major determinant of adverse consequences, as is clearly illustrated by the following report of an actual railway accident in 1883:

> This singular accident took place at Vroig cutting on the Cambrian railway on the evening of New Year's Day. The scene of the occurrence was a point between Llyngwril and Barmouth, where the rails skirt, at a considerable eminence from the water, the shore of Cardigan Bay. About eight feet above the railway line in the cliff side is the turnpike road, which a retaining wall protects. This wall with a portion of the road gave way, and fell on the railway. The 5.30 train from Machynlleth to Pwllheli was advancing at its ordinary speed when the engine dashed into the obstruction. The engine and tender rolled over the precipice to the sea-shore, a distance of about fifty feet. The engine-driver and the stoker were instantly killed, their bodies being shockingly mutilated on the jagged rocks. The four carriages and van, which with the engine and tender made up the train, did not go over the precipice. The first carriage turned over on its side, and lay partly overhanging the cliff, the coupling between it and the tender having fortunately broken. The second carriage turned on its side among the rubbish, while the remaining two did not leave the rails. The extent of the disaster was lessened by a second landslip, which took place as the train arrived at the spot, and this prevented the carriages from following the engine and tender by partly burying them. Only a few passengers were in the train. Captain Pryce of Cyffeonydd, Welshpool, Vice-Chairman of the Cambrian Railway, was in the overturned carriage, but he, as well as the other passengers, marvellously escaped without injury. (*The Graphic*, 13 January 1883)

In order to include a range of possible outcomes in a risk assessment it is necessary to develop *multiple consequence models* (consequence scenarios) and evaluate their likelihood. Thus, the risk associated with a particular outcome associated with a particular landslide (i.e. the event) can be represented by:

Risk = Prob. (Event) × Prob. (Consequence Scenario 1)

$\qquad$ × Adverse Consequences (Consequence Scenario 1)

The risk associated with all possible outcomes becomes

Total Risk = Prob. (Event) × $\sum$ (Prob. × Adverse Consequences

$\qquad$ × (Consequence Scenario 1 to n))

309

The probabilistic view of consequences is particularly useful for the *back analysis* of a particular consequence sequence after it has happened. For example, it may be useful to demonstrate that, in the case of the simple example used earlier, 'it was a ten million to one chance that the two trains collided following the landslide'.

This type of problem lends itself to the use of *event trees* (see Section 'Estimating probability through expert judgement' in Chapter 4) to establish the range of outcomes that could be generated by a particular event. Expert judgement is required to identify all the significant consequence scenario components and then develop estimates at each node along the event tree. For example, suppose an event (E), such as a rockfall that blocks a road, has a probability $P(E)$. Given that this event occurs, the *exposure* (L) of assets has the conditional probability $P(L|E)$. Likewise, the *impact* (I) has a conditional probability $P(I|L)$, given this level of exposure. The probability of this consequence scenario, or chain of events, occurring is:

$$\text{Consequence Scenario Probability} = P(E) \times P(L|E) \times P(I|L)$$

The event tree approach uses *forward logic* in that it starts with a single failure event and traces the possible consequences of the event. An alternative approach is to identify a particular consequence scenario (e.g. the two trains colliding) and use so-called *backward logic* to establish how the accident could happen, that is the causal relationships that lead to the accident or *top event*. The combinations of factors needed to generate a particular 'top event' can be identified through the use of *fault trees* (e.g. Fussell, 1973, 1976; Kumamoto and Henley, 1996).

Working back from the top of the tree, the required preconditions are set down in progressively more detail. Individual factors are joined by an *or gate* or an *and gate* which demonstrate whether one factor or another will cause the accident sequence to progress to the next stage, or whether the occurrence of both at the same time is necessary. The probability of a precondition being met can be derived from historical statistics (e.g. the failure frequency of early warning systems) or through the use of expert judgement. Fault tree analysis has the capability of providing useful information concerning the likelihood of a top event or accident and the means by which such an event could occur. Efforts to improve slope safety can be focused and refined using the results of the analysis.

Example 5.7
The Lei Yue Mun squatter villages, Hong Kong, are situated at the foot of a 20–40 m high slope, standing at 65–80°, which had been used for quarrying granular fill in the early 20th century (Ho *et al.*, 2000). The natural terrain above the quarried face rises some 200 m above the squatter huts. Both

Table 5.25. Example 5.7: Angles of friction for debris slides of different volumes (after Smallwood et al., 1997)

Debris slide	Small	Medium	Large	Very large	Extremely large
Volume: m^3	<50	50–500	500–5000	5000–50 000	>50 000
Apparent angle of friction: degrees	47	43	35	32	30

the abandoned quarry faces and the hillside have a history of instability. A number of large landslide events occurred during or shortly after a major rainstorm in August 1995, causing severe damage to the squatter dwellings. Loss of life was narrowly avoided by evacuation during the storm.

An event tree approach was used to model the consequences of landslide activity at the site, as part of a risk assessment to support decision-making about the need for re-housing of the squatters (Smallwood *et al.*, 1997). The approach adopted involved the following steps:

1 Estimating the frequency of landsliding at the site, from a review of aerial photographs covering a 55 year period (1940–95). A total of 115 events were identified.
2 Estimating the potential run-out of landslides of different magnitudes using the travel distance model of Wong and Ho (1996), who determined an *apparent friction angle* based on a line projected from the failure scar crest to the distal end of the debris (i.e. the travel angle). Apparent angles of friction used at the site are presented in Table 5.25.
3 Classifying the hazard posed by landslide events with different run-out potential into three levels or groupings – major, intermediate and minor, based on judgement. The hazard associated with an event of a particular volume reduces with increasing distance away from the slope foot. With reference to Fig. 5.7, small slides presented only what were considered to be minor hazards to buildings and population close to the slope foot (Zones A and B), whereas large slides presented a major hazard in these zones, an intermediate hazard in Zone C and a minor hazard in Zone D.
4 Sub-division of the squatter village area into 149 separate 20 m × 20 m grid cells (reference blocks). The number of people and the temporal presence in each block (i.e. the exposure) were determined from a population survey.
5 Establishment of an event tree for each reference block. Individual event trees considered landslides from the three hazard groupings (minor, intermediate and major).

311

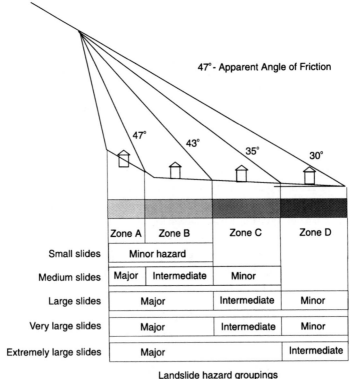

Fig. 5.7. Example 5.7: landslide hazard groupings for landslide events of different sizes and assets at different distances from the slope base (after Smallwood et al., 1997)

Figure 5.8 presents an extract of an event tree developed for one of the reference blocks and illustrates the multiple consequence scenarios associated with a *minor* debris slide. Each branch of the event tree represents a distinct consequence scenario, with the consequences (i.e. fatalities, minor and major injuries) being determined by the unique combinations of:

- the proximity of the affected buildings and population (i.e. the hazard grouping of the landslide event);
- the temporal exposure of the population (i.e. whether it was day or night when the event occurred);
- whether landslide warnings were likely to be heeded or not;
- the efficiency of the response of the emergency services;
- the occurrence of secondary hazards, such as fire.

Rule sets, based on expert judgement, were used to estimate the number of minor injuries, major injuries and fatalities resulting from events from each of the three hazard groupings. An *equivalent fatality* statistic was

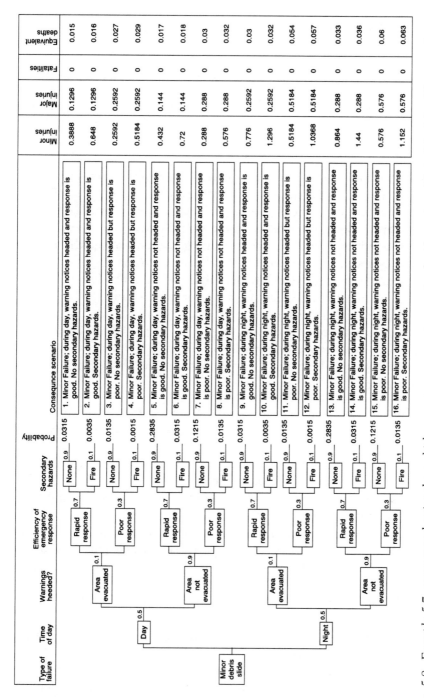

Fig. 5.8. Example 5.7: event tree showing the multiple consequence scenarios associated with a minor debris slide (after Wong et al., 1997). Note that numbers of minor and major injuries were derived by using site specific 'Rule sets' (see text).

determined as

$$\text{Equivalent Fatality} = \text{Number Fatalities}$$
$$+ (0.1 \times \text{Number Major Injuries})$$
$$+ (0.005 \times \text{Number Minor Injuries})$$

The results were used to support an assessment of individual and societal risk (see Sections 'Individual risk' and 'Societal risk' in Chapter 6).

Example 5.8
A small industrial estate has been located in a former chalk quarry. The estate comprises a combination of light structures, used by small firms making electronic components, car repair businesses and service companies, and adjacent parking lots. Following the severe wet winter of 2000–2001 there was significant rockfall activity, raising concerns about the risks to the businesses and, in particular, their staff and customers.

A series of event trees were developed to model the multiple consequence scenarios associated with rockfalls from the quarry face. Individual trees were developed for rockfall events of particular magnitude and frequency. Figure 5.9 presents one of the event trees, illustrating the consequences of a fall (5000 m^3) with an annual probability of 0.01 (established from historical records and expert judgement). The complexity of the tree reflects the need to take account of:

- the influence of natural obstacles, such as trees or boulders at the base of the quarry face, which would prevent the rockfall reaching the boundary of the developed area;
- the presence or absence of man-made protective barriers (part of the estate is protected by a wire rock fence);
- the assets at risk on the quarry floor, including open space, parking lot or light structure;
- the timing of the event. A simple population model was developed to account for the variable occupancy of the site at different times of the day. The day was split into six 4-hour periods, each of which had a different average occupancy value for the different assets at risk;
- the chance that the rockfall would miss all the assets and come to rest without causing any damage or injury;
- the likelihood that the vehicles in the parking lot would be occupied at the time of the rockfall event;
- the possibility that a boulder would pass through a window or the walls of a structure, together with the potential for partial or complete building failure.

314

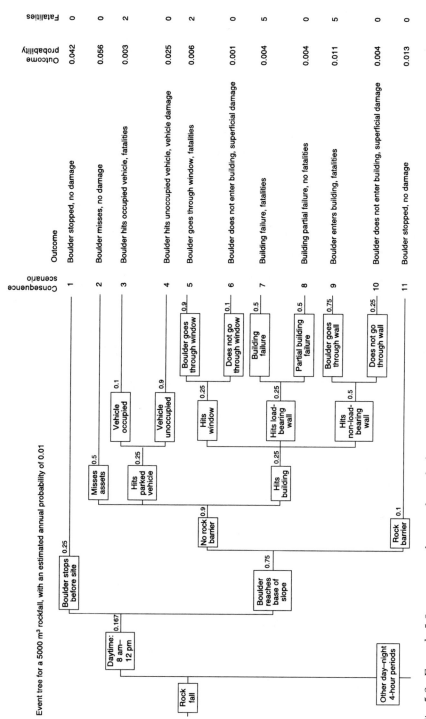

Fig. 5.9. Example 5.8: event tree showing the multiple consequence scenarios associated with a small rock fall event

315

Each branch of the event tree represents a different consequence scenario. The consequences were determined by an expert panel (see Section 'Estimating probability through expert judgement' in Chapter 4) who used their judgment and experience of rockfall incidents elsewhere to assign expected fatality statistics to each of the consequence scenarios, taking account of the population model.

The overall risk at the site, expressed in terms of potential for fatalities, was calculated as the product of the probability of the rockfall event and the sum of the fatalities associated with each of the consequence scenarios:

$$\text{Risk (Event } r) = \text{Prob. (Event)} \times \sum (\text{Prob. Adverse Consequences}$$

$$\times \text{(Consequence Scenarios 1 to } n))$$

As a range of rockfall event magnitudes/probabilities were considered, the overall risk at the site was calculated as

$$\text{Total Risk} = \sum (\text{Risk, Events 1 to } r)$$

Example 5.9
A young child was killed by a gas explosion in a house built on a pre-existing landslide. The explosion took place when an automatic timer switch controlling a washing machine in the basement caused a spark. A gas pipeline had failed as a result of a landslide reactivation event and the escaping gas had built up in the basement. Figure 5.10 presents a fault tree developed to trace the causal factors involved in the accident and estimate the probability of this particular consequence scenario. Note that it is one of numerous scenarios that could have arisen following the landslide reactivation event, the overwhelming majority of which would not have resulted in a fatality.

The key stages in the accident sequence are as follows:

- The child being present when the gas explodes. The child was playing with a wooden train set that had been laid out on the floor of the basement. The basement was used by the child every day in the morning (8 a.m. to 12 noon) for around one hour, so the probability that it was occupied at the time of the explosion was estimated to be:

$$\text{Prob. (Occupied)} = \text{Prob. (Morning Explosion)}$$

$$\times \text{Prob. (Child Present)}$$

$$= 4/24 \times 1/4 = 0.0416$$

It was assumed that if the child was caught in the explosion the likelihood of death was very high (probability $= 1.0$).

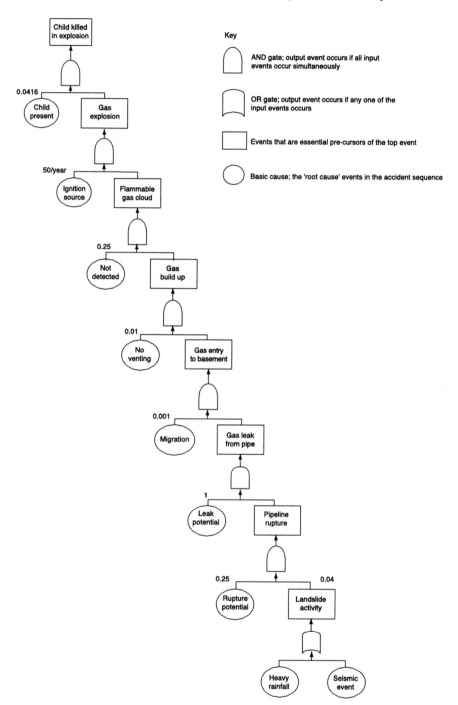

Fig. 5.10. Example 5.9: fault tree showing the sequence of events that led to the child being killed in the gas explosion (with acknowledgements to O'Riordan and Milloy, 1995)

317

- A spark in the automatic switch igniting the gas cloud. The switch was set to come on once a week (except when the family was away for two weeks on their annual holidays) and, hence, the frequency of the ignition source was 50 times a year.
- The gas build up was not detected as there were no gas detectors in the basement. It was assumed that although the child was not familiar with the smell of leaking gas, the parents might have raised the alarm. Hence, the probability that the gas was not detected was assumed to be 0.25.
- The gas built up in the basement. There was no mechanical ventilation in the basement, but the doors do not provide a perfect seal. The cloud built up to explosive limits overnight. However, the build up might have been partly dissipated when the basement door was opened in the morning. The probability that opening of the basement doors had failed to prevent the build up of the gas was judged to be 0.01.
- The entry of gas into the basement. The concrete slab on the basement floor had a number of cracks, probably as a result of poor workmanship or the effects of previous episodes of ground movement. The probability that the gas would have been able to migrate from the rupture point and entered the basement through the cracks was judged to be 0.001.
- The leakage of gas from the pipe. If the pipe had ruptured then, provided the pipe was carrying gas, there would be a leak. The probability of there being gas in the pipe at the time of rupture was assumed to be 1.0.
- The rupture of the pipe. The probability that the landslide movements would have been of sufficient intensity to cause rupture of the pipe was judged to be 0.25 (i.e. 25% of reactivation events would be of insufficient intensity to cause pipe rupture).

From the fault tree analysis, the probability of the fatal accident scenario was estimated to be

$$\text{Prob. (Consequence Scenario)} = 0.0416 \times 50 \times 0.25 \times 0.01 \times 0.001$$

$$\times 1 \times 0.25$$

$$= 0.000\,001\,3 \text{ or } (1.3 \times 10^{-4})$$

Given that the annual probability of landslide reactivation was estimated to be 0.04 (based on historical records of landslide activity in the area), the overall probability of the accident was

$$\text{Prob. (Fatal Accident)} = \text{Prob. (Event)} \times \text{Prob. (Consequence Scenario)}$$

$$= 0.04 \times 0.000\,001\,3$$

$$= 0.000\,000\,052 \text{ or } (5.2 \times 10^{-8})$$

Complex outcomes and uncertain futures

This chapter has emphasised that the potential for adverse consequences is the product of the specific hazard and exposure conditions at a particular location at a specific time. Uncertainties regarding the magnitude, character and timing of a future landslide event are compounded by uncertainties as to exactly what will suffer impact and how severe that impact will be. Slight variations in the magnitude or timing of a slope failure can result in quite dramatic variations in the level of detriment, especially where human lives are concerned. As a result, there might be need for an enormous number of possible consequence models or scenarios, in order to define and quantify the possible outcomes, especially for compound, multiple or complex events.

A balance needs to be found between the effort and resources involved in evaluating all possible outcomes and the generalisations involved in attempting to characterise a 'best estimate' scenario. On the one hand, it may be the scale of a 'worst case' scenario (e.g. the train collision described in Section 'Multiple outcome consequence models' of this chapter) that drives decision-making, on the other hand the analysis of such extreme or catastrophic scenarios may divert attention away from more likely, but less dramatic consequences (e.g. the impact of recurrent building damage caused by ground movement on the local economy). Carefully defined scoping can be useful in limiting any consequence assessment to the particular requirements of a specific risk assessment, but even so problems will occur in determining which of the full range of adverse consequences to focus on.

It is important to appreciate that whichever scenarios are considered, they will have a short life span in terms of their validity, because the nature and value of the assets at risk are almost always continuously changing. While it is difficult enough to foretell patterns of human activity in the short term and even to obtain agreement as to the values that should be placed on different artefacts and environmental resources, it becomes especially hard to assess what the situation will be like at some future date. Exponents of *Futurology* attempt to foretell the nature of society and how it will operate at varying times into the future, but hindsight reviews indicate that such 'predictions' have not been particularly successful to date (e.g. why are we not all dressed in aluminium foil suits and travelling in hover-cars?).

Population growth, economic growth and the progressive spread of human activity and infrastructure into increasingly remote and environmentally hostile areas will mean that the potential for landslide generated losses will grow. In addition, increasing standards of living will result in rising property values and the increasingly widespread ownership of expensive household goods such as televisions, washing machines, fridge-freezers etc., all of which are vulnerable to damage. Suleman et al. (1988) suggest that the

319

damage potential with reference to floods rose by over 50% for short duration and 100% for long duration flood events, between 1977 and 1987; a similar trend can be envisaged for landslide events.

Ideally, those scenarios considering the possibility of events in the more distant future could attempt to include development outcomes, changes in population distribution, numbers and concentrations, and changes in the values of assets, although all will become more uncertain with increasing time into the future. In reality, however, most consequence assessments use the current distribution and value of assets as a baseline and project this into the future, thereby controlling the speculation.

6

Quantifying risk

Introduction

Risk is expressed as the product of the likelihood of a hazard (e.g. a damaging landslide event) and its adverse consequences, that is:

Risk = Prob. (Landslide event) × Adverse Consequences

The main aim of quantitative risk assessment (QRA) is to reduce risk to a mathematical value, preferably a single mathematical value. This, it is often argued, facilitates better communication and aids appreciation, thereby improving decision making.

Reducing risk to a single mathematical value also makes it possible to compare landslide 'risk' with other risks, such as flooding, in order to determine their relative significance. However, it is important to bear in mind that the same, or similar, computed values of 'risk' can be the product of very different combinations of probability and consequence. For example, in a given year, a 1 in 2 chance (Prob. = 0.5) of landslide movement (slow creep) causing £1000 worth of damage to a footpath generates the same 'risk' as a 1 in 100 chance (Prob. = 0.01) of debris flow causing £50 000 of damage, or a 1 in 10 000 chance (Prob. = 0.0001) of a catastrophic large run-out landslide with losses of £5 million, that is all three have *mathematical expectation values* of £500 within the time period. Thus mathematically computed measures of 'risk' may be the same, although the 'perceived threats' are rather different (see Section 'Risk evaluation' in Chapter 1). As a result, the attitudes of decision-makers and the public to these three problems may differ significantly because of the differing scale of the potential losses, resulting in contrasting views as to their prioritisation and the way that they may be managed, thus necessitating the process of risk evaluation (see Chapter 7).

Bearing in mind the uncertainties and assumptions involved in developing hazard models, estimating the probability of landsliding and placing values on the adverse consequences, it is often misleading to present a single point estimate of risk (see Section 'Uncertainty and risk assessment' in Chapter 1). *Sensitivity testing* can be used to examine how sensitive the estimated risk value is to changes in the input parameters and underlying assumptions. Typically it will be useful to repeat the analysis using a range of landslide probabilities, from a '*worst case*' estimate to a lower bound estimate. Similarly, the values of the adverse consequences could be varied to reflect uncertainties in the price base that might have been used.

Current annual risk
In many instances it will be sufficient to provide an estimate of the landslide risk under current conditions, i.e. *what is the risk at the moment?* This can be expressed as an annual value, as illustrated in the following examples.

Example 6.1
A hilltop community is vulnerable to major first-time landslide events that could result in the sudden loss of a strip of land up to 100 m wide. Such events have occurred in the past and their historical frequency suggests an annual probability for a further event of 1 in 400 (0.0025). A total of 25 buildings situated within 100 m of the edge would be destroyed by the landslide, resulting in losses of £3.75 million at current values. It is expected that there would be further indirect losses to the local community, resulting from a decline in tourist numbers and disruption to businesses; these indirect losses are expected to total £5 million. The risk, expressed as an annual value, is (Table 6.1):

$$\text{Annual Risk} = \text{Prob. (Landslide event)} \times (\text{Direct Losses} + \text{Indirect Losses})$$

Example 6.2
A large town has grown-up within an area of deep-seated landsliding. Throughout its history it has been subjected to very slow ground movements, with less frequent episodes of more active movement. The landslide hazard is

Table 6.1. Example 6.1: Annual risk calculation

Probability of first-time landslide	Direct losses: £ million	Indirect losses: £ million	Annual risk: £ million
0.0025	3.75	5	0.022

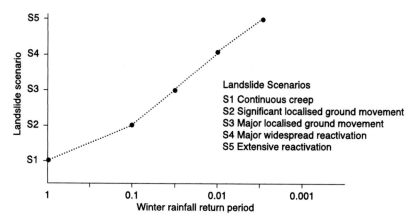

Fig. 6.1. Example 6.2: the relationship between winter rainfall and landslide reactivation

associated with the reactivation of the landslides and varies from almost continuous deep-seated movement, in the order of mm/year, to infrequent, short periods of significant ground movement that can result in widespread surface cracking and heave. Reactivation events are triggered by the high groundwater levels that coincide with wet winters. The wetter the winter, the greater the resulting ground movement, that is minor events are regular occurrences, while large events are infrequent. It is assumed, therefore, that there is a continuous range of discrete reactivation events, with each magnitude of event associated with a different return period (i.e. probability) of winter rainfall (Fig. 6.1).

The losses associated with reactivation can be identified as a combination of repairable damage and property destruction, together with traffic disruption, with the severity of losses related to the size of the reactivation event, as shown in Table 6.2.

From this information it is possible to construct a landslide reactivation damage curve that relates losses to the probability of the event (Fig. 6.2). As any of the events could occur in a given year, depending on the winter rainfall, it is necessary to calculate the average annual damage so as to take account of every possible combination of event probability and loss (Table 6.3). In Fig. 6.2 the average annual risk equates to the area below the damage curve, calculated as a series of slices between event probabilities.

Example 6.3

A mountain road between two small towns in the Scottish Highlands is regularly blocked by rockfalls and the traffic is forced to follow a lengthy diversion around the affected area. In Example 5.5 a consequence model

323

Table 6.2. Example 6.2: Relationship between winter rainfall events and landslide damage

Reactivation scenario	Winter rainfall* event (return period)	Property damage: and destruction: £ million	Traffic disruption: £ million
1	1 in 1	0.25	0.1
2	1 in 10	0.5	0.2
3	1 in 50	1	0.5
4	1 in 100	5	1
5	1 in 500	10	5

Note. *Winter rainfall is the overall total for the months of October to January.

was presented to estimate the traffic disruption costs associated with three rockfall scenarios:

- Scenario 1. A small rockfall ($<100\,m^3$) which results in the road being closed for one day, during which the debris is removed by the road maintenance crews (estimated annual probability of 0.3).

- Scenario 2. A relatively small rockfall ($100–1000\,m^3$) which results in the road being closed for five days because of the need for repairs to the road surface (estimated annual probability of 0.05).

- Scenario 3. A large rockfall ($>10\,000\,m^3$) which results in the road being closed for 50 days, during which time emergency works are undertaken

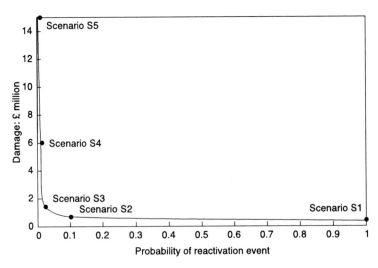

Fig. 6.2. Example 6.2: landslide damage curve

324

Table 6.3. Example 6.2: Annual risk calculation (see Fig. 6.2 for the landslide damage curve)

Damage category	Return period (years) and annual probability of reactivation event: damage (£ million)				
	1 1.000	10 0.100	50 0.020	100 0.010	500 0.002
Residential property	0.25	0.5	1	5	10
Industrial property	0	0	0	0	0
Indirect losses (e.g. tourism)	0	0	0	0	0
Traffic disruption	0.1	0.2	0.5	1	5
Emergency services	0	0	0	0	0
Other	0	0	0	0	0
Total damage	0.35	0.7	1.5	6	15
Area (damage × frequency)		0.47	0.09	0.04	0.08
Average annual risk (area beneath curve): £ million					0.68

Notes. Area (damage × frequency) for between the 1 in 1 (Scenario 1) and 1 in 10 (Scenario 2) year events is calculated as: Area (damage × frequency) = (Prob. 1 − Prob. 2) × (Total Damage 2 + Total Damage 1)/2; Average annual risk = $\sum$ Area (damage × frequency).

to stabilise the unstable rockfall backscar and repair the road surface (estimated annual probability of 0.01).

As explained in Example 5.5, a potential loss of £11 900 occurs for each day that the road is blocked.

Ignoring any threat to the road users (see Section 'Individual risk', below) and clean-up or repair costs, the risk associated with the traffic disruption resulting from each individual scenario is:

Risk (Scenario s) = Prob. (Event) × Total Disruption Cost

= Prob. (Event) × Daily Disruption Cost

× Length of Delay

Risk (Scenario 1) = 0.3 × 11 900 = £3570

Risk (Scenario 2) = 0.05 × (11 900 × 5) = £2975

Risk (Scenario 3) = 0.01 × (11 900 × 50) = £5950

Simply adding the annual risk associated with each of these three scenarios (£12 495) would underestimate the overall risk because, in reality, there is a continuous range of discrete rockfall events with a characteristic magnitude

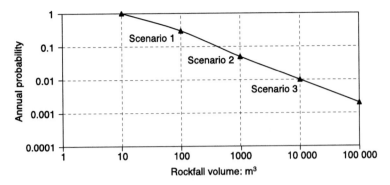

Fig. 6.3. *Example 6.3: rockfall magnitude/probability distribution*

frequency relationship (Fig. 6.3; see Example 2.11). However, by using the three scenarios it is possible to construct a landslide damage curve that relates losses to the probability of the event (Fig. 6.4). As in the previous example, any of the rockfall events could occur in a given year. By calculating the *average annual damage*, account is taken of every possible combination of rockfall probability and loss (Table 6.4). The resulting value of £26 190 turns out to be double the value obtained by simply adding the risk associated with the three scenarios.

It is important to note that three scenarios probably represent the minimum number of points required to generate a landslide damage curve. However, if there is a good correlation between event probability and damage, then it is generally not necessary to model every conceivable scenario in order to generate a reasonable indication of the average annual risk.

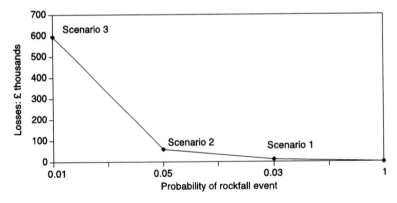

Fig. 6.4. *Example 6.3: landslide damage curve*

326

Table 6.4. Example 6.3: Annual risk calculation (see Fig. 6.4 for the landslide damage curve)

Damage category	Return period (years) and annual probability of landslide event: damage (£ thousands)			
	1 1.000	3.3 0.3	20 0.05	100 0.01
Residential property	0	0	0	0
Industrial property	0	0	0	0
Indirect losses (e.g. tourism)	0	0	0	0
Traffic disruption	0	11.9	59.5	595
Emergency services	0	0	0	0
Other	0	0	0	0
Total damage	0	11.9	59.5	595
Area (damage × frequency)		4.16	8.94	13.09
Average annual risk (area beneath curve): £ thousands				26.19

Notes. Area (damage × frequency) for between the 1 in 1 (Scenario 1) and 1 in 3.3 (Scenario 2) year events is calculated as: Area (damage × frequency) = (Prob. 1 − Prob. 2) × (Total Damage 2 + Total Damage 1)/2; Average annual risk = $\sum$ Area (damage × frequency).

Cliff recession risk

Where assets are threatened by retreating clifflines, as on the coast or some river meander scars, losses are inevitable at some point in the future because of the on-going recession. Decision-makers will tend to view the severity of the recession problem in terms of when the assets are likely to be lost and the value of the expected losses. Risk, therefore, needs to be expressed in terms of which year the losses will occur or the chance of loss in a particular year in the future.

The approach is different from the assessment of current annual risk as outlined in Section 'Current annual risk' in this chapter. An important concept that must be appreciated is that the economic value of any asset does not remain constant for the timescale over which the risk is being considered. If an asset is lost in Year 1 it will normally have a higher economic value than if it were to be lost in Year 10. One reason for this is because people discount the future, preferring their benefits now rather than sometime in the future; a phenomenon known as *time preference* ('Goods now or goods in the future').

The second reason is the *productivity of capital*. If landslide management were not undertaken, then the resources could be diverted elsewhere and, over time, would show a return. For example, if the return from investing the resources in industry were 5% in real terms, then a £100 sum invested now would yield the equivalent of £105 in one year's time. Therefore,

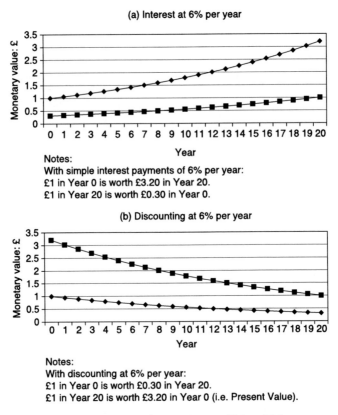

(a) Interest at 6% per year

Notes:
With simple interest payments of 6% per year:
£1 in Year 0 is worth £3.20 in Year 20.
£1 in Year 20 is worth £0.30 in Year 0.

(b) Discounting at 6% per year

Notes:
With discounting at 6% per year:
£1 in Year 0 is worth £0.30 in Year 20.
£1 in Year 20 is worth £3.20 in Year 0 (i.e. Present Value).

Fig. 6.5. The effect of discounting to achieve a Present Value (PV)

having £100 now and £105 in one year are equivalent. For an asset that gives no return, £100 in one year is worth less than £100 now; in other words, it has depreciated (see Fig. 6.5).

To achieve this modification of future asset values, it is necessary to express all future losses in terms of their Present Value (PV) by discounting. The discount factor applicable to a particular year n can be calculated as follows:

$$D_n = (1 + r)^{-n} \quad \text{or} \quad D_n = \frac{1}{(1 + r)^n}$$

where r is the discount rate, expressed as a decimal, and D_n the discount factor applicable for year n.

The sum of the discounted flows (i.e. Present Value) is expressed at the mid-point of Year 0.

At the time of writing (2002) the UK Government discount rate was 6% (MAFF, 1999a). The effect of using a discount rate is to reduce the value of predicted future losses to their value as seen from the present day. Thus if the present value of an asset is £1 million in Year 0, it will decline to £0.56 million in Year 10 and £0.06 million in Year 50 if a 6% discount rate is used.

Two different examples are presented below which show the potential application of this approach. The first can be described as a deterministic method, in that the asset is assumed to be lost in a particular year. The second is a probabilistic method, developed by Hall *et al.* (2000), which recognises that there is a chance that the assets may be lost in any year.

Example 6.4
A small coastal community is threatened by cliff recession (see Example 5.1). Inspection of historical topographic maps of different dates has indicated that the cliffline has been retreating at an average annual rate of 1.25 m/year over the last 150 years. Continued recession over the next 50 years will threaten a number of properties and important services, including a gas main and sewer pump station. It is assumed that the future pattern of recession can be determined by simply projecting the historic average annual rate into the future; that is in 10 years time 12.5 m will have been lost, while in 50 years the cliff will have retreated 62.5 m.

The risk assessment involves extrapolating the average recession rate in order to estimate the year in which particular properties will be lost. The Present Value (PV) risk of the lost assets is the summation of the discounted asset losses over the period of consideration (in this case 50 years). For each year the calculation is

Risk (Year T) = Market Value × Discount Factor (Year T)

and over 50 years:

$$\text{Risk (Years 0–49)} = \sum (\text{Market Value} \times \text{Discount Factor (Years 0–49)})$$

Note that the market value is the price for which an asset is bought or sold in an open market and is used as a measure of the economic value of the asset.

The results are presented in Table 6.5 (note that although the time period under consideration is 50 years, no losses were predicted to occur after Year 25 and so the table has been shortened to save space). Thus, although the current market value of the assets is £6 226 000, the Present Value (PV) risk is lower at £2 952 100. For example, the properties 1–5 Acacia Avenue are currently valued at £350 000, but are expected to be lost in Year 8. Applying the discount factor for Year 8 (0.6274) produces a Present Value (PV) of £219 600.

Example 6.5
The Holderness cliffs, UK, range in height from less than 3 m to around 40 m. The cliffs are formed in a sequence of glacial tills, predominantly silty clays

Table 6.5. Example 6.4: Present Value risk calculation

Year	Discount factor	Property lost due to recession	Market value: £ thousands	Present value risk: £ thousands
0	1.000			0
1	0.943			0.0
2	0.890			0.0
3	0.840			0.0
4	0.792			0.0
5	0.747			0.0
6	0.705			0.0
7	0.665	Trunk gas main	450	299.3
8	0.627	1–5 Acacia Ave	350	219.6
9	0.592			0.0
10	0.558	Café	120	67.0
11	0.527	Sunnyview and Dunswimmin	180	94.8
12	0.497	Sewage Pump Station	2300	1143.0
13	0.469	Hightrees House	1500	703.3
14	0.442			0.0
15	0.417			0.0
16	0.394			0.0
17	0.371	2–8 Acacia Ave	400	148.5
18	0.350	The Saltings	76	26.6
19	0.331			0.0
20	0.312			0.0
21	0.294	14–20 Rocco Blvrd	850	250.0
22	0.278			0.0
23	0.262			0.0
24	0.247			0.0
25	0.233			0.0
Total			6226	2952

Notes. Discount rate $= 6\%$; Present Value Risk (Year T) $=$ Discount Rate (Year T) $\times$ Market Value (Property Lost, Year T).

with chalky debris and lenses of sand and gravel. They are subject to severe marine erosion, but remain unprotected for most of their length. Long term recession rates are in the order of 1.2–1.8 m/year (Valentin, 1954; Pethick and Leggett, 1993). At the example site, ongoing recession threatens cliff top land and property, including a caravan site.

A probabilistic model was developed by Jim Hall, and presented in Lee and Clark (2002), to simulate the episodic cliff recession process on the cliffline (see Example 2.10; Chapter 2). The statistical model was based on generating random sequences of landslide event sizes and of durations between events, with statistics that conform to the measured values at the sites. From each random sequence, the annual recession distance between

Years 1 and 49 was extracted. A large number of simulations (in this case 10 000) were used to generate a histogram or annual recession distance. A kernel density estimation method was then used to obtain a smooth probability density estimate from the histogram.

The analysis involves estimating both the levels of damages/losses that could result from a particular event (i.e. the loss of cliff-top land) and the probability that such an event occurs in a particular year. The potential losses were calculated for each 1 m-wide strip of cliff-top land between the present cliff edge and 200 m further inland. As soon as the cliff edge encroaches within a strip containing part of an asset, the asset is assumed to be completely lost.

For example, considering the strip of land between the cliff edge and 1 m inland (*Strip 1*), the present value (PV) of the losses associated with recession in a particular year (Year *T*) was calculated as follows:

PV losses (Strip 1; Year *T*) = Prob. (Loss Strip 1; Year *T*)

$$\times \text{ asset value (within Strip 1)}$$

$$\times \text{ discount factor (Year } T)$$

As Strip 1 could be lost (or not) in any year over a 50-year period, the PV of losses associated with the event (i.e. loss of Strip 1) is the sum of the annual losses (Year 0–49) for Strip 1.

For another 1 m-wide strip of land, located say 9–10 m from the present cliff edge (i.e. Strip 10):

PV losses (Strip 10; Year *T*) = Prob. (Loss Strip 10; Year *T*)

$$\times \text{ asset value (within Strip 10)}$$

$$\times \text{ discount factor (Year } T)$$

The overall risk is the sum of the PV losses for each 1 m-wide strip of cliff-top land over a 50-year period, that is the risk, per unit of market value (MV), can be calculated directly from:

$$PV(\text{damage risk}) = MV \sum_{i=0}^{j} \frac{f_{XT}(i \mid X = x)}{(1+r)^i}$$

where j is the appraisal period, r is the discount rate and $f_{XT}(i \mid X = x)$ is the time to recede a given distance.

The results are presented in Table 6.6 (note that for ease of reproduction the timescale has been reduced to 25 years and that the recession distance limited to 15 m). The sum of the risk for that part of the example presented in Table 6.6 is about £523 990.

331

Table 6.6. Example 6.5: Present Value risk calculation

	Recession distance: m															
	0	1	2	3	4	5	6	7	8	9	10	11	12	13	14	15
A Asset values at given distance: £ thousands	0	0	0	0	0	0	0	0	0	0	140	0	200	0	350	140
B Sum of annual probabilities times discount rate	0.91	0.86	0.84	0.82	0.80	0.78	0.76	0.74	0.72	0.70	0.68	0.67	0.65	0.63	0.61	0.60
C Risk at given distances: £ thousands, i.e. A × B	0.00	0.00	0.00	0.00	0.00	0.00	0.00	0.00	0.00	0.00	95.58	0.00	129.7	0.00	214.9	83.79

		Simulated probability of receding given distance in given year: m															
Year	Discount factor	0	1	2	3	4	5	6	7	8	9	10	11	12	13	14	15
1	0.94	0.65	0.24	0.23	0.16	0.12	0.08	0.07	0.05	0.03	0.02	0.02	0.02	0.01	0.01	0.01	0.01
2	0.89	0.20	0.34	0.25	0.22	0.18	0.15	0.12	0.10	0.08	0.06	0.05	0.04	0.03	0.03	0.02	0.02
3	0.84	0.08	0.20	0.20	0.20	0.19	0.18	0.15	0.13	0.12	0.10	0.08	0.07	0.06	0.05	0.04	0.03
4	0.79	0.04	0.11	0.13	0.15	0.16	0.16	0.15	0.14	0.13	0.12	0.11	0.10	0.08	0.07	0.06	0.06
5	0.75	0.02	0.05	0.08	0.10	0.12	0.13	0.14	0.14	0.14	0.13	0.12	0.11	0.10	0.09	0.08	0.07
6	0.70	0.01	0.03	0.05	0.07	0.08	0.10	0.11	0.12	0.13	0.13	0.12	0.12	0.11	0.10	0.10	0.09
7	0.67	0.01	0.02	0.03	0.04	0.06	0.07	0.08	0.09	0.10	0.11	0.12	0.12	0.11	0.11	0.10	0.10
8	0.63	0.00	0.01	0.02	0.02	0.04	0.05	0.06	0.07	0.08	0.09	0.10	0.10	0.10	0.10	0.10	0.10
9	0.59	0.00	0.00	0.01	0.01	0.02	0.03	0.04	0.05	0.06	0.07	0.08	0.08	0.09	0.09	0.10	0.09
10	0.56	0.00	0.00	0.00	0.01	0.01	0.02	0.03	0.03	0.04	0.05	0.06	0.07	0.08	0.08	0.08	0.09
11	0.53	0.00	0.00	0.00	0.00	0.01	0.01	0.01	0.02	0.03	0.04	0.04	0.05	0.06	0.07	0.07	0.08
12	0.50	0.00	0.00	0.00	0.00	0.01	0.01	0.01	0.02	0.02	0.03	0.03	0.04	0.05	0.05	0.06	0.06
13	0.47	0.00	0.00	0.00	0.00	0.00	0.00	0.01	0.01	0.01	0.02	0.02	0.03	0.03	0.04	0.05	0.05

Year																				
14	0.44	0.00	0.00	0.00	0.00	0.00	0.00	0.00	0.00	0.00	0.00	0.01	0.01	0.01	0.02	0.02	0.02	0.03	0.04	0.04
15	0.42	0.00	0.00	0.00	0.00	0.00	0.00	0.00	0.00	0.00	0.00	0.00	0.01	0.01	0.01	0.01	0.02	0.02	0.03	0.03
16	0.39	0.00	0.00	0.00	0.00	0.00	0.00	0.00	0.00	0.00	0.00	0.00	0.00	0.00	0.01	0.01	0.01	0.02	0.02	0.02
17	0.37	0.00	0.00	0.00	0.00	0.00	0.00	0.00	0.00	0.00	0.00	0.00	0.00	0.00	0.00	0.01	0.01	0.01	0.01	0.02
18	0.35	0.00	0.00	0.00	0.00	0.00	0.00	0.00	0.00	0.00	0.00	0.00	0.00	0.00	0.00	0.00	0.01	0.01	0.01	0.01
19	0.33	0.00	0.00	0.00	0.00	0.00	0.00	0.00	0.00	0.00	0.00	0.00	0.00	0.00	0.00	0.00	0.00	0.00	0.01	0.01
20	0.31	0.00	0.00	0.00	0.00	0.00	0.00	0.00	0.00	0.00	0.00	0.00	0.00	0.00	0.00	0.00	0.00	0.00	0.00	0.01
21	0.29	0.00	0.00	0.00	0.00	0.00	0.00	0.00	0.00	0.00	0.00	0.00	0.00	0.00	0.00	0.00	0.00	0.00	0.00	0.01
22	0.28	0.00	0.00	0.00	0.00	0.00	0.00	0.00	0.00	0.00	0.00	0.00	0.00	0.00	0.00	0.00	0.00	0.00	0.00	0.00
23	0.26	0.00	0.00	0.00	0.00	0.00	0.00	0.00	0.00	0.00	0.00	0.00	0.00	0.00	0.00	0.00	0.00	0.00	0.00	0.00
24	0.25	0.00	0.00	0.00	0.00	0.00	0.00	0.00	0.00	0.00	0.00	0.00	0.00	0.00	0.00	0.00	0.00	0.00	0.00	0.00
25	0.23	0.00	0.00	0.00	0.00	0.00	0.00	0.00	0.00	0.00	0.00	0.00	0.00	0.00	0.00	0.00	0.00	0.00	0.00	0.00
Overall probability	1.00	1.00	1.00	1.00	1.00	1.00	1.00	1.00	1.00	1.00	1.00	1.00	1.00	1.00	1.00	1.00	1.00	1.00	1.00	1.00

Note. The sum of the risk is the sum of Row C = £523,990.

A: This row contains the market value of properties within each 1 m-wide strip of cliff-top land (i.e. a £140 000 house lies 10 m from the cliff edge)

B: This row contains values for the product of the discount rate × probability of loss for each year from 1 to 25.

C: This row contains risk values calculated as: Risk (Strip 10) = Row A (Strip 10) × Row B (Strip 10) = 140 × 0.6827 = 95.58.

Note that:

- Row A (*asset values £ thousands at given distance*) has been determined from estate agent valuations.
- Row B (*sum of annual probability times discount rate*) is calculated from

$$\text{Sum (1 m recession)} = \sum \text{Prob. Receding 1 metre}$$
$$\times \text{Discount Rate (Years 1–25)}$$

- Row C (*risk at a given distance*) is calculated for each 1 m-wide strip from

$$\text{Risk (Strip 1)} = \text{Row A} \times \text{Row B (Strip 1)}$$

- *The simulated probabilities of receding a given distance in a given year (m) have been derived from the probabilistic model* (see Example 2.10 for details).
- The *discount rate* used is 6%.

Comparing the risks associated with different management options

Risk-based methods can also be used to compare the level of risk (i.e. mathematical expectation value) associated with different management strategies. By using risk-based methods to compare strategies they act as a *project* or *options appraisal* tool. One approach could be simply to compare the annual risk between the current situation and that expected to occur with the proposed management strategy in place:

$$\text{Risk (Current Situation)} = \text{Prob. (Landslide)} \times \text{Losses}$$

compared with

$$\text{Risk (with Management Strategy)} = \text{Prob. (Landslide)} \times \text{Losses}$$

This would identify which is the most effective strategy. However, the purpose of making such a comparison is usually to address the question 'is it worth it?' to make the investment in a particular management strategy (i.e. does the risk reduction achieved justify the cost):

$$\text{Risk Reduction Benefits} = \text{'Without Project' Risk} - \text{'With Project' Risk}$$

The decision to invest in landslide management activity should depend on a thorough appraisal of the benefits of risk reduction over the *expected lifetime* of the scheme/project. The simple comparison between the annual risk associated with the current situation and that achieved following a stabilisation project can significantly underestimate the risk reduction achieved. This is because landslide stabilisation reduces risk in every year until the end of its design life (and often beyond), although major costs are only incurred at the time of construction (Year 0). Maintenance and repair costs will, of course, be spread over the scheme life. The benefits of management are the difference between the value of the losses that could be expected to be incurred without a scheme/project and the value of the losses that would be incurred when the scheme fails and landslide activity is renewed. An alternative way of looking at this is that the management scheme reduces the probability of ground movement and thereby reduces the risk over its expected lifetime, *not just in a single year*.

For example, a stabilisation scheme with a design life of 50 years will reduce the risk over a period of 50 years. Thus the benefits of this management option are

Landslide Management Benefits = 'Without Project' Risk (Years 0–49)

– 'With Project' Risk (Years 0–49)

or

Landslide Management Benefits = 'Without Project' Losses (Years 0–49)

–'With Project' Losses (Years 0–49)

As highlighted in the previous section, the economic value of risk reduction in future years is worth less than that achieved at present. It is, therefore, necessary to express all future risks in terms of their Present Value (PV), by discounting

'Without Project' Risk (Years 0–49)

$$= \sum (\text{Prob. Event} \times \text{Losses} \times \text{Discount Factor (Years 0–49)})$$

'With Project' Risk (Years 0–49)

$$= \sum (\text{Prob. Event} \times \text{Losses} \times \text{Discount Factor (Years 0–49)})$$

A variety of options can be compared by decision-makers in order to select the most 'desirable'. To make this comparison it is necessary to establish a baseline against which the various options, including continuing with the current management practice, can be assessed. This baseline is the so-called *do nothing* option, which should involve no active landslide management whatsoever, simply walking away and abandoning all maintenance, repair or management activity. It should be easy to demonstrate that continuing the current practices are better than the *do nothing* case.

Example 6.6
A cliff top community is vulnerable to a landslide event with an annual probability 0.01. The event could result in estimated total losses of

£25 million. A combined seawall and slope stabilisation scheme is proposed that would prevent further cliff foot erosion and reduce the probability of a major event to an estimated 1 in 1000 (0.001).

The comparison of the 'Without Project' case (i.e. 'do nothing') and the 'With Project' case (i.e. the scheme) is presented in Table 6.7. Note that to reduce the space required for this table, calculations for only 21 of the 50 year design life are shown and that:

PV (Without Project) Risk (Year *T*)

  = Discount Factor (Year *T*)

    × Prob. (Landslide occurring in year) × Losses (£25 million)

PV (With Project) Risk (Year *T*)

  = Discount Factor (Year *T*)

    × Prob. (Landslide occurring in year) × Losses (£25 million)

PV Risk Reduction = Without Project Risk (Year 0–20)

      − With Project Risk (Years 0–20)

It should be noted that for events which can be assumed to occur only once at a particular site (i.e. *one-off events*), it is necessary to take account of the fact that the event may have already occurred in Year 1 and, hence, could not occur in Year 2 (and so on). Thus, the annual probability for Year 2 (and subsequent years) is modified, as follows:

Prob. (Event; Year *T*)

  = Annual Prob.

    × (Prob. Event has not already occurred by Year *T* − 1)

From Table 6.7, the risk reduction achieved by the scheme is:

PV without project risk	PV with project risk	PV risk reduction
£2.88 million	£0.31 million	£2.57 million

Example 6.7
Periodic reactivation of an extensive area of deep-seated landsliding causes a combination of repairable damage and property destruction, with the severity of damage related to the size of the reactivation event. It is proposed to undertake extensive stabilisation works to reduce the frequency of reactivations. It is assumed that the scheme will reduce the event probability by a

Table 6.7. *Example 6.6: Present Value risk calculation (without project and with project cases)*

Year	Discount factor	Without project case				With project case				PV risk	
		Prob. of a landslide	Prob. landslide occurring in year	Prob. landslide has not occurred	Cumulative probability of landslide	Prob. of a landslide	Prob. landslide occurring in year	Prob. landslide has not occurred	Cumulative probability of landslide	PV without project losses: £ million	PV with project losses: £ million
0	1.000	0.010	0.010	0.990	0.010	0.001	0.001	0.999	0.001	0.25	0.03
1	0.943	0.010	0.010	0.980	0.020	0.001	0.001	0.998	0.002	0.23	0.02
2	0.890	0.010	0.010	0.970	0.030	0.001	0.001	0.997	0.003	0.22	0.02
3	0.840	0.010	0.010	0.961	0.039	0.001	0.001	0.996	0.004	0.20	0.02
4	0.792	0.010	0.010	0.951	0.049	0.001	0.001	0.995	0.005	0.19	0.02
5	0.747	0.010	0.010	0.941	0.059	0.001	0.001	0.994	0.006	0.18	0.02
6	0.705	0.010	0.009	0.932	0.068	0.001	0.001	0.993	0.007	0.17	0.02
7	0.665	0.010	0.009	0.923	0.077	0.001	0.001	0.992	0.008	0.15	0.02
8	0.627	0.010	0.009	0.914	0.086	0.001	0.001	0.991	0.009	0.14	0.02
9	0.592	0.010	0.009	0.904	0.096	0.001	0.001	0.990	0.010	0.14	0.01
10	0.558	0.010	0.009	0.895	0.105	0.001	0.001	0.989	0.011	0.13	0.01
11	0.527	0.010	0.009	0.886	0.114	0.001	0.001	0.988	0.012	0.12	0.01
12	0.497	0.010	0.009	0.878	0.122	0.001	0.001	0.987	0.013	0.11	0.01
13	0.469	0.010	0.009	0.869	0.131	0.001	0.001	0.986	0.014	0.10	0.01
14	0.442	0.010	0.009	0.860	0.140	0.001	0.001	0.985	0.015	0.10	0.01
15	0.417	0.010	0.009	0.851	0.149	0.001	0.001	0.984	0.016	0.09	0.01
16	0.394	0.010	0.009	0.843	0.157	0.001	0.001	0.983	0.017	0.08	0.01
17	0.371	0.010	0.008	0.835	0.165	0.001	0.001	0.982	0.018	0.08	0.01
18	0.350	0.010	0.008	0.826	0.174	0.001	0.001	0.981	0.019	0.07	0.01
19	0.331	0.010	0.008	0.818	0.182	0.001	0.001	0.980	0.020	0.07	0.01
20	0.312	0.010	0.008	0.810	0.190	0.001	0.001	0.979	0.021	0.06	0.01
									Total	2.88	0.31

Note: PV Risk (Year T) = Discount factor × Prob. (landslide occurring in Year T) × Losses (£25 million)

337

factor of 10; thus, a 1 in 15 year event, for example, would become a 1 in 150 year event with the scheme in place.

The average annual risk can be calculated from the area beneath reactivation damage curves developed for both the 'Without Project' and 'With Project' cases. The PV risks for both cases can be calculated from:

Risk (Years 0–49)

$$= \sum (\text{Average annual risk} \times \text{Discount Factor (Years 0–49)})$$

From Table 6.8, the risk reduction achieved by the scheme is:

PV without project risk	PV with project risk	PV risk reduction
£2 090 000	£355 000	£1 735 000

Example 6.8

Cliff recession is threatening a small cliff-top community. A proposed seawall scheme has been designed to safeguard the properties and delay any losses for the design life of the scheme which is 50 years (see Examples 5.1 and 6.4).

The risk reduction achieved by the scheme can be assessed using the deterministic approach developed by MAFF (1993). This considers the risk in terms of an annual value of the assets for each year until they are lost by cliff retreat. The logic of this approach becomes clear when the method is used to compare options for reducing the retreat rate, whereby the benefits of intervention are the resulting *increase* in asset value.

The benefit associated with each Year T that a given cliff-top asset remains usable is calculated by considering the cliff-top asset's risk-free market value. The risk-free market value, MV, can be thought of as being equivalent to the present value (PV) of n equal annual payments A (the equivalent annual value), where n is the life of the asset. If the annual payments occur from Year 1 to Year n, then:

$$A = \frac{MV \cdot r}{1 - D_n}$$

where r is the discount rate (6% in the UK, at the time of writing) and D_n the discount factor (see Section 'Cliff recession risk' in this chapter).

More usually the value of a cliff-top asset is thought to extend from Year 0 to Year $n - 1$, in which case:

$$A = \frac{MV}{1 - D_n} \frac{r}{1 + r}$$

Table 6.8. Example 6.7: Present Value risk calculation (without project and with project cases)

Damage category	Return period and annual probability of reactivation event									Total PV risk: £ thousands
	5 0.200	10 0.100	15 0.067	25 0.040	50 0.020	100 0.010	150 0.007	250 0.004	Infinity 0	
Without project										
Residential property			600	750	1000	2500	6000	7000	8500	2090
Industrial property									0	0
Indirect losses (e.g. tourism)									0	0
Total damage: £ thousands	0	0	600	750	1000	2500	6000	7000	8500	
Area (damage × frequency)		0.00	10.00	18.00	17.50	17.50	14.17	17.33	31.00	

Average annual risk (area beneath curve): £125 500; PV Factor, 16.650; Present Value risk £2 090 000

	5 0.200	10 0.100	15 0.067	25 0.040	50 0.020	100 0.010	150 0.007	250 0.004	Infinity 0	
With project										
Residential property							600	750	8500	355
Industrial property									0	0
Indirect losses (e.g. tourism)									0	0
Total damage: £ thousands	0	0	0	0	0	0	600	750	8500	
Area (damage × frequency)		0.00	0.00	0.00	0.00	0.00	1.00	1.80	18.50	

Average annual risk (area beneath curve): £21 300; PV Factor, 16.650; Present Value risk £355 000

Notes. Area (damage × frequency) for between the 1 in 5 (Scenario 1) and 1 in 10 (Scenario 2) year events is calculated as: Area (damage × frequency) = (Prob. 1 − Prob. 2) × (Total Damage 2 + Total Damage 1/2; Average annual risk = ∑ Area (damage × frequency). PV Factor = ∑ Discount factor (Years 0–49).

A can be approximated by

$$A = MV \times r$$

for annual payments starting in Year 1.

With reference to Table 6.9, the row of houses 1–5 Acacia Avenue has a total market value of £350 000 and an equivalent annual value (A) of

$$A = MV \times r = 350\,000 \times 0.06 = £21\,000$$

For the 'Without Project' case, it is assumed that these houses would be lost in Year 8. The asset value that would be lost is equivalent to eight years of an annual payment equal to the equivalent annual value (A) of £21 000. As each of these payments has to be brought back to a Present Value (PV), the 'Without Project' risk is

'Without Project' Risk

$$= \sum (A \times \text{Discount Factor}) \text{ for Years } 1–8 \text{ (excluding Year 0)}$$

$$= \sum (21\,000 \times \text{Discount Factor}) \text{ for Years } 1–8 = £130\,000$$

The overall 'Without Project' risk is the sum of the PV of assets lost in different years over the time period under consideration (in this case 50 years, although in order to save space, only 25 years are presented in Table 6.9).

Assuming that the scheme is implemented in Year 0 (i.e. immediately), a seawall with a 50 year design life would increase the life of the asset by 50 years (i.e. the life of 1–5 Acacia Avenue would be extended to 58 years). The 'With Project' risk is calculated as

'With Project' Risk

$$= \sum (A \times \text{Discount Factor}) \text{ for Years } 1–58 \text{ (excluding Year 0)}$$

$$= \sum (21\,000 \times \text{Discount Factor}) \text{ for Years } 1–58$$

$$= £338\,000$$

The overall 'With Project' risk is the sum of the PV of assets over the time period under consideration (Table 6.9).

The reduction in risk (i.e. the scheme benefits) achieved by the seawall is expressed in terms of the increase in asset value generated by extending the asset life:

Risk Reduction (PV) = PV 'With Project' asset value

$$- \text{PV 'Without Project' asset value}$$

Table 6.9. Example 6.8: Present value risk calculation (without project and with project cases)

Year	Property	MV: £ thousands	Utilities	MV: £ thousands	Equivalent annual value: A	Asset value: £ thousands	
						Without project	With project
0		0		0	0	0	0
1		0		0	0	0	0
2		0		0	0	0	0
3		0		0	0	0	0
4		0		0	0	0	0
5		0		0	0	0	0
6		0		0	0	0	0
7		0	Trunk gas main	450	27	151	434
8	1–5 Acacia Ave	350		0	21	130	338
9		0		0	0	0	0
10	Café	120		0	7	53	116
11	Sunnyview and Dunswimmin	180		0	11	85	175
12		0	Sewage PS	2300	138	1157	2238
13	Hightrees House	1500		0	90	797	1462
14		0		0	0	0	0
15		0		0	0	0	0
16		0		0	0	0	0
17	2–8 Acacia Ave	400		0	24	251	392
18	The Saltings	76		0	5	49	75
19		0		0	0	0	0
20		0		0	0	0	0
21	14–20 Rocco Blvrd	850		0	51	600	836
22		0		0	0	0	0
23		0		0	0	0	0
24		0		0	0	0	0
25		0		0	0	0	0
Totals		3476		2750		3274	6066

Notes. Equivalent Annual Value $(A) = MV \times$ Discount Factor (0.06); Asset Value (Without Project) $= \sum A \times$ Discount Factor (Years 1 to T), where T is the expected year of loss; Asset Value (With Project) $= \sum A \times$ Discount Factor (Years 1 to $T + 50$), where T is the expected year of loss

For 1–5 Acacia Avenue:

Risk Reduction (PV) $= £338\,000 - £130\,000 = £208\,000$

For the whole scheme the risk reduction is:

PV asset value – with project	PV asset value – without project	PV risk reduction
£6 066 000	£3 274 000	£2 792 000

Example 6.9
Probabilistic appraisal of cliff recession and the potential benefits of landslide management can be viewed as preferable to the use of deterministic methods because it takes a more 'realistic' account of the large uncertainties associated with the recession process. A probabilistic methodology for economic evaluation has been developed by Hall *et al.* (2000). The method takes the write-off value of the threatened assets and evaluates the loss associated with the probability that cliff retreat will result in the assets being written off in a given year.

The probability of property loss varies with time and location. The probability density function (p.d.f.) $f_{XT}(x, i)$ is the probability of damage at distance x from the cliff edge during Year i. The function f is a function of a distance random variable X and a time random variable T. For the purposes of benefit assessment, T is considered to be a discrete random variable measured in years.

For example, a single house is at threat from gradual cliff recession (albeit at an irregular rate). The setting is shown diagrammatically in Fig. 6.6. Profile *a* shows the current cliff position; three of the many possible future locations are labelled as *u*, *v* and *w*. If erosion has advanced to profile *w* then the asset will have been destroyed, whereas if it has proceeded only as far as profile *u* or *v*, it will not.

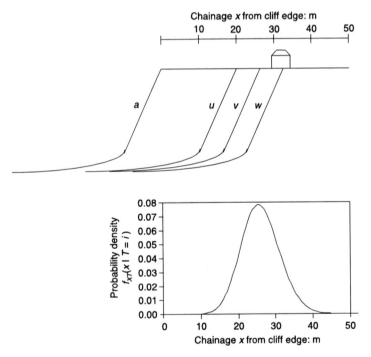

Fig. 6.6. Example 6.9: a schematic representation of the probability of cliff recession reaching a single property (after Hall et al., 2000)

The p.d.f. of cliff location in a particular year $(T = i)$, $f_{XT}(x \mid T = i)$ is shown on the same horizontal scale as the cliff cross-section. In a probabilistic analysis, there is no single conventional erosion contour for Year i but instead a band of potential erosion limits or risk. The contour of predicted average recession lies at the mean of the distribution. The probability p_i of damage to the house before or during Year i is given by:

$$p_i = \int_0^d f_{XT}(x \mid T = i)\, dx$$

where d is the distance of the house from the cliff edge.

The more convenient representation for the purposes of benefit assessment is the distribution of the predicted year of loss for assets at a given distance x from the cliff edge. The p.d.f. $f_{XT}(i \mid X = x)$ can be entered directly into the discounting table to obtain the probability weighted sum of the damage risk, that is:

$$\text{PV(damage)} = MV \sum_{i=0}^{j} \frac{f_{XT}(i \mid X = x)}{(1 + r)^i}$$

where j is the appraisal period.

Thus, the present value (PV) of the loss in any year is calculated as follows:

$$\text{PV(damage Year } i) = \text{Probability of loss} \times \text{Asset value}$$

$$\times \text{Discount factor (Year } i)$$

The overall present value losses are:

'Without Project' PV(damage Years 0–49)

$$= \sum \text{Probability of loss} \times \text{Asset value} \times \text{Discount factor (Years 0–49)}$$

Table 6.10 shows an example of the probabilistic discounting procedure for a house with a risk-free market value of £100 000 situated 10 m from the edge of an eroding cliff, using an illustrative probability distribution (obtained using the 2-distribution probabilistic method described in Example 4.10). According to the probabilistic discounting procedure the PV damage risk is £29 620.

A coast protection scheme would delay the recession scenario by a length of time equivalent to the scheme design life (e.g. 50 years). The 'With Project' losses can be calculated as follows:

'With Project' PV(damage Years 50–99)

$$= \sum \text{Probability of loss} \times \text{Asset value} \times \text{Discount factor (Years 50–99)}$$

Note that between Years 0 and 49 the probability of loss would be zero, that is the scheme had delayed the recession losses by 50 years.

Table 6.10. Example 6.9: Probabilistic discounting for a single cliff-top asset (from Hall et al., 2000)

Year i	Discount factor	Probability of damage in Year i	PV damage: £ thousands
0	1.00	0.000	0.00
1	0.94	0.002	0.18
2	0.89	0.002	0.19
3	0.84	0.005	0.39
4	0.79	0.006	0.47
5	0.75	0.010	0.73
6	0.70	0.011	0.75
7	0.67	0.013	0.89
8	0.63	0.018	1.11
9	0.59	0.019	1.10
10	0.56	0.024	1.34
11	0.53	0.023	1.20
12	0.50	0.024	1.20
13	0.47	0.028	1.32
14	0.44	0.032	1.42
15	0.42	0.032	1.34
16	0.39	0.033	1.30
17	0.37	0.035	1.29
18	0.35	0.035	1.22
19	0.33	0.035	1.14
20	0.31	0.037	1.15
21	0.29	0.033	0.98
22	0.28	0.035	0.97
23	0.26	0.033	0.87
24	0.25	0.032	0.79
25	0.23	0.030	0.70
26	0.22	0.031	0.68
27	0.21	0.029	0.60
28	0.20	0.030	0.58
29	0.18	0.027	0.50
30	0.17	0.025	0.44
31	0.16	0.022	0.36
32	0.15	0.021	0.33
33	0.15	0.021	0.30
34	0.14	0.019	0.26
35	0.13	0.018	0.23
36	0.12	0.017	0.21
37	0.12	0.016	0.19
38	0.11	0.013	0.14
39	0.10	0.013	0.13
40	0.10	0.012	0.12
41	0.09	0.009	0.08
42	0.09	0.011	0.10

Table 6.10. Continued

Year i	Discount factor	Probability of damage in Year i	PV damage: £ thousands
43	0.08	0.008	0.06
44	0.08	0.009	0.07
45	0.07	0.008	0.06
46	0.07	0.007	0.04
47	0.06	0.006	0.04
48	0.06	0.006	0.04
49	0.06	0.004	0.02

Total PV risk: £ thousands   29.62

Notes. Risk-free market value of asset = £100 000. Discount rate = 6%. Distance of asset from cliff edge = 10 m. PV damage (Year i) = discount factor (Year i) × probability of damage (Year i) × asset value.

An alternative approach would be to consider the scheme as reducing the probability of loss in each year up to the end of its design life:

'With Project' PV(damage Years 0–49)

$$= \sum \text{Revised probability of loss} \times \text{Asset value}$$

$$\times \text{Discount factor (Years 0–49)}$$

Table 6.11. Example 6.9: Summary of discounted asset values for multiple assets and different erosion control options (from Hall et al., 2000). See Example 6.8 for background

Property/utility	Distance from cliff edge: m	Market value: £ thousands	Damage risk value with different protection options: £ thousands				
			Without project	Option 1	Option 2	Option 3	Option 4
Trunk gas main	7	450	252	183	152	84	12
1–5 Acacia Ave	8	350	184	132	111	62	7
Café	10	120	62	43	30	26	1
Sunnyview B	11	180	81	61	49	29	4
Sewage PS	12	2300	940	675	555	315	47
Hightrees House	13	1500	585	417	343	195	25
2–8 Acacia Ave	17	400	136	99	75	44	7
The Saltings	18	76	26	17	14	8	0
14–20 Rocco Blvrd	21	850	165	122	100	67	9
Total PV risk: £ thousands			2431	1749	1429	830	112
Erosion control benefit: £ thousands				682	1002	1601	2319

In either case, the reduction in risk achieved by the scheme is:

Risk Reduction (PV) = 'Without Project' PV (damage)

— 'With Project' PV (damage)

Where multiple cliff-top assets are at risk, the above methodology should be repeated for each of the assets. Table 6.11 summarises the analysis for a hypothetical case in a similar format to the deterministic approach illustrated in Table 6.9. It shows how predictions of risk for various erosion control scenarios, ranging from the 'Without Project' option to a high standard of protection in Option 4, could be summarised.

Individual Risk

The risk specific to humans, such as the general public, road users or particular activity groups, is the frequency with which individuals within such groupings are expected to suffer harm (Health and Safety Executive, 1992; Royal Society, 1992).

This notion is usually referred to as *Individual Risk*. It is a somewhat abstract concept and must not be confused with the risk faced by a specific individual, which is best termed *Personal Risk*. A formal definition of Individual Risk is the frequency with which an individual within a specific group or population may be expected to sustain a given level of harm from the realisation of a specific hazard or particular combination of hazards (IChemE, 1992). In the case of landsliding, it can be considered to be the risk of fatality or injury to individuals who live within a zone liable to be impacted by a landslide or who follow a particular pattern of life that might subject them to the adverse consequences of a landslide (Fell and Hartford, 1997). Individual Risk has traditionally been taken to be the risk of death, expressed as the probability per year, although in the case of the transport sector, it is normally expressed as 'per journey' or 'per passenger mile or km'. It is the risk deemed to be experienced by a notional single individual in a given time period. It generally reflects the amount of time the individual is exposed to the hazard and its severity.

Individual Risk is a problematic concept and discussions in the literature are often contradictory. It is essential, therefore, when computing values for individual risk, to specify at the outset whether the calculated frequencies relate to those most at risk from a given activity (say as a result of their location, work, recreation or time periods for which they remain vulnerable) or whether they relate to an 'average' or 'shared value' representative of all potentially affected individuals. The use of an average value is only strictly appropriate where the risk is relatively uniformly distributed over the affected population. Otherwise this measure can be highly misleading, in that where a few individuals are exposed to high risk levels, this could be concealed when averaged over a large number of people at relatively low risk.

There are typically three different types of Individual Risk (e.g. IChemE, 1992; Kauer *et al.*, 2002):

1 *Location-Specific Individual Risk (LSIR)*. The risk for an individual who is present at a particular location for the entire period under consideration, which may be 24 hours per day, 365 days per year or during the entire time that a risk generating plant, process or activity is in operation. The LSIR can be a misleading risk measure as few individuals remain at the same location all the time or are constantly exposed to the same risk. However, it is useful in spatial planning as it is a property of the location in question, rather than the behaviour patterns of the population (e.g. Bottelberghs, 2000):

$$\text{LSIR} = \text{Prob. (Landslide)} \times \text{Vulnerability}$$

2 *Individual-Specific Individual Risk (ISIR)*. The risk for an individual who is present at different locations during different periods. The ISIR can be a more realistic measure than LSIR:

$$\text{ISIR} = \text{Prob. (Landslide)} \times \text{Prob. (Wrong Place)}$$
$$\times \text{Prob. (Wrong Time)} \times \text{Vulnerability}$$

Prob. (Wrong Place) is the probability that the path of the landslide intersects the location where the individual could be (i.e. the spatial probability of impact).

Prob. (Wrong Time) is the probability that the individual is in the landslide danger zone during the landslide occurrence (i.e. the temporal probability of impact).

3 *Average Individual Risk (AIR)*. The AIR can be calculated from historical data of the number of fatalities per year divided by the number of people at risk. Alternatively, a measure of the average individual risk can be derived from the societal risk (see Section 'Societal Risk' in this chapter) divided by the number of people at risk:

$$\text{AIR} = \frac{\text{Number of Fatalities/Year}}{\text{Exposed Population}}$$

$$\text{AIR} = \frac{\text{Societal Risk}}{\text{Exposed Population}}$$

Note that the *exposed population* could be either the total population or the average population expected to be within the danger zone during a given time period, such as an hour or a day (see Section 'Exposure' in Chapter 5).

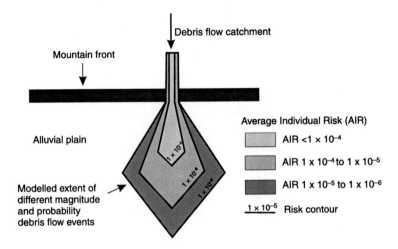

Fig. 6.7. *Individual risk contours associated with a series of debris flow events*

In order to estimate ISIR it is necessary to determine:

- The probability of an individual being in the *wrong place*, that is in the danger zone affected by the landslide (see Section 'Exposure' in Chapter 5);
- the probability of an individual being present at the *wrong time*, that is in the danger zone when the landslide event occurs (see Section 'Exposure' in Chapter 5);
- the vulnerability of an individual to the landslide event (see Section 'Vulnerability' in Chapter 5).

Locations with equal individual risk can be shown on a map by so-called *risk contours* (Fig. 6.7). These contours can be constructed using a commercial package, although terrain unit boundaries can provide a more realistic framework for distinguishing between different zones than simply relying on mathematical contouring.

Individual Risk can also be expressed by means of the *Fatal Accident Rate* (FAR), which is the number of fatalities per 1000 hours of exposure (e.g. Bedford and Cooke, 2001). The FARs are more convenient and more readily understandable than Individual Risk per year. A variant is the *death per unit activity*, where the time unit is replaced by a unit measuring the amount of activity. The risks of travel by car, train or aeroplane are often expressed in the form of the number of deaths per kilometre travelled.

As discussed in Section 'Consequence models' in Chapter 5, a different approach to estimating potential loss of life and individual risk has been developed for use in Hong Kong. In this instance, the potential loss of life was considered in relation to the loss of life expected from a *reference landslide* (see Example 5.6).

Example 6.10
Where beaches are backed by cliffs, rockfalls can present a significant risk to the public. For example, in 1986 a lifeguard was killed when a beach hut was hit by falling mud and rocks at Newquay on the Cornish coast. The 10 km length of cliff between Hayle and Portreath is of particular concern because of the combination of frequent rockfalls and high visitor numbers to the beaches during the summer months.

The risk to beach users and individuals was calculated by Coggan *et al.* (2001) using the following methodology (Fig. 6.8), which involved several simplifying assumptions:

1 *Developing a simple rockfall hazard model.* The threat was found to be posed by individual small blocks which fall with great frequency (1000 events per year).
2 *Establishing the extent of the hazard zone.* The potential for harmful impacts was considered to be confined to a 5 m-wide *impact danger zone* at the base of the cliffs. This was divided into 100 000 cells 'equivalent in volume to a person's head' (i.e. circa 0.25 m sides). It was assumed that each falling rock would impact 20 cells within the danger zone.
3 *Establishing the threat to the beach users.* The assumption was made that the beach would only be used during relatively good weather conditions over the warmer part of the year, and certainly not during storms or high tide conditions. The beach-use season was, therefore, estimated to be 200 days long with an average occupancy of six hours per day (the *potential occupancy time*). During the course of a year, the overall period of beach use (i.e. *danger time*) would be:

$$\text{Danger Time} = \text{Beach Season} \times \text{Daily Use} = 200 \times 6/24 = 50 \text{ days}$$

Estimating the number of rockfalls expected during the 'danger time' required the removal of all rockfall events that occurred during storms, when the beach was not in use (estimated at 90%) and averaging the remainder over the year:

Rockfalls during Danger Time

$$= (\text{Total Rockfalls} - \text{Storm Rockfalls}) \times \text{Danger Time}$$

$$= (1000 - 900) \times 50/365 = 100 \times 0.137 = 13.7$$

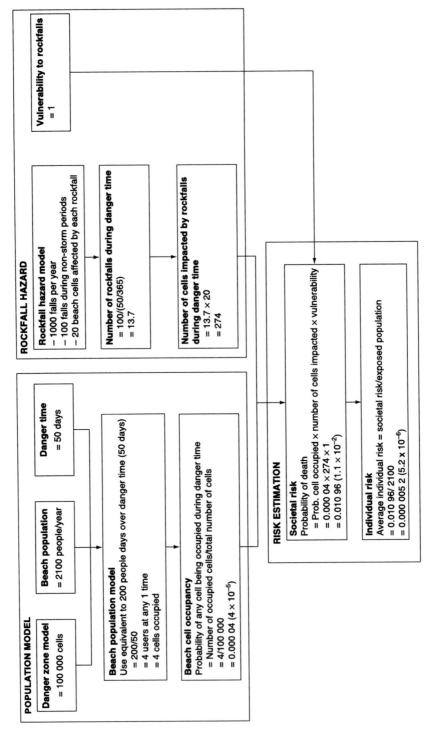

Fig. 6.8. Example 6.10: risk estimation procedure (adapted from Coggan et al., 2001)

As each rockfall was assumed to impact 20 cells, the average number of cells impacted per year during the 'danger time' was calculated as:

Average Number of Impacted Cells/Year

= Rockfalls during Danger Time

× Number of Cells Impacted by a Rockfall

$= 13.7 \times 20 = 274$

4 *Establishing the risk to beach users.* This required an estimate of the likelihood of a cell being occupied by a person at the moment it is impacted by a falling block. Clearly the risk varies with the length of time an individual spends on the beach, the nature of recreational activity and the size of the population. In this instance, the occupancy of the beach during the 'danger time' was estimated to be the equivalent of 200 people days, which translates as an average of four people on the beach at any one time, or four cells occupied. The probability of any cell being occupied during the 'danger time' was, therefore:

$$\text{Prob. Cell Occupied} = \frac{\text{Number of Occupied Cells}}{\text{Total Number of Cells}}$$

$$= 4/100\,000 = 0.000\,04 \quad \text{or} \quad 4 \times 10^{-5}$$

As 274 cells per year were impacted by rockfalls, the annual probability of a single fatality from the whole population of beach users, assuming that if a person is hit by a rockfall then they are killed (i.e. vulnerability $= 1$), was calculated as:

Risk = Prob. Cell Occupied × Number of Cells Impacted

$$= 0.000\,04 \times 274 = 0.010\,96 \quad \text{or} \quad 1.1 \times 10^{-2}$$

This measure of risk is the *societal risk*, that is the risk to the beach users (see Section 'Societal Risk' in this chapter), rather than the individual risk.

5 *Calculation of individual risk.* A number of measures of individual risk can be derived:

- the *Average Individual Risk (AIR)*. Assuming that the number of beach users over the 200 day 'danger time' was 2100, the risk becomes:

$$\text{Average Individual Risk} = \frac{\text{Societal Risk}}{\text{Exposed Population}}$$

$$= \frac{0.010\,96}{2100} = 0.000\,005\,2 \quad \text{or} \quad 5.2 \times 10^{-6}$$

- the *Individual-Specific Individual Risk (ISIR)*. Assuming that a specific individual remains on the beach over the entire 'danger time', occupying a single cell the risk becomes:

Individual-Specific Individual Risk

$$= \text{Prob. Cell Occupied} \times \text{Number of Cells Impacted}$$

$$= \frac{\text{Number of Occupied Cells}}{\text{Total Number of Cells}} \times \text{Number of Cells Impacted}$$

$$= 1/100\,000 \times 274 = 0.002\,74 \quad (2.7 \times 10^{-3})$$

- the *Location-Specific Individual Risk (LSIR)*. Assuming that a specific individual remains within a single cell over the entire year, the 'danger time' is 365 days, the number of rockfalls during the 'danger time' becomes 1000 and the number of impacted cells becomes 20 000, that is 1000×20. The risk is calculated as:

Location-Specific Individual Risk

$$= \text{Prob. Cell Occupied} \times \text{Number of Cells Impacted}$$

$$= \frac{\text{Number of Occupied Cells}}{\text{Total Number of Cells}} \times \text{Number of Cells Impacted}$$

$$= 1/100\,000 \times 20\,000 = 0.2 \quad \text{or} \quad 2 \times 10^{-1}$$

A different approach would be to employ the notions of *wrong place* and *wrong time* by calculating the likelihood of a cell being impacted at the same time as it is occupied by a person. The probability of a particular cell being impacted during the 'danger time' is:

$$\text{Prob. (Wrong Place)} = \frac{\text{Number of Impacted Cells/Year}}{\text{Total Number of Cells}}$$

$$= \frac{274}{100\,000} = 0.002\,74 \quad \text{or} \quad 2.74 \times 10^{-3}$$

If an individual occupied the same cell for the entire duration of the 'danger time', the probability of being there at the *wrong time* is:

$$\text{Prob. (Wrong Time)} = \frac{\text{Exposure Time}}{\text{Danger Time}} = 50/50 = 1$$

Assuming a vulnerability of 1, the risk of death is:

$$\text{Prob. (Death)} = \text{'Wrong Place'} \times \text{'Wrong Time'} \times \text{Vulnerability}$$

$$= (2.74 \times 10^{-3}) \times 1 \times 1 = 2.74 \times 10^{-3}$$

These are the same as the *Individual-Specific Individual Risk* calculated earlier. However, if the individual only spends 50% of the period in the specific cell then the risk of death becomes:

Prob. (Death) = 'Wrong Place' × 'Wrong Time' × Vulnerability

$$= (2.74 \times 10^{-3}) \times 0.5 \times 1 = 1.37 \times 10^{-3}$$

The risk continues to reduce with the decrease in time spent in the 'danger zone', so that a specific person who sits in a particular cell for one hour during the 'danger time' has a risk of death (personal risk) of:

Prob. (Death) = 'Wrong Place' × 'Wrong Time' × Vulnerability

$$= (2.74 \times 10^{-3}) \times (1/(50 \times 24)) \times 1$$

$$= (2.74 \times 10^{-3}) \times (8.33 \times 10^{-4}) \times 1 = 2.28 \times 10^{-6}$$

In the case of a fossil hunter who spends the entire 'danger time' walking up and down the beach in the 'danger zone', then the risk can be calculated as follows:

$$\text{Prob. (Wrong Place)} = \frac{\text{Number of Impacted Cells/Year}}{\text{Total Number of Cells}}$$

$$= 0.002\,74 \quad (2.74 \times 10^{-3})$$

This represents the probability of rockfall activity during the danger time in *any one cell*:

$$\text{Prob. (Wrong Time)} = \frac{\text{Exposure Time}}{\text{Danger Time (50 days)}}$$

If the fossil hunter stays in the same cell for the entire duration of the 'danger time':

Prob. (Wrong Time) = 50/50 = 1

However, if the fossil hunter moves randomly from cell to cell while remaining in the danger zone throughout the danger time, each cell will have the same overall occupancy (i.e. Prob. Wrong Time) over the course of the danger time:

$$\text{Prob. (Wrong Time)} = \frac{\text{Exposure Time/Total Number of Cells}}{\text{Danger Time}}$$

$$= \frac{50/100\,000}{50} = 0.000\,01 \quad (1 \times 10^{-5})$$

The probability of death in a particular cell becomes:

$$\text{Prob. (Death)} = \text{'Wrong Place'} \times \text{'Wrong Time'} \times \text{Vulnerability}$$

$$= 0.002\,74 \times 0.000\,01 \times 1$$

$$= 0.000\,000\,027\,4 \quad (2.74 \times 10^{-8})$$

This represents the probability of death given rockfall activity during the danger time in *any one cell*.

As the fossil hunter is exposed to this level of risk in each one of the 100 000 cells he/she visits during the danger time, the overall risk becomes:

$$\text{Prob. (Death)} = \sum \text{Prob. (Death) (Cells 1 to 100\,000)}$$

$$= 0.000\,000\,027\,4 \times 100\,000 = 0.002\,74 \quad (2.74 \times 10^{-3})$$

This example clearly illustrates that the risk to individuals can vary dramatically, depending on their activity patterns and exposure. Risks faced by particular individuals (personal risk) should not be confused with individual risk, which can be either a measure of the risk at a particular location or an average value shared by all individuals within a particular population.

Example 6.11
A major highway through mountainous terrain in British Columbia is susceptible to rockfalls that can cause delays, damage, injury and death to road users. Along a particular section, where the road passes through a deep rock cutting, maintenance records and rockfall impact marks on the carriageway suggest a minimum rockfall frequency of 2.2 incidents per year (Bunce *et al.*, 1997). The road carries an average of 4800 vehicles per day (on average, 200 per hour), at speeds of around 80 km per hour.

The binomial distribution can be used to model the probability of a vehicle being hit by a falling rock (see Example 4.3 for details). Each rockfall is represented by a separate trial with two possible outcomes, collision or no collision. The probability of one or more collisions is related to the probability of the rockfall in a specific trial hitting a vehicle (i.e. a vehicle being in the 'wrong place') and the number of falls per year:

$$\text{Prob. (Collision)}$$

$$= 1 - (1 - \text{Prob. (Vehicle in the 'Wrong Place')})^{\text{Number of Falls/Year}}$$

To estimate the probability of a *specific vehicle* being hit while stationary in traffic, say for half an hour, it is necessary to calculate the probability of it being in the 'wrong place' at the 'wrong time', as well as the probability of the trial (i.e. the rockfall) resulting in a collision outcome:

$$\text{Prob. (Wrong Place)} = \frac{\text{Length of Vehicle}}{\text{Length of Road Cutting}} = \frac{5.4}{476} = 0.011$$

$$\text{Prob. (Wrong Time)} = \frac{\text{Length of Stay (hours)}}{\text{Length of Year}} = \frac{0.5}{8760} = 5.7 \times 10^{-5}$$

As there is more than one rockfall in a single year, the probability of a collision between a rockfall and the *specific vehicle* (Prob. Collision) can be calculated from the binomial model:

$$\text{Prob. (Collision)} = 1 - (1 - \text{Prob. (Wrong Place)})^{\text{Number of Falls/Year}}$$

$$= 1 - (1 - 0.011)^{2.2} = 0.025$$

$$\text{Annual Prob. (Collision)} = \text{Prob. (Wrong Time)} \times \text{Prob. (Collision)}$$

$$= 5.7 \times 10^{-5} \times 0.025 = 1.4 \times 10^{-6}$$

Thus, the annual probability of a specific vehicle being hit by a falling rock is 1.4×10^{-6}, that is 0.000 001 4.

To calculate the probability of *any vehicle* in the line of stationary traffic being hit, it is assumed that the jam extends for the full length of the cutting with vehicles 'bumper to bumper', in which case:

$$\text{Prob. (Wrong Place)} = \frac{476}{476} = 1$$

$$\text{Prob. (Wrong Time)} = \frac{\text{Length of Stay (hours)}}{\text{Length of Year}} = \frac{0.5}{8760} = 5.7 \times 10^{-5}$$

Prob. (Collision) = 1 (i.e. if a rockfall occurs then it will hit a vehicle)

Annual Prob. (Collision)

$$= \text{Prob. (Wrong Time)} \times \text{Frequency of Rockfalls}$$

$$= 5.7 \times 10^{-5} \times 2.2 = 0.000\,125\,4 \quad \text{or} \quad 1.25 \times 10^{-4}$$

Note that the 'individual risk' to all vehicles is the same, as can be shown by calculating the number of vehicles in the cutting and dividing by the above number, as follows:

$$\text{Number of Vehicles} = \frac{\text{Length of Cutting}}{\text{Length of Vehicle}} = \frac{476}{5.4} = 88.15$$

$$\text{'Individual Risk' (Vehicle)} = \frac{1.25 \times 10^{-4}}{88.15} = 0.000\,001\,4 \quad \text{or} \quad 1.4 \times 10^{-6}$$

The same principles apply when estimating the probability of a *moving vehicle* being hit. When vehicles are in motion, the proportion of time that a part of the highway is occupied by a vehicle is:

$$\text{Prob. (Wrong Place)} = \frac{\text{Number of Vehicles/hour} \times \text{Length of Vehicle}}{\text{Speed (in metres per hour)}}$$

$$= (200 \times 5.4)/80\,000 = 1.35 \times 10^{-2}$$

$$\text{Prob. (Collision)} = 1 - (1 - \text{Prob. (Wrong Place)})^{\text{Number of Falls/Year}}$$

$$= 1 - (1 - 0.0135)^{2.2} = 1 - (0.9865)^{2.2} = 2.95 \times 10^{-2}$$

Assuming that, throughout the year, the vehicles are evenly distributed, then:

$$\text{Prob. (Wrong Time)} = 1.0$$

The annual probability of a moving vehicle being hit becomes:

$$\text{Annual Prob. (Collision)} = \text{Prob. (Wrong Time)} \times \text{Prob. (Collision)}$$

$$= (2.95 \times 10^{-2}) \times 1 = 2.95 \times 10^{-2}$$

The probability of an accident on a single trip through the road cutting can be approximated by the annual probability of a collision, divided by the total number of trips per year (4800×365):

$$\text{Annual Prob. (Single Trip Collision)} = 2.95 \times 10^{-2}/(4800 \times 365)$$

$$= 1.7 \times 10^{-8}$$

This is equivalent to the 'individual risk' shared between all vehicles. Bunce *et al.* (1997) assumed that the probability of a fatality following a collision (i.e. boulder impact) is 0.125 and 0.2 for stationary vehicles and moving vehicles, respectively. The probability of one or more deaths is:

$$\text{Prob. (Death)} = \text{Annual Prob. (Collision)} \times \text{Prob. (Fatality)}$$

	Stationary traffic for 30 minutes (specific vehicle)	Stationary traffic for 30 minutes (any vehicle)	Moving traffic (any vehicle)	Moving traffic (single trip)
Prob. (Death)	1.75×10^{-7}	1.56×10^{-5}	5.9×10^{-3}	3.4×10^{-9}

Societal Risk

Societal Risk is the frequency and the number of people suffering a given level of harm from the realisation of specified hazards (IChemE, 1992). It usually refers to the risk of death, and is expressed as risk per year.

Potential loss of life (PLL), equivalent to the *expected value of the number of deaths per year* (see Vrijing and van Gelder, 1997), is used as a measure of the risk to all individuals exposed to the full range of landslide events that might occur in an area (i.e. the *societal risk from landsliding*). To calculate PLL it is necessary to estimate, for each event and its possible outcome, the frequency per year (f) and the associated number of fatalities (N). The PLL is the sum of the outcome of multiplying f and N for each event:

$$PLL = \sum f_1 N_1 + f_2 N_2 + \cdots + f_n N_n$$

Societal risk can also be calculated from individual risk (see Section 'Individual Risk' in this chapter):

PLL = Individual Risk × Exposed Population

Frequency and number of fatalities data are usually presented as so-called F–N curves, which show the cumulative frequency (F) of all event outcomes with N or more fatalities. The advantage of F–N curves is that they provide a framework for comparing the societal risk associated with landsliding, or other sources of risk, against *risk criteria* (see Chapter 7). These risk criteria can be the means by which the results of a risk assessment exercise can be compared to assist decision-making.

Carter and Riley (1998) have suggested that societal risk can be measured using a *scaled risk index* (SRI), which takes account of the individual risk level (IR; expressed as deaths per million per year) and other characteristics of a site or area:

$$SRI = \frac{\text{Population factor} \times \text{Individual Risk} \times \text{Occupancy Time}}{\text{Area in hectares}}$$

The Population factor (P) is a function of the number of persons (n) in the area:

$$P = \frac{n + n^2}{2}$$

The Occupancy Time is the proportion of time the area is occupied by n persons.

Example 6.12
Natural slopes in Hong Kong are often strewn with large boulders, especially when mantled with colluvium or beneath rock cliffs. Boulder falls are a common occurrence, especially during or following intense rainstorms, and can result in property damage and fatalities. Between 1984 and 1995, there were 169 reported rockfall and boulder fall incidents, causing three injuries but no fatalities (ERM-Hong Kong, 1998b). Over the 69 years between 1926 and 1995 there were, however, three fatal incidents:

Table 6.12. Example 6.12: F–N data for historical landslide incidents and boulder falls in Hong Kong (modified from ERM-Hong Kong, 1998b)

Number of fatalities, N	Boulder falls 1926–1995 (69 years)		Landslides on man-made slopes 1917–1995 (78 years)	
	Number of events with N or more fatalities	Frequency F of N or more fatalities	Number of events with N or more fatalities	Frequency F of N or more fatalities
1	3	0.0435	117	0.974
2			31	0.397
3	2	0.0290	19	0.244
4			11	0.141
5	1	0.0145	9	0.115
6			8	0.103
8			7	0.0897
16			5	0.0641
18			4	0.0513
67			3	0.0385
71			2	0.0256
73			1	0.0128

- Elliot Pumping Station, Pok Fu Lam, in 1926 (5 deaths)
- Shau Kei Wan squatter area in 1976 (3 deaths)
- Kings Road in 1981 (1 death).

By way of contrast, there have been 117 fatal landslide incidents associated with man-made slopes, with the 'worst case' event being the collapse of a low-rise building in Po Hing Fong during 1917, which caused 73 deaths.

These incidents, along with fatal landslide events on man-made slopes, are presented as F–N data in Table 6.12. The average historical PLL associated with boulder falls is 0.13 fatalities per year (i.e. 9 deaths in 69 years). Figure 6.9 presents the F–N curves derived from these statistics. To express this societal risk as an economic value, it is necessary to assign a *value of life*, typically assumed to be around £2 million, as discussed in Section 'Loss of life and injury' in Chapter 5:

Annual Societal Risk = Value of Life × PLL

$$= £2 \text{ million} \times 0.13 = £260\,000$$

Example 6.13
During the early hours of 13 August 1995 a cutting failed along the Fei Tsui Road, Hong Kong, triggered by heavy rainfall. The road in front of the slope

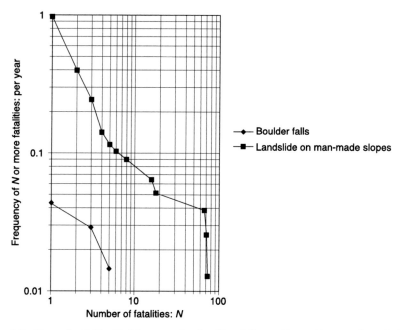

Fig. 6.9. Example 6.12: F–N curve for boulder falls in Hong Kong (adapted from ERM-Hong Kong, 1998a)

was totally engulfed by landslide debris up to about 6 m thick; a man was slightly injured, but his son was trapped in the debris and killed. An assessment of the societal risk was undertaken as part of the post-event investigations, in order to establish a reliable indication of the risk posed to the affected community (Wong *et al.*, 1997; Ho *et al.*, 2000). The objective of the assessment was to consider what might have happened rather than simply focusing on what did happen. Indeed, the loss of life could have been higher if the landslide had occurred during daytime, rather than at 1.15 a.m., when traffic flows were very low.

The assessment involved estimating the potential loss of life (PLL) associated with a number of consequence scenarios (Table 6.13) and the development of an *F–N* curve for the event (Fig. 6.10). The following example only considers the threat to road users. Note, however, that the societal risk assessment presented in Wong *et al.* (1997) also considers the threat to an area of open space, a playground, a Baptist Church and a kindergarten.

The risk assessment involved the following steps:

1 *Estimating the probability of the landslide event.* Analysis of rainfall records suggested that the storm that preceded the event had a return period of around 100 years. The probability of the landslide was assumed, therefore, to be 0.01 (1 in 100).

Table 6.13. *Example 6.13: Societal Risk associated with the 1995 Fei Tsui landslide (adapted from Wong et al., 1997)*

Prob. landslide event	Consequence scenario	Number of people exposed	Prob. consequence scenario	Vulnerability factor	Probable fatalities N	PLL	Risk	Frequency of event: F	Frequency of N or more fatalities: $\geqslant N$
0.01	1	0	0.4989	0.85	0	0.000 000	0.000 000	4.99E-03	1.00E-02
0.01	2	1	0.1875	0.85	0.85	0.159 375	0.001 594	1.88E-03	5.01E-03
0.01	3	5	0.2225	0.85	4.25	0.945 625	0.009 456	2.23E-03	3.14E-03
0.01	4	10	0.0875	0.85	8.5	0.743 750	0.007 438	8.75E-04	9.11E-04
0.01	5	30	0.0033	0.85	25.5	0.084 150	0.000 842	3.30E-05	3.60E-05
0.01	6	100	0.000 28	0.85	85	0.023 800	0.000 238	2.80E-06	3.00E-06
0.01	7	200	0.000 02	0.85	170	0.003 400	0.000 034	2.00E-07	2.00E-07
					Total	1.960 100	0.019 601		

Notes:

Probable fatalities N = Exposed population $\times$ Vulnerability factor

PLL = Prob. consequence scenario $\times$ Probable fatalities

Frequency F = Prob. landslide event $\times$ Prob. consequence scenario

Frequency $\geqslant N$ = Frequency N + Frequency $> N$

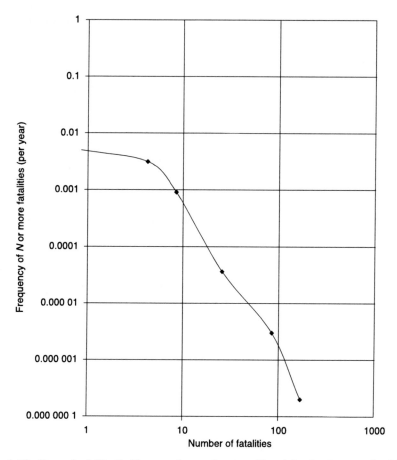

Fig. 6.10. Example 6.13: F–N curve for road users affected by the Fei Tsui landslide, Hong Kong (after Wong et al., 1997)

2 *Development of consequence scenarios.* A range of scenarios were considered, each with different numbers of people exposed along the road through the landslide area. The extreme scenario was for 200 people to be within the cutting; this might occur if a traffic jam were to be caused by a road accident. The probability of the consequence scenarios was estimated from available traffic data and analogy with traffic flow conditions on similar roads.

3 *Estimating the vulnerability of the people exposed to the landslide.* A vulnerability factor of 0.85 was used, based on an assessment of the proximity of the road to the cut face and the travel distance of the landslide debris (see Section 'Consequence models' in Chapter 5 and Example 5.6 for further details of the general approach to defining vulnerability factors).

4 *Estimating the probable number of fatalities for each consequence scenario.* This was calculated as follows:

Probable Fatalities (Scenario *s*) = Exposed Population (Scenario *s*)

$$\times \text{Vulnerability Factor}$$

So for Scenario 7:

Probable Fatalities (Scenario 7) = Exposed Population (Scenario 7)

$$\times \text{Vulnerability Factor}$$

$$= 200 \times 0.85 = 170$$

5 *Calculating the potential loss of life associated with each consequence scenario.* This was calculated as follows:

PLL (Scenario *s*) = Prob. (Scenario *s*)

$$\times \text{Probable Fatalities (Scenario } s)$$

PLL (Scenario 7) = Prob. (Scenario 7)

$$\times \text{Probable Fatalities (Scenario 7)}$$

$$= 0.000\,02 \times 170 = 0.0034$$

The overall potential loss of life is:

$$\text{PLL (Scenarios 1–7)} = \sum \text{PLL (Scenarios 1–7)} = 1.96$$

This provided a measure of the consequences given that the landslide has occurred. To establish the risk associated with each scenario it was necessary to take account of the probability of the landslide event:

Risk (Scenario *s*) = Prob. (Event) × Prob. (Scenario *s*)

$$\times \text{Probable Fatalities (Scenario } s)$$

Risk (Scenario 7) = Prob. (Event) × Prob. (Scenario 7)

$$\times \text{Probable Fatalities (Scenario 7)}$$

$$= 0.01 \times 0.000\,02 \times 170$$

$$= 0.000\,034 \quad (\text{i.e. } 3.4 \times 10^{-5})$$

The overall risk was:

$$\text{Risk (Scenarios 1–7)} = \sum \text{Risk (Scenarios 1–7)}$$

$$= 0.0196 \quad \text{or} \quad 1.96 \times 10^{-2}$$

6 *Compiling an F–N curve for the landslide event.* This involved calculating the event frequency (F) and the frequency of N or more fatalities ($>N$):

Frequency (F, Scenario s) = Prob. (Event) × Prob. (Scenario s)

Frequency (F, Scenario 7) = Prob. (Event) × Prob. (Scenario 7)

$$= 0.01 × 0.00002$$

$$= 0.0000002 \quad \text{or} \quad 2 × 10^{-7}$$

The frequency of N or more fatalities was calculated as follows, using the results presented in Table 6.13:

Frequency ($\geqslant 85$) = Frequency (85 fatalities)

$$+ \text{Frequency (170 fatalities)}$$

$$= 0.0000028 + 0.0000002$$

$$= 0.000003 \quad \text{or} \quad 3.0 × 10^{-6}$$

The *F–N* curve presented in Fig. 6.10 was compiled by plotting the calculated values of F and $>N$ for each consequence scenario.

Example 6.14
Landslides present a risk to highway traffic in Hong Kong. For example, historical data for the 14 km stretch of Castle Peak Road indicates that between 1984 and 1996, there were 32 recorded landslide incidents (2.38/year). The risk to life was estimated by ERM-Hong Kong (1999), using the Landslide Consequence model described in Section 'Consequence models' in Chapter 5 and Example 5.6, which relates potential fatalities to the landslide volume. Considering those situations where the road is in cutting and threatened by debris flows and landslide run-out (87% of slope sections along the road are in cuttings, the remainder are fill slopes), the PLL was estimated as

$$\text{PLL} = \sum \text{Event Frequency} × \text{Consequence (for all events)}$$

1 *Event frequency.* The historical frequency of past incidents along Castle Peak Road was used to generate a probability distribution for the height/volume of recorded slides (the *slope height–volume distribution*; see Table 6.14).

The annual frequency of events of a particular size was calculated as follows:

Frequency = Total Incident Frequency

$$× \text{Proportion of Road in Cutting}$$

$$× \text{Height/Volume Probability}$$

Table 6.14. Example 6.14: Slope height–volume distribution for landslides on cut slopes along Castle Peak Road, Hong Kong (based on ERM-Hong Kong, 1999)

Volume: m^3/Height: m	<20	20–50	50–500	500–2000	>2000
<10	0.23	0.1	0.05	0.05	0
10–20	0.15	0.1	0.05	0.01	0
>20	0.05	0.08	0.08	0.05	0

Thus, for an event of between 500 and 2000 m^3 on a slope higher than 20 m:

$$\text{Frequency} = 2.38 \times 0.87 \times 0.05 = 0.104$$

2 *Consequences.* As explained in Example 5.6, the consequence is the product of the *expected fatalities* for a reference landslide, given the volume of traffic and the number of lanes, the *vulnerability* for each lane and a *scale factor* (with respect to the reference landslide).

The results are presented in Table 6.15, and indicate a PLL from upslope failures along this section of Castle Peak Road of 0.98 fatalities per year.

Statistics are signs from God?

The reduction of a wide range of landslide hazard and multiple consequence scenarios to a mathematical expectation value has considerable advantages. On one level it provides a rational framework for a decision-making process in which risk levels are compared against a pre-determined set of criteria (e.g. benefit:cost ratios, risk acceptance criteria; see Chapter 7). At another level it supports the view that the future is manageable, thereby reducing any feelings of helplessness in the face of capricious nature: 'constituting something as a statistically describable risk makes possible the ordering of the future through the use of mathematical probability calculus' (Knights and Vurdubakis, 1993).

However, it is important not to lose sight of the fact that quantitative risk assessment is neither a neutral nor objective process. It is judgemental and as a result the results can be value-laden and biased. Individuals and groups who do not share the judgements and assumptions of the assessors may see the results of the risk assessment process as invalid, flawed or irrelevant (Stern and Fineberg, 1996). Judgements that can be the source of conflict include:

- the way in which hazard models are framed can influence which adverse consequences are analysed or ignored. For example, a landslide hazard model that focuses on rainfall or basal erosion as the prime cause of instability can direct attention away from the significance of leaking swimming pools or excavation of building plots;

Table 6.15. Example 6.14: Assessment of potential loss of life for cut slopes along Castle Peak Road, Hong Kong (based on ERM-Hong Kong, 1999)

Slope height: m	Slide volume: m³	Landslide frequency/year	Slope proportion	Height/volume factor	Vulnerability factor	Expected fatality	Scale factor	Consequence factor	PLL
<10	<20	2.3800	0.8700	0.2300	0.0455	1	0.4000	0.0182	0.0087
10–20	<20	2.3800	0.8700	0.1500	0.0685	1	0.4000	0.0274	0.0085
>20	<20	2.3800	0.8700	0.0500	0.0835	1	0.4000	0.0334	0.0035
<10	20–50	2.3800	0.8700	0.1000	0.3500	1	0.7000	0.2450	0.0507
10–20	20–50	2.3800	0.8700	0.1000	0.4350	1	0.7000	0.3045	0.0630
>20	20–50	2.3800	0.8700	0.0800	0.4850	1	0.7000	0.3395	0.0562
<10	50–500	2.3800	0.8700	0.0500	0.5850	1	1.5000	0.8775	0.0908
10–20	50–500	2.3800	0.8700	0.0500	0.6650	1	1.5000	0.9975	0.1033
>20	50–500	2.3800	0.8700	0.0800	0.6950	1	1.5000	1.0425	0.1727
<10	500–2000	2.3800	0.8700	0.0500	0.8900	1	2.0000	1.7800	0.1843
10–20	500–2000	2.3800	0.8700	0.0100	0.9500	1	2.0000	1.9000	0.0393
>20	500–2000	2.3800	0.8700	0.0500	0.9500	1	2.0000	1.9000	0.1967
<10	>2000	2.3800	0.8700	0.0000	0.9500	1	2.5000	2.3750	0.0000
10–20	>2000	2.3800	0.8700	0.0000	0.9500	1	2.5000	2.3750	0.0000
>20	>2000	2.3800	0.8700	0.0000	0.9500	1	2.5000	2.3750	0.0000
Total			1.0						0.9778

Notes. Landslide frequency: recorded events/year; slope proportion: proportion of failure from cut slopes (0.87) and fill slopes (0.13); height–volume factor: see Table 6.14; vulnerability factor: see Table 5.14 and explanation in Section 'Consequence models' in Chapter 5; expected fatality: see Table 5.13, for road with heavy vehicular or pedestrian traffic density; scale factor: see Table 5.16; consequence factor = vulnerability factor × expected fatality × scale factor; PLL = landslide frequency × slope proportion × height/volume factor × consequence factor.

- the focus on readily measurable or easily valued adverse consequences can lead to other consequences being excluded, especially those that are *close to home* to many of the affected community; for example the effect of the risk assessment on property values and the availability of insurance cover, disruption of the social framework and adverse impact on the character of a neighbourhood. When confronted by a statistical risk assessment, people often reframe the question in terms of *what does it mean for me or my family* (Plough and Krimsky, 1987; Siegal and Gibson, 1988). By ignoring such direct and personal questions, risk assessment can end up being misunderstood and mistrusted by the local community;

- the use of discounting techniques to analyse future risks is very contentious and confusing to many. This practice can have the effect of reducing the significance of risks that lie more than a generation or two in the future almost to zero. Many people consider that notions of sustainability deem it appropriate to use a low or zero discount rate so as to ensure that future risks or environmental damage are given sufficient weight in any analysis;

- the use of loss of life statistics as a measure of risk can be a source of controversy. Treating all fatalities as equal involves a judgement. It is assumed that the deaths of the old and the young are the same; deaths that occur during an event are treated the same as deaths that follow a protracted and painful period of hospitalisation. No value is placed on people who were exposed to the event and spent many years in constant fear of another incident. Few realise that individual risk is an abstract statistic and does not provide a realistic measure of the risk to *me or my family* (see Example 6.10). The situation is even worse when 'value of life' statistics are used, for there is widespread misunderstanding of this abstract measure (see Section 'Loss of life and injury' in Chapter 5). Once again the controversy arises because of the widespread misinterpretation of a risk measurement as representing the actual worth or value of an individual person.

Many people see risk in a completely different manner than the risk analyst. As discussed in Section 'Risk evaluation' in Chapter 1, research has shown that the way in which people react to risk can be described in terms of (e.g. Slovic *et al.*, 1980):

- *dread*, that is the horror of the hazard and its outcomes, the feeling of lack of control, fatal consequences, catastrophe potential;
- *the unknown* nature of the hazard and the resulting adverse consequences.

The quantitative risk assessment process generates results that are an expression of probability and loss of life or monetary value. It is important to appreciate, therefore, that such risk assessments do not deliver results that are directly relevant to many peoples' perception of risk. As Stern and Fineberg (1996) state:

> Conflicts over 'risk' may reflect differences between specialists in risk analysis and others on their definitions of the concept. In this light, it is not surprising that citations about 'actual risks' often do little to change most people's attitudes and perceptions. Nonspecialists factor complex, qualitative considerations into their estimates of risk, including judgements about uncertainty, dread, catastrophic potential, controllability, equity, and risk to future generations.

The solution is not to weight risks to conform to the majority values of the affected community, but rather to recognise that the quantitative risk assessment process produces one type of risk measure, and not to be so presumptuous as to suggest that it delivers the only valid measure:

> When lay and expert values differ, reducing different kinds of hazard to a common metric (such as number of fatalities per year) and presenting comparisons only on that metric have great potential to produce misunderstanding and conflict and to engender mistrust of expertise. (National Research Council, 1989).

7

From risk estimation to landslide management strategy

Introduction to landslide risk management

The decisions whether or not to reduce the risk posed by landsliding, and how best to reduce the risk, involve consideration of a range of views, interests and factors. The results obtained from the risk assessment process (see Section 'Risk estimation' in Chapter 1), irrespective of their form and how they may have been achieved, provide an indication of the level of risk and the likelihood of differing adverse outcomes. However, it is then necessary to ask the question *how much does it matter?* before going on to address the second question *what should be done about it?*

Risk evaluation addresses the first of these questions (see Section 'Risk evaluation' in Chapter 1). It is here that estimations of *threat*, as depicted by measures of the likelihood and severity of future adverse outcomes, including worst-case scenarios involving predictions of catastrophic events, are considered by relevant bodies and individuals, including the potentially affected population (stakeholders). The risk estimations have to be critically reviewed in terms of the *assumptions* that may have been made and the *levels of uncertainty* that are involved, in order to establish *confidence* in the results. The risks have also to be compared with other prevailing risks of concern, for the purposes of prioritisation. Stakeholders (individuals, groups or the public more generally) have to be consulted in order to find out their views on the risks that they are, or could be, exposed to and it is here that significant problems may be encountered. Many people are not well versed in the use of probabilities and are not able to appreciate the meaning of 'return periods' or 'recurrence intervals'. Each and every person has different formulations of risk based on their individual perceptions (see Section 'Risk evaluation' in Chapter 1), with the result that their toleration of different risks often shows little or no relationship with statistically-based risk estimations, especially where deaths or catastrophic outcomes are possible. However, good communications with the general public, including consultation from an early stage of a project or exercise, can significantly reduce potential tensions by limiting the extent to which risk management strategies are perceived to have been 'imposed' on stakeholders (see Chapter 1).

The results of risk estimation feed into the final risk assessment stage where the second question is addressed and decisions are taken regarding the most appropriate risk management strategy. A wide range of factors must be considered at this stage and options evaluated in terms of their technical feasibility, economic viability, environmental acceptability and political desirability (see later). However, before examining some of these issues in greater detail, it is important to emphasise that just because there is a physical problem (landsliding) does not necessarily mean that there has to be a physical solution involving engineering and the application of technology.

Humans have three main options when faced by a geohazard, such as landsliding. They can do the following:

- Accept the consequences and *bear* the costs (*loss bearing* and *do nothing*).
- Respond by abandoning a site, relocating elsewhere to safer ground or changing the use of a site so as to reduce risk (*choose change* and *risk avoidance*).
- *Take* active steps to reduce risk by limiting hazard potential and/or the potential to suffer loss (*adjustment*).

Only in the case of *adjustment* is landslide management involving engineering works an option and even here it is but one of the three main approaches outlined by Smith (2001), which are as follows:

- *Modification of loss burden*, which involves spreading the potential losses as widely as possible, through such measures as insurance. This is essentially a *loss-sharing* approach with limited emphasis on *loss-reduction*, so total risk remains roughly the same but the financial exposure of individuals, groups etc. is reduced because it is shared between a large number of participants.
- *Modification of hazard events*, which involves reducing the potential for loss by the use of hazard-resistant designs and engineered structures so as to safeguard lives and property and, if possible, to physically suppress the hazard potential of the geohazard concerned.
- *Modification of human vulnerability*, which focuses on reducing losses through land use planning programmes that seek to relate *land use zonation, building codes* to *hazard zonation*, together with the development of preparedness programmes that aim to limit losses, especially human

369

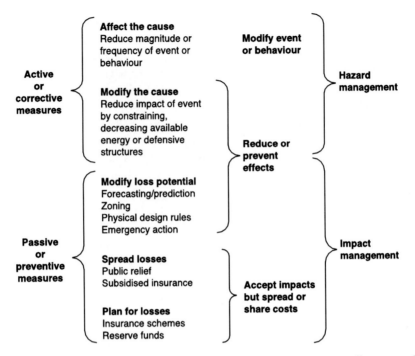

Fig. 7.1. *Classification of adjustment choices or management options, illustrating the differences between 'hazard management', 'vulnerability management' and 'risk management' (developed from Burton* et al. *(1978) and Jones (1996))*

casualties, through the installation of monitoring networks linked to forecasting and warning systems that translate into emergency actions.

A more detailed division of these landslide management approaches is shown in Fig. 7.1, which is based on the work of Burton *et al.* (1978) but subsequently modified for use in Jones (1991, 1996) and Royal Society (1992). The five categories of adjustment recognised are as follows, with specific reference to landsliding:

1 Actions designed to *affect the cause* of risk from landsliding by limiting the potential for slope failure through the use of land use management (e.g. soil conservation and afforestation; Sidle *et al.*, 1985), slope stabilisation measures, slope drainage (e.g. Hutchinson, 1977; Holz and Schuster, 1996; Wyllie and Norrish, 1996), and erosion control structures in rivers and along the coast (e.g. Lee and Clark, 2002). The objective is to *reduce the likelihood of landsliding.*

2 Actions designed to reduce risk by *modifying or constraining* slope failure so as to limit adverse impacts. Such measures assume that landsliding will

occur but seek to limit its ability to cause detriment by pre-determining pathways and run-out areas and, if possible, reducing landslide frequency, volume and velocity. Technical measures include using barriers to stop rolling rocks, nets to catch falling rocks, rock shelters and chutes over roads and railways, specially constructed debris flow channels/chutes around villages and under transport routes, check dams in gullies to inhibit debris flow development, storage basins etc. (e.g. Costa and Wieczorek, 1987; Wyllie and Norrish, 1996).

3 Actions designed to *modify loss potential* through improved forecasting and prediction (*prognostication*), better education about the nature of hazard and possible adverse consequences, together with the development of warning systems (*risk communication*), the establishment of emergency action plans and procedures, the development of building codes/building ordinances designed to improve the resistance of structures to slope movements and thereby limit damage, and the creation of a planning framework that seeks to relate geographical patterns of hazardousness (*hazard zonation*) with patterns of land use (*land use zonation*) so as to limit vulnerability (e.g. Clark *et al.*, 1996; Schuster and Kockelman, 1996).

4 Actions designed to *spread the losses*, usually taken in the wake of a serious impact when humanitarian concern makes it inevitable that aid be given to those that either could not or did not protect themselves against loss. This type of *passive loss sharing* takes the form of emergency aid or disaster relief provided both within a nation as well as between nations. National support is obtained from private donations, the work of charitable non-governmental organisations (NGOs) and expenditure from central funds via governmental organisations and agencies (e.g. the National Disaster Relief Arrangements (NDRAs) in Australia and the work of the Federal Emergency Management Agency (FEMA) in the USA which may be overridden by a Presidential Disaster Declaration). International relief and aid is also sometimes obtained from private donations, but the vast majority is in the form of *bilateral aid* (from government to government or indirectly through NGOs) and *multilateral aid* (through international bodies such as the EU, World Bank, Asian Bank and UN agencies, most especially the Disaster Relief Organisation (UNDRO)). With the passage of time it is becoming increasingly difficult to distinguish between such disaster aid and longer-term development aid.

5 Actions designed to *plan for losses* through insurance and the establishment of reserve funds (e.g. the New Zealand Earthquake Commission which provides natural disaster insurance cover to residential property owners; Murray, 2000). These measures can be termed *active loss sharing*, for the potential victims take deliberate actions in advance of an impact in order to protect themselves financially and ensure the potential for

recovery should an impact occur; a process even carried-out by the insurance industry through the process of re-insurance.

From a different perspective, the groupings of management options shown in Fig. 7.1 into *active/corrective* measures and *passive/preventative* measures emphasises whether the focus of activity is on the threat posed by hazard (i.e. landsliding) or on the potential for impact on human society, and provides the basis of the fundamental division into *hazard management* and *impact or vulnerability management* (Jones, 1996), both of which form major components within the broad field of risk management. In this context it is important to recognise the significance of the measures listed under *modify risk potential* in reducing risk, most especially risk communication and the crucial role of prognostication, without which the human population would continually be 'surprised' by hazardous events.

The five-fold division shown in Fig. 7.1 has to be recognised as imperfect because not all actions and activities fit neatly into only one of the categories and because of evolving management practices. For example, insurance is no longer a purely loss-sharing activity but can be used to actively encourage the establishment of land zonation policies and the adoption of building codes. In other instances, properties adversely affected by hazards may be purchased by the State rather than reconstructed or restored, so that loss sharing is used to achieve a reduction in future vulnerability. This is the case in the USA where the Hazard Mitigation Grant Program (HMGP) was created in November 1988, by Section 404 of the Robert T. Stafford Disaster Relief and Emergency Assistance Act. The HMGP assists States and local communities in implementing long-term hazard mitigation measures following a major disaster declaration. In 1998 the US Federal Emergency Management Agency (FEMA) and the California Governor's Office of Emergency Services provided a $1.3 million grant to the property owners in Humboldt County, California. The money was for the purchase of 17 residential properties in the Big Lagoon landslide area that had been threatened by erosion as a consequence of the El Niño storms. The grant represented 75% of the appraised value. Any structures on the properties were demolished and the land, to be maintained by the County, is to be kept as open space.

Similarly in France, the Law Barnier (2 February 1995) authorises the appropriation of and compensation by the Government for all property threatened by natural risks when the remedial works are too expensive to undertake. Compensation is funded from a State Surcharge of 9% which is added to all property insurance premiums. A Risk Prevention Plan (PPR) determines the areas where a natural risk is foreseeable. The PPR is intended to allow action to be taken in advance by the proprietor and the local authority.

Landslide risk management is the broad field of activities that covers all the strategies outlined above, including the passive acceptance of loss (e.g. the repeated repair of roads distorted by minor movements) and 'giving-up' or abandoning a site (e.g. the abandonment of the A625 road across the Mam Tor landslide, UK, in 1979; Jones and Lee, 1994); indeed the latter is now seen to be an increasingly preferred policy option for some eroding coastlines in the context of progressively rising sea-levels due to global warming (e.g. Lee and Clark, 2002). While landslide risk estimations inform all of these strategies, it has to be emphasised that the role of science and engineering figures prominently only in reducing the likelihood of landsliding, modifying or constraining slope failures when they do occur and, to a lesser extent, modifying the loss potential.

It is, however, beyond the scope of this book to examine the landslide risk management process in detail or to debate the contentious nature of risk management (see Royal Society, 1992). Instead attention will be focused on the important issues that determine whether or not landslide management is likely to be the preferred choice and how choices are made between different management strategies.

Assessment criteria

In the event that the risk evaluation process reveals that the risk from land-sliding is significant and needs to be reduced, then a choice has to be made between various management options. The nature and scale of the problem, together with the value of the elements at risk, will greatly influence the decision, although the level at which the decision is made is also a crucial factor. Individuals tend to view the adverse consequences associated with geohazards as *imposed risks* and are, therefore, generally less tolerant of them than they are of *chosen risks* (see Section 'Risk evaluation' in Chapter 1). In the Developed World there is also a well-established and growing view that the application of science, technology and engineering should protect people from the harmful aspects of the physical environment and that people should be fully informed of the risks they face. As a consequence, a property owner whose house is threatened by a developing landslide will usually insist on slope stabilisation (landslide management) funded from elsewhere (it must be someone else's fault/responsibility so they should pay); the Local Authority may well conclude that abandonment is the best option on the grounds of cost. However, where a large number of properties are involved, the pressure for tangible evidence of protection is greatly increased.

Irrespective of the precise details, it follows from the above that the selection process requires that consideration should only be given to those strategies that can deliver an *acceptable level of risk reduction*, while being

373

both *economically viable* and *environmentally acceptable*, in the broadest sense. In many instances, the decision will also be influenced by social and political pressures.

Landslide management often involves the planning of *public* expenditure to increase social welfare by reducing land instability losses. As only a minority of the tax-paying community (i.e. the nation) is affected, the use of public funds can be seen as a subsidy (e.g. extending the property life and safeguarding investments). Investment in landslide management can, therefore, be viewed as a means of safeguarding the vulnerable within society and helping towards the redistribution of wealth. There are, of course, other mechanisms for delivering improved social welfare (e.g. education, health and efficient infrastructure), all of which compete for resources.

Allocation of *public resources* for landslide management is, therefore, influenced by the need to find an acceptable balance between investments in a wide range of competing public services. Three tests are usually applied to decision-making about the allocation of public expenditure:

- the *scarcity* of resources requires that investments give the *highest returns* from the relevant perspective (i.e. national, regional or local);
- decisions to invest public funds must be *accountable and justifiable*;
- decisions must be based on a *rational comparison* between the available options.

Economic evaluation provides a mechanism for comparing the benefits of landslide management with the costs incurred, to determine:

- whether the benefits exceed the costs;
- the strategy that is expected to deliver the greatest economic return, that is the most efficient use of resources;
- the anticipated 'loss' to be incurred if it is decided to proceed with an 'uneconomic' strategy.

Not all landslide management activity is funded by National or Local Government, as individuals or organisations may wish to undertake works to protect their own property or assets. There are important differences between economic evaluation, which seeks to examine the returns to the community at large, and financial appraisal which examines whether the investment is worthwhile to an individual or organisation. For example, for an individual developer, the decision whether to protect a proposed hotel site from debris flow activity will be influenced by the additional profits to be generated after the implementation of mitigation measures. However, from a national perspective, the new hotel may simply divert visitors from other hotels in the country or even in the neighbourhood. From this perspective, the national benefits of landslide mitigation might be minimal.

374

Managing landslide risk cannot be viewed as solely an economic or financial issue, as the environment has become an increasingly important factor in determining the preferred option and the level of risk that is acceptable. This is because landslide mitigation works may result in environmental losses. To the individuals directly affected by landsliding or the threat of landsliding, the benefits of mitigation may far outweigh these losses. To others, the losses can represent an unacceptable price to pay for subsidising the lifestyle of a few.

Acceptable or tolerable risks

A cornerstone of risk management is the concept that there is a degree of risk that is acceptable. During the 1980s, however, the term 'acceptable' came to be progressively replaced by 'tolerable' as research revealed that in many instances people do not accept risks but merely tolerate them (see below and Royal Society, 1992). Central to this shift was the work of the UK Health and Safety Executive (HSE, 1988, 1992) into the *Tolerability of Risk* (TOR), when a framework was developed for making decisions on the tolerability of risk arising from any practice, activity, action or location (Fig. 7.2) based on the recognition of three levels of risk. Above a certain threshold the risks might be considered intolerable or unacceptable. Below another threshold, the risk might be considered to be so small that it is acceptable. It is widely accepted that between these two conditions the

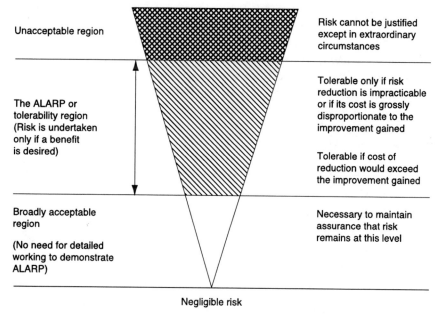

Fig. 7.2. Risk tolerability and the ALARP concept

375

level of risk should be reduced to a level which is *as low as reasonably practicable* (the so-called ALARP principle).

In England, the definition of reasonably practicable has been established by case law:

> Reasonably practicable is a narrower term than 'physically possible' and seems to me to imply that a computation must be made by the owner in which the quantum of risk is placed on one scale and the sacrifice involved in the measures necessary for averting the risk (whether in money, time or trouble) is placed in the other, and that, if it be shown that there is a gross disproportion between them – the risk being insignificant in relation to the sacrifice – the defendants discharge the onus on them. (Judge Asquith, Edwards v National Coal Board, All England Law Reports Vol. 1, 747 (1949))

The ALARP principle forms the basis of the approach used by the UK Health and Safety Executive in its regulation of the major hazardous industries, such as the nuclear, chemical and offshore oil and gas industries. The concept, as illustrated in Fig. 7.2, implies that:

- if the risk is unacceptable it must be avoided or reduced, irrespective of the benefits, except in extraordinary circumstances;
- if the risk falls within the ALARP or tolerability region, then cost may be taken into account when determining how far to pursue the goal of minimising risk or achieving safety. Beyond a certain point investment in risk reduction may be an inefficient use of resources. Thus risk does not have to be reduced to *as low as possible* employing *best available techniques* (BAT), as this will almost certainly involve excessive cost. The benefits to be gained from a reduction in risk are normally expected to exceed the costs of achieving such a reduction. This comparison leads to the important concepts of *as low as reasonably achievable* (ALARA) and *best available technique not entailing excessive cost* (BATNEEC) which underpin the ALARP principle.

Within the ALARP region, risks may be tolerated; however, tolerability does not mean 'acceptability':

> To tolerate a risk means that we do not regard it as negligible or something we might ignore, but rather as something we need to keep under review and reduce still further if we can. For a risk to be 'acceptable' on the other hand means that for purposes of life or work, we are prepared to take it pretty well as it is. (HSE, 1988)

The UK HSE has suggested that, in terms of individual risk, the upper and lower boundaries of the ALARP region are 10^{-4} and 10^{-6} fatalities per

Table 7.1. Possible tolerable individual risk criteria for landslides (from Fell and Hartford, 1997)

Situation	Tolerable individual risks (fatalities per year)
Natural slopes	10^{-3}
Existing engineered slopes	10^{-4} to 10^{-6}
New engineered slopes	10^{-5} to 10^{-6}

year, respectively (i.e. 1 in 10 000 and 1 in 1 000 000), for people living close to hazardous industrial sites. Fell and Hartford (1997) present possible acceptability criteria for landslides, based on historical evidence of what people appear willing to tolerate (Table 7.1).

Societal risk criteria are often presented on $F–N$ curves. In the absence of agreed risk criteria for landslides, Fell and Hartford (1997) considered dam safety to be a good analogy to landsliding. Figures 7.3 and 7.4 present $F–N$ curves for dams developed by:

1 British Columbia Hydro (BC Hydro; a major dam owner and operator), who recognise 'tolerable' and 'intolerable regions' (Fig. 7.3).
2 Australian National Committee on Large Dams (ANCOLD), who identify an ALARP region between 'acceptable' and 'unacceptable' risk regions (Fig. 7.4). Of interest, the ALARP region is truncated horizontally at a 10^{-6} per year failure probability, because ANCOLD felt that it was unrealistic to design a dam with a failure probability lower than this figure. However, it tends to imply that it is no more unacceptable for 10 000 or more to die in a failure incident than 100 people.

Recently the Hong Kong Government has published interim risk guidelines for natural terrain landslide hazards for trial use (ERM, 1998a; Reeves *et al.*, 1999; Ho *et al.*, 2000). These criteria involve both individual and societal risks (in the form of $F–N$ curves). The limits on individual risk for the most vulnerable person affected by the landslide hazard are shown in Table 7.2.

In terms of societal risk, two options are being tested (Fig. 7.5). The first option involves a conventional 3-tier system incorporating an unacceptable region, a broadly acceptable region and an intervening ALARP region. The second option involves a 2-tier system comprising an unacceptable region and an ALARP region. When the risk level is assessed to be within the ALARP region, cost–benefit calculations need to be carried out to demonstrate that all cost-effective and practicable risk mitigation measures are being undertaken. An *intense scrutiny* zone has been included in both options and is intended to reflect society's aversion to events with 1000 or more fatalities.

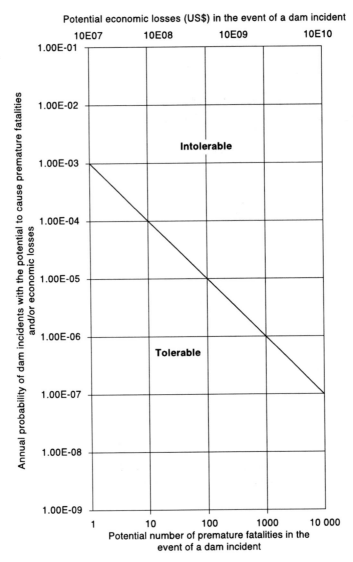

Fig. 7.3. *Societal risk criteria for dam failures: BC Hydro*

It is less clear, however, as to what the limits of the ALARP region represent in terms of economic risk, where the losses are property, services, productivity and infrastructure rather than loss of life. In Britain, for example, landowners are responsible for protecting their own property. However, they do not have to exercise their rights, as there is no process of law by which this responsibility might be enforced (i.e. it is a duty of *imperfect obligation*). Local authorities have powers that can be used to manage landslide risk, but these powers are

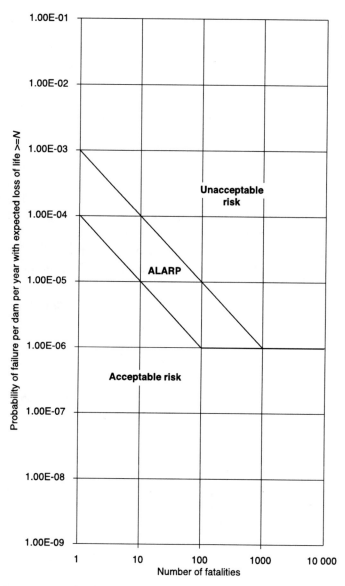

Fig. 7.4. Interim societal risk criteria: ANCOLD

permissive rather than *mandatory*. This clearly limits the role of the State to only providing works that are:

- deemed to be in the national interest;
- provide a sound economic return on the investment, and
- are environmentally acceptable.

Table 7.2. *Interim individual landslide risk criteria: (ERM-Hong Kong 1998a; Reeves et al., 1999)*

Type of development	Maximum allowable individual risk
New	1×10^{-5}
Existing	1×10^{-4}

It is possible to speculate that, from the British Government's viewpoint, there is a continuous ALARP region in which economic efficiency and environmental acceptability dictate whether the risk *to the nation* has been reduced to a reasonably practicable level. It follows that, in Britain, the concept of an unacceptable region, where landslide risk *must* be avoided or reduced, is only applicable to individuals or organisations. The situation is, however, more complicated as is testified to by the UK Government having recently published so-called *indicative standards of protection* for

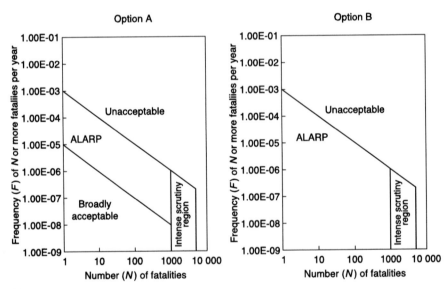

Notes.
1. The above societal risk criteria are to be used in conjunction with a reference toe length of the natural hillside of 500 m (Reeves *et al.*, 1999).
2. If a development is affected by more than 500 m toe length of natural terrain, an appropriate linear scaling factor should be used to scale up the risk criteria. For example, in the case of a large development affected by natural terrain with a toe length of 5 km, then the above societal risk criteria should be increased by one order of magnitude.
3. If the development is affected by less than 500 m toe length of natural terrain, then the same criteria as proposed above are taken to apply (i.e. the criteria will not be scaled down).
4. The societal risk criteria are intended to aid decision-making and not intended to be mandatory.

Fig. 7.5. *Proposed societal risk criteria for landslides and boulder falls from natural terrain in Hong Kong (from Ho et al., 2000)*

Table 7.3. Indicative standards of coast protection used in England (from MAFF, 1999a)

Land use band	Annual probability of failure	Return period: years
A Intensively developed urban areas	0.003–0.01	100–300
B Less intense urban areas with some high grade agricultural land or environmental assets	0.005–0.02	50–200
C Large areas of high grade agricultural land and/or environmental assets; some property at risk	0.01–0.10	10–100
D Mixed agricultural land with occasional properties at risk	0.05–0.40	2.5–20
E Low grade agricultural land with isolated properties	>0.20	<5

coast protection works (Table 7.3). These standards are intended to convey a broad target level of protection for different land use bands, provided the works prove to be cost-effective. They can also be viewed as representing what the Government considers to be reasonable and practicable levels of risk. For example:

- on an urban cliffline (Land use band A, Table 7.3) where cliff-top property is under threat from rare major landslide events, such as on the Scarborough coast (see Example 4.2), Government funding might be limited to undertake slope stabilisation works that are designed to reduce the annual probability of a major landslide to 0.003 (i.e. around 1 in 300). Any additional reduction in the risk might require funding from alternative sources;
- on a rural cliffline (Land use band C, Table 7.3) with isolated cliff-top properties at risk from similar rare landslide events, there might only be Government funding for works that reduce the annual probability of failure to around 0.01 (1 in 100).

Economic risks

The *benefit* of a landslide management strategy is the reduction in risk, expressed in monetary terms, compared with a 'do nothing' case (see Section 'Comparing the risks associated with different management options' in Chapter 6). The *costs* should include all the expenditure incurred during the investigation, planning and design, construction and operation of the strategy. Both benefits and costs should be considered over the strategy lifetime and, hence, need to be brought back to their Present Value by discounting.

A range of strategy options should be evaluated, including:

- the 'do nothing' case involving no active landslide management, simply walking away and abandoning all maintenance, repair or management activity;
- a 'do minimum' case which might involve limited intervention aimed at attempting to reduce, rather than control, the problems and provide a minimum level of protection; for example, promoting the build-up of a beach in front of an unprotected cliff, preventing water leakage on unstable slopes, or the provision of early warning systems for cliff instability;
- a variety of combinations of mitigation works that provide different levels of risk reduction.

The Benefit:Cost ratio (BCR) is a widely used measure of economic cost-effectiveness. Often decision-makers seek to identify strategies that maximise the BCR while ensuring that the level of residual risk would be acceptable, given the current or proposed land use.

Other useful measures include the Net Present Value (NVP) and the Incremental Benefit:Cost ratio which represents the change in Present Value (PV; see Section 'Cliff recession risk' in Chapter 6) costs and benefits between options:

NVP = Risk Reduction − Costs

Incremental BCR

$$= \frac{\text{Risk Reduction (Option B)} - \text{Risk Reduction (Option A)}}{\text{Costs (Option B)} - \text{Costs (Option A)}}$$

Maximisation of the BCR is often the aim of project appraisal for landslide management works. However, it is common for the option with the greatest BCR to fall short of providing an acceptable standard of protection (Table 7.3) or risk reduction. In Britain, a *decision rule* is used to help identify the most economic option and involves the following steps (MAFF, 1993):

1 Examine the BCR of all options. If none is above 1.0, then the project is uneconomic.
2 Identify the option with the greatest BCR that is at least 1.0. If this option delivers an acceptable risk reduction, it should be the final choice. If not, then it is necessary to examine other options (Step 3).

3 Determine whether an increase in standard of protection would be economically efficient. If the incremental BCR of the next option exceeds 1.0, then this option will be economic and should be chosen.
4 If the choice under Steps 2 and 3 falls short of delivering an acceptable standard of protection, then the option that approaches the standard should be chosen, provided the BCR is at least 1.0 and its incremental BCR exceeds 1.0.

Example 7.1
A major public building has started to show signs of cracking and settlement. Investigations have shown that it had been built within an ancient landslide complex, prone to periodic reactivation. A range of landslide management options have been proposed, including:

● 'do nothing';
● Option A; 'do minimum', ensuring that water supply and sewerage pipes are monitored and repaired to prevent leakage into the landslide;
● Option B; the installation of a network of surface and deep drains;
● Option C; the construction of a combined toe weighting and drainage scheme.

Table 7.4 sets out the present value (PV; see Section 'Cliff recession risk' in Chapter 6) costs and benefits (risk reduction) associated with each option, for a 50-year design life.

The risk reduction is the 'do nothing' risk minus the residual risk associated with a particular option.

The option with the greatest NPV is C – the construction of a combined toe weighting and drainage scheme – and might be considered to be the most cost effective. However, it has a lower BCR and Incremental BCR than Option B. If the *decision rule* is applied to this example, the first preference would be Option B – surface and deep drains – as it has the highest BCR. This option should be selected, provided it is expected to deliver an acceptable reduction in the risk. Were this not to be the case, Option C could be selected as it is also economically efficient, with a BCR and incremental BCR above 1.0.

Example 7.2
The lower slopes of a mountain range are prone to debris flow activity. A variety of diversion and control works have been proposed to reduce the risk to downslope properties. Each of the eight options would provide a particular level of protection and risk reduction, corresponding to the return period event it is designed to provide protection against. For example, Option A will only provide protection against a 1 in 1-year debris flow (Table 7.5).

Table 7.4. *Example 7.1: costs and benefits*

	Do nothing	Option A	Option B	Option C
PV costs: £ thousands		100	250	2500
PV risk: £ thousands	5250	5000	4000	500
PV risk reduction		250	1250	4750
Net Present Value		150	1000	2250
Average BCR		2.5	5	1.9
Incremental BCR			6.67	1.56

Notes. PV cost is the present day value of the option costs, that is discounted at a rate of 6% per year (see Section 'Cliff recession risk' in Chapter 6). For example, for a £1 million scheme which will be implemented in Year 5:

PV Cost = Scheme Cost × Discount Factor (Year 5) = 1 × 0.747 = £747 000

PV Risk is the present value of the risk associated with each option, over a 50-year period:

Do Nothing Risk (Years 0–49)

$$= \sum (\text{Prob. Event} \times \text{Losses} \times \text{Discount Factor (Year 0–49)})$$

With Project Risk (Years 0–49)

$$= \sum (\text{Prob. Event} \times \text{Losses} \times \text{Discount Factor (Year 0–49)})$$

PV Risk Reduction is the difference in risk between the With Project risk (i.e. Options A–C) and the Do Nothing risk:

PV Risk Reduction = Do Nothing Risk (Year 0–49) − With Project Risk (Year 0–49)

Net Present Value is the difference between the option benefits (i.e. risk reduction) and the option costs:

NVP = PV Risk Reduction − PV Costs

Average BCR is the ratio of the option benefits (i.e. risk reduction) and the costs:

BCR = PV Risk Reduction/PV Costs

Incremental BCR represents the change in PV costs and PV benefits between options:

$$\text{Incremental BCR} = \frac{\text{Risk Reduction (Option B)} - \text{Risk Reduction (Option A)}}{\text{Costs (Option B)} - \text{Costs (Option A)}}$$

The choice of option will reflect both the level of risk that can be accepted *and* the economic efficiency of the option. The cost of improving the defences from 1 in 100 years to 1 in 200 years would be an extra £1.125 million, but would only reduce the risk by a further £0.5 million. In economic terms, this would not be an efficient use of resources.

Figure 7.6 illustrates how the interplay between benefits and costs can lead to difficult choices. In this example, the 'acceptable' standard of protection to the semi-urban developments at the mountain foot corresponds

384

Table 7.5. Example 7.2: Costs and benefits (adapted from MAFF, 1999a)

Option	Standard of protection (maximum return period)	Benefits: £ million	Costs: £ million	BCR	Incremental BCR
A	1	4.5	1.5	3.0	
B	2.5	7.25	1.95	3.7	6.1
C	10	11.75	2.5	4.7	8.2
D	25	14.5	2.875	5.0	7.3
E	50	16	3.250	4.9	4.0
F	100	16.75	3.625	4.6	2.0
G	200	17.25	4.750	3.6	0.4
H	300	17.5	5.5	3.2	0.2

to the 1 in 100 year event. However, the maximum BCR coincides with the option that delivers protection against the 1 in 25 years event (Option D, Table 7.5). An increase in the standard of protection beyond this level can be justified by the incremental BCR of 2.0 for the option that delivers protection against the 1 in 100 year event (Option F). Further increases in the standards of protection would not be justifiable because of the low incremental BCRs.

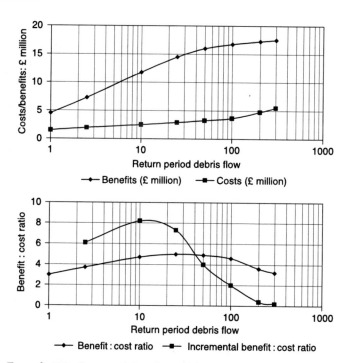

Fig. 7.6. Example 7.2: Costs and benefits and benefit:cost ratios for different schemes (each for protection against different return period debris flow events)

Loss of life

If the risk to people at a particular site or area falls within the ALARP region (Fig. 7.2) it will be necessary to carry out an economic evaluation to establish the cost-effectiveness of any proposed mitigation measures. As with economic risk, a comparison needs to be made between the *costs* of an option and the *benefits* it is expected to deliver. The risk reduction associated with each option is usually expressed in terms of potential loss of life (PLL; see Section 'Societal Risk' in Chapter 6). For discrete events, such as landslides, PLL can be calculated as follows:

$$\text{Potential Loss of Life (PLL)} = \text{Event Probability}$$
$$\times \text{Expected Number of Deaths}$$

Therefore, reducing the likelihood of landsliding reduces the PLL, as does reducing the magnitude or suddenness of events. Thus, if the implementation of mitigation measures upslope of a resort hotel would reduce the annual probability of debris avalanches likely to kill 10 people from 0.01 to 0.005, then the PLL Risk Reduction would be

$$\text{PLL Risk Reduction} = \text{PLL ('do nothing')} - \text{PLL (post stabilisation)}$$
$$= (10 \times 0.01) - (10 \times 0.005) = 0.1 - 0.05 = 0.05$$

However, in order to make direct comparisons with the costs it is necessary to establish a monetary value to the reduction in PLL. This involves determining the *value of life* or the *value of a statistical life* (VOSL), from the amount that people would pay for a very small change in risk (i.e. their *willingness to pay*; Marin, 1992; see Section 'Loss of life and injury' in Chapter 5 for a full discussion):

$$\text{Value of a Statistical Life (VSOL)} = \frac{\text{Willingness to Pay}}{\text{Risk Reduction}}$$

For example, a 'value of a statistical life' of £2 million is equivalent to saying that people would pay £200 for a reduction in the risk of death of 1 in 10 000 (0.0001):

$$\text{Value of a Statistical Life} = \frac{200}{0.0001} = \text{£2 million}$$

The 'value of life' estimates used by a range of countries are listed in Table 7.6.

Society tends to be more adverse to a single large multiple death event than to a series of events that yield a comparable cumulative death toll (Horowitz and Carson, 1993). For example, a landslide that left 2000 dead would almost certainly be viewed as much worse than 1000 separate

386

Table 7.6. Typical 'value of life' figures (from ERM-Hong Kong, 1998b)

Sector	Country	Value of life: £ million	Year applicable
Transport	USA	1.67	1993
	New Zealand	0.75	1993
Railway industry	UK	2–5	1993
	France	4	1993
	Germany	1.3–2.1	1993
	Netherlands	0.3	1993
Dangerous goods	UK	2	1991
transportation	Hong Kong	2	1991

incidents that each caused two deaths, even if the latter were the product of a 'landslide generating event' (see Chapter 1 and Section 'Using the historical record' in Chapter 5). This has led some organisations, such as the Hong Kong Government, to employ *aversion factors* that result in the use of higher 'value of life' figures when the potential for large multiple death events is being considered. In essence, the 'value of life' is envisaged to increase with the size of the risk, above a certain threshold. For example, the railway industry in Switzerland uses a threshold level of 10 deaths. Below this threshold the 'value of life' is considered to be £0.7 million, whereas above it the value immediately rises to £50 million. Such *disaster* or *catastrophe weightings* remain rather arbitrary in terms of the factors used and the thresholds at which they come into play.

In Hong Kong, an aversion factor of 20 is used for sites where landsliding has the potential to cause over 1000 deaths (i.e. sites that fall within the *intense scrutiny* zone; Fig. 7.5; ERM-Hong Kong, 1998a). Note, however, that Kong (2002) suggests that 20 fatalities would be regarded as an unacceptable event and should be the threshold for the application of the aversion factor.

In addition to the Benefit:Cost ratio (BCR), useful indicators of the economic efficiency of measures to reduce loss of life are:

1 *The maximum justifiable expenditure.* This provides a guide to the upper limit of annual investment in risk reduction measures:

$$\text{Maximum Expenditure} = \text{Total PLL} \times \text{Value of Life}$$

$$\times \text{Aversion Factor}$$

2 *The implied cost of averting a fatality (ICAF).* This provides a means of comparing options according to the benefits delivered over the lifetime

of the measures:

$$ICAF = \frac{\text{Cost of Option}}{(\text{PLL Risk Reduction}) \times (\text{Lifetime of Measures})}$$

It is similar to the BCR, but provides a monetary value that can be compared to the 'value of life'. If the ICAF is less than the 'value of life', then the option may be considered to be cost-effective.

Example 7.3

A small town lies within an ancient landslide complex. Periodic reactivation causes significant economic damage, but poses little threat to public health and safety. However, there remains a potential for more dramatic first-time failure of the landslide backscar area. Such an event could be sudden and be accompanied by rapid movements. A number of fatalities could be expected, mainly due to falling masonry.

An expert panel has estimated that the first-time failure has an annual probability in the order of 0.001 (1 in 1000). Analogues suggest that this type of first-time failure in an urban area could cause up to 10 deaths. The Potential Loss of Life (PLL) is:

$$PLL = \text{Event Probability} \times \text{Number of Deaths}$$

$$= 0.001 \times 10 = 0.01$$

Using a 'value of life' of £2 million (no aversion factor has been applied because of the limited number of anticipated fatalities), the maximum justifiable expenditure per year for this event is:

$$\text{Maximum Annual Expenditure} = PLL \times \text{Value of Life}$$

$$= 0.01 \times 2 = £0.02 \text{ million}$$

Any proposed mitigation measures would involve a 'one-off' cost, but would remain effective over a 50-year design lifetime. Therefore, the maximum 'one-off' expenditure would be

$$\text{Maximum 'One-off' Expenditure} = 0.02 \times 50 = £1 \text{ million}$$

Example 7.4

A new road is under construction through landslide-prone, mountainous terrain. The risk to potential users has been established and, in places, lies within the ALARP region but close to the unacceptable threshold. A variety of landslide mitigation measures have been proposed, including the use of boulder fences, check dams and retaining walls.

Table 7.7 sets out the costs and benefits of each of the various options, with a 'do nothing' option providing a baseline for the comparison. For

Table 7.7. *Example 7.4: The BCR and implied cost of averting a fatality for a landslide mitigation scheme (adapted from Kong, 2002)*

Option	Scheme type	Scheme life	PLL	PLL reduction	Value of life: £ thousands	Scheme lifetime benefits: £ thousands	Scheme costs: £ thousands	BCR	ICAF: £ thousands	ICAF < Value of Life
Do nothing			0.008		2000					
A	Boulder fence	40	0.0073	0.0007	2000	56	25	2.2	893	Yes
B	Check dam	40	0.007	0.001	2000	80	100	0.8	2500	No
C	Retaining wall	40	0.006	0.002	2000	160	250	0.6	3125	No

Notes. Scheme Benefits = PLL Reduction × Value of Life × Scheme Life; ICAF = Cost of Option/((PLL Risk Reduction) × (Scheme Life)).

each option, the scheme benefits are the risk reduction, expressed in monetary terms:

Benefits = PLL Reduction × Value of Life

For a 'value of life' of £2 million, the only option with a BCR above 1.0 and with an ICAF less than £2 million is Option A, that is the use of boulder fences. The other two options are not cost-effective. Had the site fallen within the area of 'intense scrutiny' on Fig. 7.5, and an aversion factor of 20 been applicable, then all three options would have been cost-effective.

Environmental risk

Landslides can be important in creating and sustaining internationally important environmental resources (e.g. Clark *et al.*, 1996). In Britain, for example, some landslides have been designated Sites of Special Scientific Interest (SSSI) for their geomorphological importance for earth science research and training (e.g. Alport Castles in the Peak District). Landslides can also create unique landscapes and habitats, as in the Landslip Nature Reserve on the east Devon coast. Most cliffs are shaped by and dependent on landslide processes. Coastal cliffs can also figure prominently among a country's assets, as is the case in Britain where many stretches are safeguarded by the protection afforded by their inclusion in National Parks and AONBs (Areas of Outstanding Natural Beauty). Coastal erosion also maintains exposures of geological features, some of which may be internationally important stratigraphic or fossil reference sites. Cliff recession can have an important role in supplying sediment to beaches, sand dunes and mudflats on neighbouring stretches of coastline. These landforms absorb wave and tidal energy arriving at the coast and can form important components of flood defence or coast protection solutions elsewhere (Lee, 1995). Disruption of the supply of sediment from eroding cliffs will invariably lead to the starvation of some coastal landforms and, hence, may lead to increased risks elsewhere.

Some approaches to landslide management can present significant threats to these environmental resources. For example, many coast protection schemes have had significant impacts on the environment (Fig. 7.7; Lee *et al.*, 2001a; Lee and Clark, 2002). Seawalls or rock revetments have been built which stop the recession process. Cliff faces have been stabilised by drainage works, regraded and landscaped. As a result, geological exposures have become obscured, hardy grasses of little or no conservation value have replaced bare soil and early pioneer stages, and wet areas have dried out. In Britain, a significant proportion of the soft rock cliff habitat resource

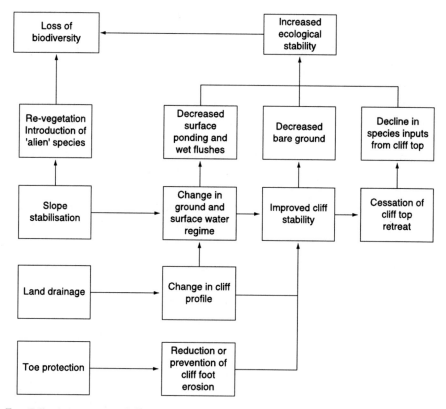

Fig. 7.7. *A summary of the impacts of coast protection schemes on biodiversity (from Lee* et al., *2001a)*

has been adversely affected, with consequent loss or degradation of biological sites of national and international conservation value.

The risks to the environment need to be identified and taken into account in the appraisal of landslide management options, as the environmental losses or mitigation measures are additional project costs over and above the costs associated with a particular option. Environmental impact assessment (EIA) is a widely used approach for identifying the likely significance of the various options being considered. The key factors that will need to be investigated in an EIA include the potential impacts on flora, fauna, population, amenity cultural heritage, property and the built environment, landscape and geological/geomorphological features. Other effects might include:

- the impact of construction traffic;
- impact on access;
- impacts due to construction noise and emissions;
- health and safety;
- water quality implications.

It is important to carry out a scoping study at an early stage of the appraisal of possible management options, in order to pin-point the key environmental issues and concerns that will need to be addressed in a more formal EIA. In many countries an EIA will need to be undertaken where landslide management is likely to have a significant effect. The EIA should help identify which options are best from an environmental perspective. Its key findings should be incorporated in the design and construction process.

In order to fully quantify the risks associated with a management option it is important that the environmental costs, such as a reduction in habitat area or quality, are included in the assessment of risks and economic evaluation. The need to value environmental resources in monetary terms, in order that they can be included in an economic evaluation, presents major difficulties. A number of measures have been proposed, including:

- direct use values, that is the direct use of the environmental resource by humans (e.g. recreation, fishing etc.);
- indirect use values, covering the value to humans of 'background' environmental resources, such as flood regulation, soil fertility etc.;
- option or future use values, including the desire for preservation of environmental resources for possible future use as enshrined in notions of sustainability (e.g. plant communities for possible use in medicine);
- existence and non-use values, representing a human's preferences for the preservation of the environment, over and above use and option values.

Perhaps the most pragmatic and lowest-cost approach is to estimate a *proxy value* for the resource, based on:

- the cost of creating a similar site elsewhere of equivalent environmental value (e.g. a maritime woodland habitat);
- the cost of relocating a resource to another site (e.g. relocation of a specially protected species);
- the cost of local protection *in situ* (e.g. the construction of a viewing chamber for access to a geological exposure).

This would provide an indication of the minimum environmental value. Other methods, such as contingent valuation (e.g. Mitchell and Carson, 1989; Penning-Rowsell *et al.*, 1992; Bateman and Willis, 1999; Bateman *et al.*, 2002; DTLR, 2002), may be necessary if an estimate of the full environmental value is needed.

Replacement costs should include land acquisition, planning, design and implementation, and ongoing monitoring and management of the site. As for other aspects of economic valuation, the costs should be adjusted to their Present Value (PV) by discounting.

A limitation of economic evaluation is that it seeks to compare alternative options in terms of a single objective, their economic efficiency. As it can be difficult to give environmental risks their full significance in monetary terms, it is useful to also set out environmental objectives for a project. Possible options can then be compared in terms of their ability to deliver both economic efficiency and the environmental objectives.

Example 7.5
An urban development sited on mountain footslopes is vulnerable to the impact of channelised debris flows. Possible mitigation measures include the construction of concrete-lined debris chutes to carry the flows through the developed area or the use of a combination of upstream diversion barriers and storage basins. Environmental studies identified a number of important objectives:

• ensuring public access and enjoyment of the natural views along a well-used right-of-way next to the urban channel;
• ensuring that there was no net loss of nationally important orchid and insect species that occur within the proposed storage basin area.

Consultation with local residents and nature conservation bodies revealed that the loss of 'naturalness' along the urban channel banks could not be satisfactorily overcome in scheme design, but it might be acceptable to relocate the orchids and insects by re-creating a similar habitat in what was currently an area of grazing land.

From Table 7.8, it is clear that the option with the highest BCR and net present value (NPV) is Option A, the construction of debris chutes through

Table 7.8. Example 7.5: Costs and benefits

Option	PV risk: £ million	PV risk reduction: £ million	PV costs: £ million	BCR	NPV: £ million	Satisfies environmental objectives
Do nothing	2.5					
Option A: debris chutes	0.5	2	1	2	1	No
Option B1: diversion barriers and storage basin	0.4	2.1	1.25	1.7	0.85	No
Option B2: diversion barriers and storage basin – with habitat replacement	0.4	2.1	1.4	1.5	0.7	Yes

Table 7.9. Example 7.6: Costs and benefits

Option	Environmental benefits	Environmental losses	PV risk: £ million	PV risk reduction: £ million	PV costs: £ million	BCR	NPV: £ million	Satisfies environmental objectives
Do nothing	Ongoing geological exposure and habitats		3.5					
Option A: concrete seawall and slope stabilisation		Geological exposure obscured Degradation of habitats	0.15	3.35	2.5	1.3	0.85	No
Option B: beach management and rock revetment	Natural processes largely preserved	Partial reduction in exposure Dynamic nature of habitats reduced, but not lost	1.25	2.25	2	1.1	0.25	Yes

the urban area. However, this option fails to satisfy the environmental objectives of ensuring access and unspoilt naturalness along the channel banks. The favoured option then becomes Option B2, involving the re-creation of the orchid and insect habitats at an additional cost of £150 000. This achieves the environmental objectives and would still be economically efficient, with a BCR greater than 1.0.

Example 7.6
Over the next 50 years a small group of houses located on a cliff-top will be destroyed as a result of cliff recession. Environmental scoping studies have identified that proposals to prevent recession by constructing a seawall at the cliff foot, together with slope stabilisation measures, would lead to degradation of an internationally important geological exposure and lead to the loss of maritime cliff habitats. As a result, the proposals were considered to be unacceptable from an environmental perspective. It was recognised that the geological and habitat value of the cliffs could be maintained if the rate of recession was reduced from its present value, rather than prevented.

It was agreed that the objectives of the management scheme would be to reduce the level of risk to the cliff-top community to an acceptable level, while maintaining the scientific quality of the geological site with unrestricted access and continuing to provide high biodiversity habitats on the coastal slopes.

Table 7.9 summarises the economic evaluation for three alternative options, along with their environmental benefits and losses. On economic grounds, the most efficient option would be Option A, as it has the highest BCR and NPV. However, Option B is preferred, as it delivers a reduction in the environmental losses while remaining economically efficient.

Environmental acceptance criteria
In the past, it has often proved possible to find a compromise solution that delivers what is perceived to be an acceptable balance between risk reduction and environmental impacts. Hence, mitigation measures have been put forward and then modified to address environmental objections. However, the value placed on environmental resources in some countries, such as Britain, has grown to the point where further degradation is increasingly considered unacceptable. Thus, there has been a shift from a position of limiting environmental damage to one which seeks to ensure that there is no loss of environmental resource.

Lee (2000) describes how the environment has become an increasingly important factor in determining the way in which coastal landslide risks are managed in England. Of particular importance are:

1 *The EC Habitats and Species Directive* (the 'Habitats Directive'; Council Directive 92/43/EEC). This requires member states to designate areas of importance for particular habitats and species as Special Areas of Conservation (SACs). Together with Special Protection Areas (SPAs) designated under the Conservation of Birds Directive (the 'Birds Directive'; Council Directive 79/409/EEC), these areas form a Europe-wide series of sites known as 'Natura 2000'. In Great Britain the Habitats Directive is implemented through the Conservation (Natural Habitats Etc.) Regulations 1994 (SI 2716), which employs the term 'European Site' to encompass SACs and SPAs.

The Regulations set out measures intended to maintain at, or restore to, a 'favourable conservation status' those habitats and species designated as SAC/SPA. The conservation status of a habitat is considered to be favourable when:

- its natural range and areas it covers within that range are stable or increasing;
- the specific structure and functions which are necessary for its long-term maintenance exist and are likely to continue to exist for the foreseeable future; and
- the conservation status of its typical species is favourable.

The Directive identifies 'Vegetated Sea Cliff of the Atlantic and Baltic coasts' as requiring the designation of SAC. The UK coast supports a significant proportion of the EC sea cliff resource and, to date, ten lengths of cliffline have been put forward as candidate SACs, including the eroding, soft rock cliffs of Suffolk, East Devon, West Dorset and the Isle of Wight.

Along these clifflines, the Government is required to take appropriate steps to avoid the deterioration of the natural habitats and the habitats of species, as well as the significant disturbance of species.

2 *Biodiversity Targets.* The UK Government has set out its commitments to the Convention on Biological Diversity (the Rio Convention) in the document 'Biodiversity: the UK Action Plan'. The overall goal is 'to conserve and enhance the biological diversity within the UK and to contribute to the conservation of global biodiversity through all appropriate mechanisms'. In pursuit of this objective, the Government has published a series of Habitat Action Plans which contain habitat creation and rehabilitation targets. Coast protection authorities have specific High Level Targets in relation to biodiversity. When carrying out works they must aim to ensure that there is no net loss to habitats covered by biodiversity action plans (MAFF, 1999b).

The Maritime Cliff and Slope Habitat Action Plan includes five targets, three of which are directly related to coast protection (UK Biodiversity Group, 1999):

- to maintain the existing maritime cliff resource of cliff top and slope habitat;
- to maintain, wherever possible, free functioning of coastal physical processes acting on maritime cliff and slope habitats;
- to retain and where possible increase the amount of maritime cliff and slope habitats unaffected by coastal defence and other engineering works.

Included within the habitat action plan are a number of proposed actions agreed by various agencies and local government. These proposed actions include:

(a) encouraging a presumption against the stabilisation of any cliff face except where human life, or important natural or man-made assets, are at risk;

(b) where stabilisation of a cliff face is necessary, ensuring that adequate mitigation and/or compensation measures are provided in order to maintain the overall quantity and quality of maritime cliff and slopes habitat;

(c) encouraging the increased use of soft (e.g. foreshore recharge) rather than hard engineering techniques where some degree of cliff stabilisation is necessary;

(d) giving consideration to the non-replacement of defences which have come to the end of their useful life.

The Habitats Directive and habitat action plans have introduced a 'no net loss' policy for maritime cliff and slope habitats, with the aspiration of actually achieving, over time, a 'net gain'. These commitments and aspirations could lead to the situation where some existing coast protection and slope stabilisation works will have to be removed in the future in order to ensure that biodiversity targets are met. Indeed, a coast protection scheme that might affect the integrity of the habitats would only be approved if there were imperative reasons for overriding public interest. In such circumstances, compensation measures would be required as part of the scheme; that is the creation of replacement vegetated sea cliff habitat. As the only viable option for recreating vegetated sea cliffs of international quality is to 'restore' natural habitats at previously protected cliffs, for every length of new defences there would have to be an equivalent abandonment of existing defences (e.g. Lee *et al.*, 2001a).

A major obstacle to achieving biodiversity targets are the contrasting attitudes to coastal erosion within British society. Many feel that loss of land is unacceptable and needs to be resisted by public investment in coast protection. As John Gummer (2000), formerly the Minister at the Government department responsible for coastal defence policy, wrote:

Of course it has happened before. It's just that the last time our shores were successfully invaded was 1066. Now the east coast is being crossed again, and more effectively than by Norman or Dane. East Anglia is threatened as far inland as Bedfordshire. Already, more than 70 per cent of its beaches are in retreat. If erosion goes on at its present rate, my own constituency of Suffolk Coastal will simply continue falling into the sea.

Britain is proud of being an island, and her people of being an island race. The time has now come for us to pay the price of defending this blessed plot from the very sea which has been our defence so often in the past.

This *fortress Britain* attitude is in marked contrast to the view that the erosion process is necessary for maintaining the natural beauty of the coastline (the *living coast* view). To this latter group, coastal defence leads to environmental degradation and should only be contemplated where there is an over-riding national need. The 'living coast' view reflects current legislation and Government policies. However, the 'fortress Britain' view has considerable popular support, ensuring that there will be a continued demand for coast protection schemes and a resistance to the abandonment of current defences (e.g. Lee, 2000; Lee et al., 2001a; Lee, 2002).

As described earlier in this chapter, conventional approaches to risk reduction generally involve seeking an acceptable balance between the costs of the management measures and the benefits (i.e. prevention of loss of life, injury or economic losses). Risk criteria, such as the ALARP principle, are often used to indicate whether a risk should be tolerated or not. The level of expenditure on risk reduction measures should be proportional to the level of risk. However, the increasing importance of environmental legislation and international commitments appears to be placing further constraints on landslide management.

Climate change uncertainty: implications for landslide management

Global climatic changes are occurring as the result of human-induced accumulation of so-called greenhouse gasses such as CO_2 in the atmosphere (IPCC, 1990, 1995a,b, 2001). Evidence from ice cores, supplemented by direct measurements since the mid-1950s, reveals a steady rise in greenhouse gas concentrations from the late 1700s, changing to a rapid rise post 1950. Atmospheric concentrations of carbon dioxide (CO_2), the primary anthropogenic greenhouse gas, have risen from about 270 ppm in pre-industrial times to over 360 ppm. Global temperature has risen by about 0.6 °C since the beginning of the 20th century, with about 0.4 °C of this warming occurring since the 1970s. There is evidence that precipitation has increased by 0.5–1% per decade in the 20th century over the Northern Hemisphere continents, together with a 2–4% increase in the frequency of heavy precipitation events.

Climatic modelling is improving rapidly and recent results provide an indication of the scale of changes that could be expected by 2080:

- mean global temperature will rise by 1–3.5 °C (IPCC, 2001). Note that warming at the higher end of this range would shift climatic zones pole-ward by about 550 km;
- at latitudes of 45° or greater, annual precipitation will increase by 100–300 mm (i.e. northern Europe, Russia, China, northern and central USA, Canada and the southern extremes of South America);
- in lower latitudes (5–45°), annual precipitation will decrease by 100–700 mm (i.e. Australia, southern Africa, southern USA, western South America, Central America, the Caribbean, north Africa, the Mediterranean region, the Middle East and India);
- within 5° of the Equator, annual precipitation changes are expected to be complex, with a decrease of 100–600 mm predicted for the Americas and South-east Asia, but an increase of 100–300 mm expected in central Africa;
- global mean sea-levels are expected to rise by around 0.4 m, in response to thermal expansion of sea water. Global mean sea-level in 2100 is projected to have risen by between 0.09 and 0.88 metres relative to the 1990 level (IPCC, 2001). However, it is not expected that mean sea-level rise will be the same everywhere as the heating up of the oceans will not be uniform;
- the frequency of extreme high sea water levels is expected to increase dramatically. The UK Climate Impacts Programme (UKCIP98, e.g. Hulme *et al.*, 1998), for example, has predicted that by the 2050s extreme high water levels will increase in frequency from once a century to, typically, once a decade. This situation would be further exacerbated if storminess were to increase more rapidly than predicted.

It is not just the climate that could change, but also the weather. Changes in the weather are likely to be significant as it is the extremes of climate (droughts, hurricanes, intense rainstorms, periods of extreme heat and cold etc.) that can be the most damaging to society.

The predicted changes in climate and weather will lead to changes in the probability of landsliding in many areas and, hence, landslide risk. For example, the UKCIP98 climate change scenarios give estimates of future potential change to mean precipitation and mean evapo-transpiration for southern Britain. The application of the UKCIP98 change scenarios to the

Table 7.10. Predicted changes in effective rainfall, Ventnor, UK (from Halcrow, 2001)

	September–November		December–February	
	mm	% change	mm	% change
Mean effective rainfall (2000)	76.4	0	67.4	0
Mean effective rainfall: low	80.2	5	71.7	6
Mean effective rainfall: medium–low	84.4	11	74.6	11
Mean effective rainfall: medium–high	82.7	8	81.8	21
Mean effective rainfall: high	85.2	12	84.0	25

Note. Low, med–low, med–high and high estimates based on UKCIP98 climate change scenarios

Ventnor Undercliff rainfall data (see Examples 2.2 and 4.4) suggests a 5–6% increase in mean monthly effective rainfall under the *Low scenario* and a 12–25% increase for the *High scenario* (Table 7.10; Halcrow Group, 2001). This is expected to result in an increase in the frequency or probability of landslide events, assuming the distribution of events will be similar to the historic record. Figure 7.8 presents the mean monthly effective rainfall frequency distribution for the period December to February based on 1839–2000 Ventnor data. The UKCIP98 change scenarios have been applied to this distribution to derive 2080 *Low* and *High* change scenarios. The distributions indicate that the current probability of mean monthly effective rainfall of

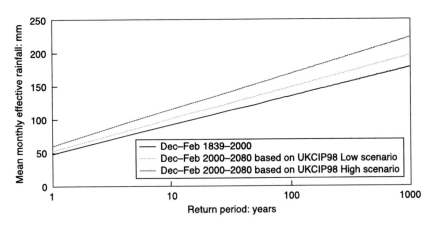

Fig. 7.8. Ventnor, Isle of Wight: the predicted changes in the mean monthly effective rainfall frequency distribution for the period December to February, for the UKCIP Low and High scenarios (based on 1839–2000 rainfall data, see Fig. 4.11; from Halcrow, 2001)

100 mm is 0.1 (or 1:10 years). Under the UKCIP98 *Low scenario* the probability will remain largely unchanged, but under the *High scenario* the probability is increased to 0.2 (1:5 years). The potential change for more extreme conditions is greater with, for example, the probability of mean monthly effective rainfall of 150 mm increasing from the current 0.005 (1:200 years) to 0.02 (1:50 years) for the UKCIP98 *High scenario*.

This potential increase in rainfall event frequency in Ventnor is likely to apply to the established relationship between effective rainfall and ground movement susceptibility (see Example 4.4). It follows, therefore, that the annual probability of landslide reactivation events will increase over time.

It should be stressed, however, that there are major uncertainties associated with the forecasting of future climate change, some of which arise from the reliability of the available Global Circulation Models, the emission scenarios (e.g. Greenhouse gas only (GHG) or Greenhouse gas plus sulphate aerosols (GHG + SUL); Conway, 1998) and the climate variables that are modelled. This uncertainty is compounded by speculations about the possible impact on slope stability and future levels of landslide hazard and, as a result, landslide risk. However, it can be argued that the changes to landslide risk due to Global Warming will probably be less than those arising from population growth, development changes and alterations in land use, except in especially sensitive areas such as urban zones developed on old and extensive landslides prone to reactivation. Nevertheless, the two sets of factors taken together show that landslide risk will change in the future, sometimes quite dramatically, and that decisions will have to be made despite this uncertainty. Decisions taken now will have implications in the future in terms of the vulnerability of society to landslide hazards.

A number of strategies are available for managing landslide problems in the face of uncertain change, including:

- the use of so-called *no regret* or *low regret* options that are worthwhile pursuing irrespective of how the climate changes; for example, improving drainage systems in urban landslide areas (e.g. McInnes, 2000);
- making sound decisions, based on a thorough analysis of the information available, while ensuring that future generations are not committed to inflexible and/or expensive landslide management practices (*adaptive management*). The approach should leave scope for amending decisions at a later date as improved information becomes available on climate change and the associated uncertainties;
- *delaying action* if the assessment of current and future risks suggest problems are not likely to become significant unless and until certain climate change thresholds are reached; for example, in mountain regions prone to

401

debris flows, the decision to replace sacrificial bridges across debris flow chutes with more permanent structures;

- *delay and buy time*, through the implementation of a short-term solution to the current slope problems and delaying the point at which significant investment or relocation decisions have to be made. For example, the A3055 road on the Isle of Wight, UK, lies immediately inland of the crest of 70 m-high Chalk cliffs and is threatened by coastal erosion. Construction of erosion control measures at the cliff foot is not considered appropriate, because of the environmental value of the site. In order to extend the life of this important tourist route, a monitoring system was installed to provide early warning of ground movement that could affect the safety of road users (Fort and Clark, 2002). A decision on the long-term future of the road was delayed until an acceptable solution could be reached between the local authority and conservation bodies;
- *changing the land use* to one where the future risks were likely to be more acceptable; for example, a change from allocation of land for housing in an area where climate change might lead to an increase in landslide activity to a lower risk use, such as playing fields or public open space. In some places, *site abandonment* might be the best option;
- *contingency planning*, involving making plans and provisions for a possible increased frequency of extreme events and climatic 'surprises'. For example, this might involve establishing a coordinated approach to disaster management, such as the US Federal Emergency Management Agency (FEMA) which supervises the Hazard Mitigation Grant Program;
- *making allowance for climate change* in the design of slope stabilisation and erosion control measures. This can be achieved through the use of design specifications that allow for uncertainty or variability in the design parameters or loadings, and by ensuring that structures can be modified at a later date if the allowances are found to be inadequate. For example, in the UK, allowances to take account of possible changes in the rate of sea-level rise have been given by the Government (MAFF, 1999a) for the design or adaption of coastal defences. These allowances range from 6 mm per year (eastern and southern England) to 4 mm per year (north west and north east England) and 5 mm per year (the remainder of England and Wales).

All of these strategies should be supported by monitoring of climate change and its impact on landslide risk. In addition to improving the understanding of changing conditions, monitoring also allows adaptive decisions themselves to be modified in the light of improved knowledge. However, monitoring should not be an excuse to delay more direct intervention where the existing or future level of risk is considered to be unacceptable.

Risk assessment, decision making and consultation

Risk assessment is part of a process that begins with the recognition and formulation of a problem – the likelihood of damage or harm – and ends with a decision about how best to manage the problem (see Fig. 1.3). The decision should take account of the results of the risk assessment process, but not be determined solely by them. There are a wide range of other factors that will usually need to be considered in reaching a decision, including environmental issues, financial constraints and the broader socio-economic and political context.

In many instances the participation in the decision-making process may extend wider than a regulatory authority or a corporate management team. There has been a long-running debate as to whether the traditional approach of restricting decision-making on risk issues to a few scientifically well informed individuals (*narrow participation*) should be replaced by broader involvement (see Section 'Risk evaluation' in Chapter 1 and Royal Society, 1992). It is now generally appreciated that broad participation is preferable (see Pidgeon, Funtowicz and Ravetz, and O'Riordan in Hood and Jones, 1996), because the information base is widened, the accountability of the technical decision-makers is increased and the acceptability of the final decision rises if those that are affected have been involved in the process. Thus pressures exerted from homeowners, landowners, interest groups, business groups and environmentalists will also influence the choice of landslide risk management strategy.

Consultation must, therefore, be seen to be an essential part of the decision-making process, despite the fact that wider involvement can increase costs and cause delay. Stern and Fineberg (1996) suggest that effective consultation should involve:

- ensuring that all interested parties are identified and invited to participate in the consultation process, preferably at an early stage of the assessment process;
- ensuring that the participants are aware of the legal status of the consultation process and how their representations will be used by the decision-makers;
- making information readily available to the consultees and, where appropriate, providing access to technical expertise and other resources for groups that lack these resources.

Getting the broad range of groups and interests that may be affected by landslide problems to accept risk-based decisions is often critical to the successful implementation of landslide risk management strategies. Such acceptance is dependent on the establishment of trust and this, in turn, is dependent on openness, involvement and good communications. It follows, therefore, that the risk assessment process should seek to inform all stakeholders, addressing *their* questions in a form that *they* can understand. Herein lies the true importance of risk communication, a process which is not simply confined to the development of forecasts and the issuing of warnings, but embraces all the diverse ways by which information on risks is disseminated among all those involved in order to improve safety and security.

Glossary of terms

Acceptable Risk The level of risk that individuals and groups are prepared to accept at a particular point in time, as further expenditure in risk reduction is not considered justifiable (e.g. the costs of risk reduction exceed the benefits gained). The term is increasingly coming to be replaced by Tolerable Risk.

Adverse Consequences The adverse effects, losses or harm from a human perspective, resulting from the realisation of a hazard. They can be expressed quantitatively (e.g. deaths, economic losses) or qualitatively (loss of amenity) and occur in the short or longer term.

Adverse Outcomes See Adverse Consequences.

Benefit:Cost Ratio The ratio of the present value of benefits to the present value of costs.

Consequences The adverse effects and benefits arising from the realisation of a hazard.

Consequence Assessment The identification and quantification of the full range of adverse consequences arising from the identified patterns and sequences of hazard.

Deterministic method A method in which precise, single-values are used for all variables and input values, giving a single value as the output.

Detriment A numerical measure of the expected harm or loss associated with an adverse event and, therefore, an important ingredient in benefit–cost and risk–benefit analyses. Increasingly used to describe adverse consequences in general.

Direct and indirect economic losses Those losses capable of being given monetary values because of the existence of a market, with all other losses classified as intangibles. Direct economic losses arise principally from the physical impact of a landslide on property, buildings, structures, services and infrastructure. Indirect economic losses are those that subsequently arise as a consequence of the destruction and damage caused by primary hazard (i.e. the landslide itself), secondary hazards or follow-on hazards.

Disaster An imprecise term which should only be applied to situations where the level of adverse consequences is sufficiently severe so that either outside assistance is required to facilitate the recovery process or the detrimental effects are long-lasting and debilitating.

Discounting The procedure used to arrive at the sum of either costs or benefits over the lifetime of a project, using a discount rate to scale down future benefits and costs. The effect of using a discount rate is to reduce the value of projected future costs or benefits to their values as seen from the present day.

Economic Risk The risk of financial loss due to potential hazards causing loss of production, damage or other adverse financial consequences.

Elements at Risk The population, buildings and engineering works, economic activities, public services utilities, artefacts, valued possessions, infrastructure and environmental features in any area that are valued by humans and potentially adversely affected by landslides.

Environmental Hazard The threat potential posed to humans or nature by events originating in, or transmitted by, the natural or built environment.

Environmental Risk An amalgam of the probability and scale of exposure to loss arising from hazards originating in, or transmitted by, the physical and built environments.

Event Tree A form of logic diagram designed to specify the steps by which the range of possible adverse outcomes can arise after the occurrence of a selected initiating event.

Exposure The proportion of time that an asset or person is exposed to the hazard. Exposure involves the notion of being in 'the wrong place' (i.e. the 'danger zone' where a landslide impacts – the spatial probability) at the 'wrong time' (i.e. when the landslide occurs – the temporal probability).

Fault Tree A form of logic diagram designed to work backwards from a particular event or outcome (known as the top event) through all the chains of possible events that could be precursors of the top event. Its purpose is to analyse why a particular outcome occurred or could occur.

F–N Curve A plot showing the frequency–magnitude relationship of adverse consequences arising from different types of hazards. Usually, the consequence referred to is the number of deaths, n, and the cumulative frequency of incidents with n or more deaths is plotted against number of fatalities (Fn).

Frequency A measure of likelihood expressed as the number of occurrences of an event in a given time.

Global risk assessment A procedure to determine the overall risk faced by a community.

Hazard A property or situation that in particular circumstances could lead to harm from a human perspective. For a hazard to exist situations have to arise or circumstances occur where human value systems might be adversely impacted. Hazards are threats to humans and what they value: life, well-being, material goods and environment. A primary hazard event (i.e. a major landslide) may generate three other types of hazard:

- **Post-event hazards** which occur after the initial sequence and are a product of the specific system returning towards stability, as is the case with the aftershocks following major earthquakes.
- **Secondary hazards** which are different geohazards generated by the main hazard event sequence; for example, destructive tsunami generated by earthquakes and major landslides and the floods caused by the failure of landslide-generated dams.
- **Follow-on hazards** generated by the primary events but arising due to failures of infrastructure and management systems; for example, fire caused by overturned stoves, electrical short-circuits and broken gas pipes, localised flooding caused by broken supply mains or sewers and disease.

Individual Risk The risk specific to humans, such as the general public, road users or particular activity groups, is the frequency with which individuals within such groupings are expected to suffer harm. Measures of individual risk include:

- **Location-Specific Individual Risk (LSIR)** The risk for an individual who is present at a particular location for the entire period under consideration, which may be 24 hours per day, 365 days per year or during the entire time that a risk generating plant, process or activity is in operation.
- **Individual-Specific Individual Risk (ISIR)** The risk for an individual who is present at different locations during different periods.

407

- **Average Individual Risk (AIR)** This can be calculated from historical data of the number of fatalities per year divided by the number of people at risk. Alternatively, a measure of the average individual risk can be derived from the societal risk divided by the number of people at risk.

Intangible losses The vague and diffuse adverse consequences that arise from an event and which cannot easily be valued in economic terms because there is no market. They include effects on the environment, nature conservation, amenity, local culture, heritage, aspects of the local economy, recreation and peoples' health, as well as their attitudes, behaviour and sense of well-being.

Likelihood Used as a qualitative description of probability or frequency.

Market Value The price for which individual goods are bought or sold in the market.

Natural Hazard Those elements of the physical environment harmful to humans and caused by forces extraneous to human society. Because of the scale of human impacts on the environment, the term is increasingly being replaced by Environmental Hazard and Geohazard.

Outcomes The range of hazards, adverse consequences and benefits that can result from an initial hazard event.

Perceived Risk The combined evaluation that is made by an individual of the likelihood of an adverse event occurring in the future and the magnitude of its likely adverse consequences.

Potential Loss of Life (PLL) A measure of the expected number of fatalities for a population exposed to the full range of landslide events that might occur in an area (i.e. the societal risk). To calculate PLL it is necessary to estimate, for each event and its possible adverse outcomes, the frequency per year (f) and the associated number of fatalities (N).

Present Value (PV) The value of a stream of benefits or costs when discounted back to the present time.

Probabilistic method A method in which the variability of input values and the sensitivity of the result are taken into account to give results in the form of a range of probabilities for different adverse outcomes.

Probability The likelihood of a specific outcome, measured by the ratio of specific outcomes to the total number of possible outcomes. Probability is expressed as a number between 0 and 1, with 0 indicating an impossible outcome, and 1 indicating that a particular outcome is certain.

Residual Risk The remaining risk after all proposed improvements in management of the system have been made.

Return Period The average length of time separating the occurrence of extreme events of a similar, or greater, magnitude. Also known as Recurrence Interval.

Risk The potential for adverse consequences, loss, harm or detriment from a human perspective. Risk is often expressed as a mathematical expectation value, the product of the probability of occurrence of a defined future hazard and the monetary value of the adverse consequences.

Risk Analysis The use of available information to estimate the risk to individuals or populations, property, or the environment, posed by identified hazards. Risk analyses generally contain the following steps: problem definition, hazard identification and risk estimation.

Risk Assessment The process by which risk is analysed, estimated and evaluated.

Risk Communication Any purposeful exchange of information about health or environmental risks between interested parties.

Risk Control or Risk Treatment The process of decision-making for managing risk, the implementation, or enforcement, of risk mitigation measures and the re-evaluation of its effectiveness from time to time, using the results of risk assessment as one input.

Risk Estimation The process used to produce a measure of the level of health, property, or environmental risks being analysed. Risk estimation contains the following steps: hazard assessment, consequence analysis and their integration.

Risk Evaluation The stage at which values and judgements enter the decision process perceived by including consideration of the importance of the estimated risks and the associated social, environmental and economic consequences, in order to identify a range of alternatives for managing the risks. Public participation and obtaining the views of stakeholders are the keys to success.

Risk Management The process whereby decisions are made to accept a known or assessed level of risk and/or the implementation of actions to reduce the consequences or probability of occurrence.

Risk Perception How people's knowledge, experience, cultural background and attitudes (i.e. their socio-cultural make-up) leads them to interpret the stimuli and information that they receive concerning risk. Risk perception involves people's beliefs, attitudes, judgements and feelings, as well as the wider social or cultural values and dispositions that people adopt towards hazards and their benefits.

Risk Register An auditable record of the project risks, their consequences and significance, together with proposed mitigation and management measures.

Risk Scoping The process by which the spatial and temporal limits of a risk assessment are defined at the outset of a project. These may be determined on purely practical grounds, such as budget constraints, time constraints, staff availability or data availability.

Risk Screening The process by which it is decided whether or not a risk assessment is required or whether it is required for a particular element, or elements, within a project.

Sensitivity Testing A method in which the impact on the output of an analysis is assessed by systematically changing the input values.

Site-specific risk assessment A procedure to determine the nature and significance of the hazards and risk levels at a particular site.

Societal Risk The number of people within a group, at a location or undertaking an activity, and the frequency at which they suffer a given level of harm from the realisation of specified hazards. It is the shared level of risk and usually refers to the risk of death, expressed as risk per year.

Tolerable Risk A risk that society is willing to live with so as to secure certain net benefits in the confidence that it is being properly controlled, kept under review and will be further reduced as and when possible. In some situations, risk may be tolerated because the individuals at risk cannot afford to reduce risk even though they recognise that it is not properly controlled.

Value of a Statistical Life (VOSL) The value of a change in the risk of death, not human life itself, that is how a person's welfare is affected by an increased mortality risk, not what his or her life is worth. If 100 000 people are exposed to an annual mortality risk of $1:100\,000$ there will, statistically, be one death incidence per year. If the risk is reduced to $1:120\,000$, then there will, on average, be 0.833 deaths per year, a saving of 0.167 of a statistical life. The cost of the risk reduction measures can be used to place a value on a statistical life. Values of £1–2 million are usually applied.

Vulnerability The potential to suffer harm, loss or detriment, from a human perspective. It can be represented by the level of potential damage, or degree of loss, of a particular asset (expressed on a scale of 0 to 1) subjected to a damaging event of a given intensity. For property, the loss will be the value of the damage relative to the value of the property; for persons, it will be the probability that a particular life (the element at risk) will be lost, given that the person(s) is affected by the landslide.

Willingness to Pay The amount an individual is prepared to pay in order to obtain a given improvement in utility, or specified reduction in risk.

References

Adams, J. (1995). *Risk*. UCL Press, London.

Adams, W. C. (1986). Whose lives count? TV coverage of natural disasters. *Journal of Communication* **36**(2), 113–122.

Agnew, C. E. (1985). Multiple probability assessments by dependent experts. *Journal of the American Statistical Society* **80**, 390, 343–347.

Aki, K. (1965). Maximum likelihood estimation of b in the formula $\log(N) = a - bM$ and its confidence limits. *Bulletin of the Earthquake Research Institute, Tokyo University* **43**, 237–239.

Alén, C. (1996). Application of a probabilistic approach in slope stability analysis. In *Landslides* (ed. K. Senneset). Balkema, Rotterdam, 1137–1148.

Aleotti, P., Baldelli, P. and Polloni, G. (2000). Hydrogeological risk assessment of the Po River Basin (Italy). In *Landslides: In Research, Theory and Practice* (eds E. N. Bromhead, N. Dixon and M-L. Ibsen). Thomas Telford, London, 13–18.

Alexander, D. E. (1986). Landslide damage to buildings. *Environmental Geology and Water Science* **8**, 147–151.

Alexander, D. E. (1989). Urban landslides. *Progress in Physical Geography* **13**, 2, 157–191.

Alexander, D. (2000). *Confronting Catastrophe*. Terra, Harpenden.

Alexander, D. (2002). *Principles of Emergency Planning and Management*. Terra, Harpenden.

Alfors, J. T., Burnett, J. L. and Gay, T. E. (1973). *Urban Geology Masterplan for California*. California Division of Mines and Geology Bulletin 198.

de Ambrosis, L. (2002). Letters to the editor: re: landslide risk management concepts and guidelines. *Australian Geomechanics* **37**, 54–55.

Ang, A. H-S. and Tang, W. H. (1984). *Probability Concepts in Engineering Planning and Design. Volume II Decision, Risk and Reliability*. John Wiley and Sons, Chichester.

Australian Geomechanics Society (2000). Landslide risk management concepts and guidelines. *Australian Geomechanics* **35**, 49–52.

Bacon, J. and Bacon, S. (1988). *Dunwich Suffolk*. Segment Publications, Colchester.

Baecher, G. (1987). Statistical quality control of engineered embankments. *US Army Engineer Waterways Experiment Station*, Vicksburg.

Baeza, C. and Corominas, J. (1996). Assessment of shallow landslide susceptibility by means of statistical techniques. In *Landslides* (ed. K. Senneset). Balkema, Rotterdam, 147–152.

Baker, Sir Richard (1674). *A Chronicle of the Kings of England etc.* (6th impression). London, Ludgate-hill. George Sawbridge, Hosier Lane, London.

Bateman, I. and Willis, K. G. (1999). *Valuing Environmental Preferences*. Oxford University Press, Oxford.

Bateman, I. et al. (2002). *Economic Valuation with Stated Preference Techniques: A Manual*. Edward Elgar, Cheltenham.

Baxter, P. J. (1990). Medical effects of volcanic eruptions. *Bulletin Volcanologique* **52**, 532–544.

Bedford, T. and Cooke, R. M. (2001). *Probabilistic Risk Analysis: Foundations and Methods*. Cambridge University Press, Cambridge.

Belsom, C., Dolan, S., Glickman, L. and Haydock, R. (1991). *Probability*. Cambridge University Press, Cambridge.

Benjamin, J. R. and Cornell, C. A. (1970). *Probability, Statistics and Decision for Civil Engineering*. McGraw-Hill Book Company, New York.

Benson, M. A. (1960). *Characteristics of Frequency Curves Based on a Theoretical 1000 Year Record*. United States Geological Survey Water Supply Paper 1543-A.

Bentley, S. and Smalley, I. J. (1984). Landslips in sensitive clays. In *Slope Instability* (eds D. Brunsden and D. B. Prior). John Wiley and Sons, Chichester, 457–490.

Besson, L., Durville, J. L., Garry, G., Graszk, Ed., Hubert, Th. and Toulemont, M. (1999). *Plans de prevention des risques naturels (PPR) – Risques de mouvements de terrain*. Guide méthodologique. La Documentation Française, Paris.

Birch, G. P. and Griffiths, J. S. (1995). Engineering geomorphology. In *Engineering Geology of the Channel Tunnel* (eds C. S. Harris, M. B. Hart, P. Varley and C. Warren). Thomas Telford, London.

Bishop, A. W. (1955). The use of the slip circle in the stability analysis of slopes. *Géotechnique* **5**, 7–17.

Bishop, A. W. (1973). Stability of tips and spoil heaps. *Quarterly Journal of Engineering Geology* **6**, 335–376.

Bishop, A. W., Hutchinson, J. N., Penman, A. D. N. and Evans, H. E. (1969). Geotechnical investigation into the causes and circumstances of the disaster of 21st October (1966). *Unpublished Report for the Aberfan Disaster Enquiry*.

Blaikie, P., Cannon, T., Davis, I. and Wisner, B. (1994). *At Risk: Natural Hazards, Peoples' Vulnerability and Disasters*. Routledge, London.

Blockley, D. I. (1980). *The Nature of Structural Design and Safety*. Ellis Horwood, Chichester.

Blockley, D. I. (1985). Reliability or responsibility? *Structural Safety* **2**, 273–280.

Blockley, D. I. (1995). Process re-engineering for safety. *Proceedings of the HSE Conference on Risk Management in Civil and Mechanical Engineering*, Institution of Civil Engineers, London.

413

Blong, R. J. (1992). Some perspectives on geological hazards. In *Geohazards: Natural and Man-made* (eds G. J. H. McCall, D. J. C. Laming and S. C. Scott). Chapman and Hall, London, 209–216.

Boggett, A. D., Mapplebeck, N. J. and Cullen, R. J. (2000). South Shore Cliffs, Whitehaven – geomorphological survey and emergency cliff stabilisation works. *Quarterly Journal of Engineering Geology and Hydrogeology* **33**, 213–226.

Borgman, L. E. (1963). Risk criteria. *ASCE Journal of the Waterways and Harbours Division, WW3* **89**, 1–35.

Bottelberghs, P. H. (2000). Risk analysis and safety policy developments in the Netherlands. *Journal of Hazardous Materials* **71**, 59–84.

Brabb, E. E. and Harrod, B. L. (eds) (1989). *Landslides: Extent and Economic Significance.* Balkema, Rotterdam, 123–126.

Brand, E. W. and Hudson, R. R. (1982). CHASE – an empirical approach to the design of cut slopes in Hong Kong soils. *Proceedings of the Seventh Southeast Asian Geotechnical Conference, Hong Kong* **1**, 1–16.

British Standards Institution (BSI) (1990). *British Standard Methods of Test for Soils for Civil Engineering Purposes.* BS 1377.

British Standards Institution (BSI) (1991). *Quality Vocabulary.* BS 4778.

British Standards Institution (BSI) (1999). *Code of Practice for Site Investigation.* BS 5930.

Bromhead, E. N. (1986). *The Stability of Slopes.* Surrey University Press, London.

Bromhead, E. N., Hopper, A. C. and Ibsen, M-L. (1998). Landslides in the Lower Greensand escarpment in south Kent. *Bulletin of Engineering Geology and the Environment* **57**, 131–144.

Brunsden, D. and Jones, D. K. C. (1972). The morphology of degraded landslide slopes in south-west Dorset. *Quarterly Journal of Engineering Geology* **3**, 205–223.

Brunsden, D. and Jones, D. K. C. (1976). The evolution of landslide slopes in Dorset. *Philosophical Transactions of the Royal Society, London* **A283**, 605–631.

Brunsden, D. and Lee, E. M. (2000). Understanding the behaviour of coastal landslide systems: an inter-disciplinary view. In *Landslides: In Research, Theory and Practice* (eds E. N. Bromhead, N. Dixon and M-L. Ibsen). Thomas Telford Keynote Papers CD-ROM. Available from: e.bromhead@kingston.ac.uk.

Brunsden, D. and Prior, D. B. (1984). *Slope Instability.* John Wiley and Sons, Chichester.

Brunsden, D., Doornkamp, J. C., Fookes, P. G., Jones, D. K. C. and Kelly, J. M. N. (1975a). Large scale geomorphological mapping and highway engineering design. *Quarterly Journal of Engineering Geology* **8**, 227–253.

Brunsden, D., Doornkamp, J. C., Hinch, L. W. and Jones, D. K. C. (1975b). Geomorphological mapping and highway design. *Sixth Regional Conference for Africa on Soil Mechanics and Foundation Engineering*, 3–9.

Brunsden, D., Ibsen, M-L., Lee, E. M. and Moore, R. (1995). The validity of temporal archive records for geomorphological purposes. *Quaestiones Geographicae Special Issue 4*, 79–92.

Bryant, E. A. (1991). *Natural Hazards.* Cambridge University Press, Cambridge.

Bucknam, R. C. (2001). *Landslides Triggered by Hurricane Mitch in Guatemala: Inventory and Discussion.* US Geological Survey Open File Report 01-443.

414

Building Research Establishment (1981). *Damage in Low-rise Buildings.* BRE Digest 251.

Bunce, C., Cruden, D. M. and Morgenstern, N. R. (1997). Assessment of the hazard from rock fall on a highway. *Canadian Geotechnical Journal* **34**, 344–356.

Burton, I. and Kates, R. W. (1964). The perception of natural hazards in resource management. *Natural Resources Journal* **3**, 412–441.

Burton, I., Kates, R. W. and White, G. F. (1978). *The Environment as Hazard.* Oxford University Press, Oxford.

Buss, E. and Heim, A. (1881). *Der bergsturz von Elm.* Wurster, Zurich.

Cannon, T. (1993). A hazard need not a disaster make: vulnerability and the causes of 'natural' disasters. In *Natural Disasters: Protecting Vulnerable Communities* (eds P. A. Merriman and C. W. A. Browitt). Thomas Telford, London, 92–105.

Cardinali, M., Reichenbach, P., Guzzetti, F., Ardizzone, F., Antonini, G., Galli, M., Cacciano, M., Castellani, M. and Salvati, P. (2002). A geomorphological approach to the estimation of landslide hazards and risks in Umbria, Central Italy. *Natural Hazards and Earth System Sciences* **2**, 57–72.

Carr, A. P. (1962). Cartographic error and historical accuracy. *Geography* **47**, 135–144.

Carr, A. P. (1980). The significance of cartographic sources in determining coastal change. In *Timescales in Geomorphology* (eds R. A. Cullingford, D. A. Davidson and J. Lewin). John Wiley and Sons, Chichester, 67–78.

Carrara, A. (1983). Multivariate models for landslide hazard evaluation. *Mathematical Geology* **15**, 403–427.

Carrara, A. (1988). Landslide hazard mapping by statistical methods: a 'black box' approach. In *Workshop on Natural Disasters in European Mediterranean Countries, Perugia, Italy*, Consiglio Nazionale delle Ricerche, Perugia, 205–224.

Carrara, A. and Merenda, L. (1976). Landslide inventory in northern Calabria, Southern Italy. *Bulletin of the Geological Society of America* **87**, 1153–1162.

Carrara, A., Cardinali, M., Detti, R., Guzzetti, F., Pasqui, V. and Reichenbach, P. (1990). Geographic information systems and multivariate models in landslide hazard evaluation. In *ALPS 90 Alpine Landslide Practical Seminar, Sixth International Conference and Field Workshop on Landslides, Aug. 31–Sept. 12, Milan, Italy*, Universitá degli Studi de Mialno, 17–28.

Carrara, A., Cardinali, M., Detti, R., Guzzetti, F., Pasqui, V. and Reichenbach, P. (1991). GIS techniques and statistical models in evaluating landslide hazard. *Earth Surface Processes and Landforms* **16**, 427–445.

Carrara, A., Cardinali, M. and Guzzetti, F. (1992). Uncertainty in assessing landslide hazard and risk. *ITC Journal* **2**, 172–183.

Carrara, A., Guzzetti, F., Cardinali, M. and Reichenbach, P. (1998). Current limitations in modeling landslide hazard. In *Proceedings of IAMG '98* (eds A. Buccianti, G. Nardi and R. Potenza), 195–203.

Carreño, R. and Bonnard, C. (1999). Rock slide at Macchupicchu, Peru. In *Landslides of the World* (ed. K. Sassa). Kyoto University Press, Kyoto, 323–326.

Carter, J. and Riley, N. (1998). The role of societal risk in land use planning near hazardous installations and in assessing the safety of the transport of hazardous materials at the national and local level. In *Quantified Societal Risk and Policy Making* (eds R. E. Jorissen and P. J. M. Stallen). Kluwer Academic Publishers, Dordrecht.

Casale, R. and Margottini, C. (eds) (1995). *Meteorological events and natural disasters: an appraisal of the Piedmont (North Italy) case history of 4–6 November 1994 by a CEC field mission*. European Commission Report.

Catenacci, V. (1992). Il dissesto geologico e geoambientale in Italia dal dopoguerra al (1990). Memorie Descrittive della Carta Geolog-ica d'Italia, *Servizio Geologico Nazionale* (in Italian) **47**, 301.

Cave, P. W. (1992). Natural hazards, risk assessment and land use planning in British Columbia: progress and problems. *Proceedings of First Canadian Symposium on Geotechnique and Natural Hazards, Vancouver, Canada*, 1–12.

Chan, R. K. S. (2000). Hong Kong slope safety management system. *Proceedings of the Symposium on Slope Hazards and Their Prevention, The Jockey Club Research and Information Centre for Landslip Prevention and Land Development, The University of Hong Kong, Hong Kong*.

Charman, J. H. (2001). Desk studies. In *Land Surface Evaluation for Engineering Practice* (ed. J. S. Griffiths). Geological Society, Engineering Geology Special Publication No. 18, 19–21.

Chicken, J. C. and Posner, T. (1998). *The Philosophy of Risk*. Thomas Telford, London.

Chowdhury, R. N. (1988). Special lecture: Analysis methods for assessing landslide risk – recent developments. In *Landslides* (ed. C. Bonnard). Balkema, Rotterdam **1**, 515–524.

Chowdhury, R. and Flentje, P. (2003). Role of slope reliability analysis in landslide risk management. *Bulletin of Engineering Geology and Environment* **62**, 41–46.

Clark, A. R. and Guest, S. (1991). The Whitby cliff stabilisation and coast protection scheme. In *Slope Stability Engineering: Developments and Applications* (ed. R. J. Chandler). Thomas Telford, London, 283–290.

Clark, A. R. and Guest, S. (1994). The design and construction of the Holbeck Hall landslide coast protection and cliff stabilisation emergency works. *Proceedings of the 29th MAFF Conference of River and Coastal Engineers, Loughborough*, 3.3.1–3.3.6.

Clark, A. R. and Johnson, D. K. (1975). Geotechnical mapping as an integral part of site investigation – two case histories. *Quarterly Journal of Engineering Geology* **8**, 211–224.

Clark, A. R., Palmer, J. S., Firth, T. P. and McIntyre, G. (1993). The management and stabilisation of weak sandstone cliffs at Shanklin, Isle of Wight. In *The Engineering Geology of Weak Rock* (eds J. C. Cripps and C. F. Moon). Engineering Group of the Geological Society Special Publication, 392–410.

Clark, A. R., Lee, E. M. and Moore, R. (1996). *Landslide Investigation and Management in Great Britain: a Guide for Planners and Developers*. HMSO, London.

Clark, A. R., Fort, D. S. and Davis, G. M. (2000). The strategy, management and investigation of coastal landslides at Lyme Regis, Dorset, UK. In *Landslides: In Research, Theory and Practice* (eds E. N. Bromhead, N. Dixon and M-L. Ibsen). Thomas Telford, London, 279–286.

Clayton, C. R. I., Simons, N. E. and Matthews, M. C. (1982). *Site Investigation: A Handbook for Engineers.* Granada, London.

Clements, M. (1994). The Scarborough experience – Holbeck landslide, 3–4 June 1993. *Proceedings of the Institution of Civil Engineers, Municipal Engineers* **103**, 63–70.

Close, U. and McCormick, E. (1922). Where the mountains walked. *National Geographic Magazine* **41**, 5: 445–464.

Coburn, A. (1994). Death tolls in earthquakes. In *Medicine in the International Decade for Natural Disaster Reduction (IDNDR).* Royal Academy of Engineering, 21–26.

Coburn, A. and Spence, R. (1992). *Earthquake Protection.* John Wiley and Sons, Chichester.

Coggan, J. S., Pine, R. J. and Stead, D. (2001). A proposed methodology for rockfall risk assessment along coastlines. *Geoscience in South-west England* **10**, 190–194.

Conway, B. W., Forster, A., Northmore, K. J. and Barclay, W. J. (1980). *South Wales Coalfield Landslip Survey.* British Geological Survey Special Surveys Division, Report No. EG/80/4.

Conway, D. (1998). Recent climate variability and future climate change scenarios for Great Britain. *Progress in Physical Geography* **22**, 350–374.

Corominas, J. (1996). The angle of reach as a mobility index for small and large landslides. *Canadian Geotechnical Journal* **33**, 260–271.

Corominas, J. and Moya, J. (1996). Historical landslides in the Eastern Pyrenees and their relation to rainy events. In *Landslides* (eds J. Chacon, C. Irigaray and T. Fernandez). Balkema, Rotterdam, 125–132.

Corominas, J. and Moya, J. (1999). Reconstructing recent landslide activity in relation to rainfall in the Llobregat River basin, Eastern Pyrenees, Spain. *Geomorphology* **30**, 79–94.

Cory, J. and Sopinka, J. (1989). John Just versus Her Majesty The Queen in right of the Province of British Columbia. *Supreme Court Report*, 2, 1228–1258.

Costa, J. E. and Wieczorek, G. F. (eds) (1987). Debris flows/avalanches: processes, recognition and mitigation. *Reviews in Engineering Geology* **VII**. Geological Society of America, Boulder, Colorado.

Cox, D. R. (1962). *Renewal Theory.* Methuen, London.

Cox, S. J. and Tait, N. R. S. (1991). *Reliability, Safety and Risk Management: An Integrated Approach.* Butterworth-Heinemann, Oxford.

Crescenti, U. (1986). La grande frana di Ancona del 13 dicembre 1982. *Studi Geologici Camerti*, Special Issue, Camerino (in Italian) **1**.

Cross, M. (1987). *An Engineering Geomorphological Investigation of Hillslope Stability in the Peak District of Derbyshire.* PhD thesis, University of Nottingham.

Cross, M. (1988). Landslide susceptibility mapping using the Matrix Assessment Approach: a Derbyshire case study. In *Geohazards and Engineering Geology* (eds J. G. Maund and M. Eddleston). Geological Society Special Publication, 15, Geological Society Publishing, Bath, 247–261.

Cross, W. (1924). *Historical Sketch of the Landslides of the Gaillard Cut.* Memoir 18, National Academy of Sciences, Washington DC, 22–43.

Crozier, M. J. (1984). Field assessment of slope instability. In *Slope Instability* (eds D. Brunsden and D. B. Prior). John Wiley and Sons, Chichester, 103–142.

Crozier, M. J. (1986). *Landslides: Causes, Consequences and Environment.* Croom Helm, London.

Crozier, M. J. and Glade, T. (1999). Frequency and magnitude of landsliding: fundamental research issues. *Zeitschrift fur Geomorphologie NF Suppl.Bd* **115**, 141–155.

Crozier, M. J. and Preston, N. J. (1998). Modelling changes in terrain resistance as a component of landform evolution in unstable hill country. In *Lecture Notes in Earth Science* (eds S. Hergarten and H. J. Neugebauer) **78**, 267–284.

Cruden, D. (1991). A simple definition of a landslide. *Bulletin of the International Association of Engineering Geology* **43**, 27–29.

Cruden, D. M. (1997). Estimating the risks from landslides using historical data. In *Landslide Risk Assessment* (eds D. Cruden and R. Fell). Balkema, 177–184.

Cruden, D. M. and Varnes, D. J. (1996). Landslide types and processes. In *Landslides: Investigation and Mitigation* (eds A. K. Turner and R. L. Schuster). Transportation Research Board, Special Report 247, National Research Council, National Academy Press, Washington DC, 36–75.

Cui, W. and Blockley, D. I. (1990). Interval probability theory for evidential support. *International Journal of Intelligent Systems* **5**, 183–192.

Cvetkovich, G. and Lofstedt, R. E. (eds) (1999). *Social Trust and the Management of Risk.* Earthscan, London.

Damgaard, J. S. and Peet, A. H. (1999). Recession of coastal soft cliffs due to waves and currents: experiments. In *Proceedings of Coastal Sediments '99. 4th International Symposium on Coastal Engineering and Science of Coastal Sediment Processes, Long Island, New York*, 1181–1191.

Davies, W. N. and Christie, H. D. (1996). The Coledale mudslide, New South Wales, Australia – a lesson for geotechnical engineers. In *Landslides* (ed. K. Senneset). Balkema, Rotterdam, 701–706.

Davis, J. P. and Blockley, D. I. (1996). On modelling uncertainty. In *Hydroinformatics '96* (ed. Muller). Balkema, Rotterdam.

Dearman, W. R. and Fookes, P. G. (1974). Engineering geological mapping for civil engineering practice in the United Kingdom. *Quarterly Journal of Engineering Geology* **7**, 223–256.

DeGraff, J. V. (1978). Regional landslide evaluation: two Utah examples. *Environmental Geology* **2**, 203–214.

Del Prete, M. and Petley, D. J. (1982). Case history of the Main Landslide at Craco, Basilicata, South Italy. *Geol. App. E Idrogeol.* **17**, 291–304.

Del Prete, M., Guadagno, F. M. and Hawkins, B. (1998). Preliminary report on the landslides of 5 May 1998, Campania, southern Italy. *Bulletin of Engineering Geology and Environment* **57**, 113–129.

Department of the Environment (DoE) (1991). *Policy Appraisal and the Environment.* HMSO, London.

Department of the Environment (DoE) (1995). *A Guide to Risk Assessment and Risk Management for Environmental Protection.* HMSO, London.

Department of the Environment, Transport and the Regions (DETR) (2000).

Guidelines for environmental risk assessment and management – revised Departmental Guidance. Prepared by Institute for Environment and Health.

Department of the Environment, Transport and the Regions (DETR) (2001). *Contaminated Land Risk Assessment: A Guide to Good Practice.* CIRIA, London.

Department for Transport, Local Government and the Regions (DTLR) (2002). *Economic Valuation with Stated Preference Techniques: A Summary Guide.* DTLR Appraisal Guidance. DTLR, London.

Dikau, R., Brunsden, D., Schrott, L. and Ibsen, M-L. (1996). *Landslide Recognition.* John Wiley and Sons, Chichester.

DNV Technica (1996). *Quantitative landslip risk assessment of Pre-GCO man-made slopes and retaining walls.* DNV Technica Report to Geotechnical Engineering Office, Hong Kong Government.

Doornkamp, J. C., Brunsden, D., Jones, D. K. C., Cooke, R. U. and Bush, P. R. (1979). Rapid geomorphological assessments for engineering. *Quarterly Journal of Engineering Geology* **12**, 189–204.

Douglas, M. and Wildavsky, A. (1983). *Risk and Culture: an Essay on the Selection of Technological and Environmental Dangers.* University of California Press, Berkeley, CA.

DRM (1990). *Les études préliminaires à la cartographie réglementaire des risques naturels majeurs.* La Documentation Française, 143 pp.

Dumbleton, M. J. (1983). *Air Photographs for Investigating Natural Changes, Past Use and Present Condition of Engineering Sites.* Transport and Road Research Laboratory Report LR 1085.

Dumbleton, M. J. and West, G. (1970). *Air Photograph Interpretation for Road Engineers in Britain.* Road Research Laboratory Report LR 369.

Dumbleton, M. J. and West, G. (1976). *Preliminary Sources of Information for Site Investigation in Britain.* Transport and Road Research Laboratory Report LR 40/3.

Duncan, J. M. (1996). Soil slope stability analysis. In *Landslides: Investigation and Mitigation* (eds A. K. Turner and R. L. Schuster). Transportation Research Board, Special Report 247, National Research Council, National Academy Press, Washington DC, 337–371.

Dunnicliff, J. and Green, G. E. (1988). *Geotechnical Instrumentation for Monitoring Field Performance.* John Wiley and Sons, Chichester.

Dussauge-Peisser, C., Helmstetter, A., Grasso, J-R., Hantz, D., Desvarreux, P., Jeannin, M. and Giraud, A. (2002). Probabilistic approach to rock fall hazard assessment: potential of historical data analysis. *Natural Hazards and Earth System Sciences* **2**, 15–26.

Edgers, L. and Karlsrud, K. (1982). Soil flows generated by sub-marine slide – case studies and consequences. *3rd International Conference on the Behaviour of Offshore Structures, Cambridge* **2**, 425–437.

Eiby, G. A. (1980). *Earthquakes.* Heinemann, London.

Einstein, H. H. (1988). Special lecture: Landslide risk assessment procedure. In *Landslides* (ed. C. Bonnard). Balkema, Rotterdam **2**, 1075–1090.

Eisbacher, G. H. and Clague, J. J. (1984). *Destructive Mass Movements in High Mountains: Hazard and Management.* Geological Survey of Canada Paper 84–16.

419

Engineering News Record (1971). *Contrived landslide kills 15 in Japan* **187**, 18.

EPOCH (1993). *Temporal Occurrence and Forecasting of Landslides in the European Community.* 3 Volumes. Contract No. 90 0025.

ERM-Hong Kong (1998a). *Landslides and Boulder Falls from Natural Terrain: Interim Risk Guidelines.* GEO Report No. 75. Report prepared for the Geotechnical Engineering Office, Hong Kong.

ERM-Hong Kong (1998b). *Quantitative Risk Assessment of Boulder Fall Hazards in Hong Kong: Phase 2 Study.* GEO Report No. 80. Report prepared for the Geotechnical Engineering Office, Hong Kong.

ERM-Hong Kong (1999). *Slope Failures along BRIL Roads: Quantitative Risk Assessment and Ranking.* GEO Report No. 81. Report prepared for the Geotechnical Engineering Office, Hong Kong.

Essex, C., Lookman, T. and Nererberg, M. R. H. (1987). The climate attractor over short time scales. *Nature* **326**, 64–66.

Eusebio, A., Grasso, P., Mahtab, A. and Morino, A. (1996). Assessment of risk and prevention of landslides in urban areas of the Italian Alps. In *Landslides* (ed. K. Senneset). Balkema, Rotterdam, 189–194.

Evans, N. C. (1997). *Natural Terrain Landslide Study: Preliminary assessment of the influence of rainfall on natural terrain landslide initiation.* GEO Discussion Note DN 1/97.

Evans, N. C. and King, J. P. (1998). *The Natural Terrain Landslide Study: Debris Avalanche Susceptibility.* GEO Technical Note TN 1/98.

Evans, N. C., Huang, S. W. and King, J. P. (1997). *The Natural Terrain Landslide Study: Phase III.* GEO Special Project Report SPR 5/97.

Evans, S. G. and Hungr, O. (1993). The assessment of rockfall hazard at the base of talus slopes. *Canadian Geotechnical Journal* **30**, 620–636.

Fantucci, R. and Sorriso-Valvo, M. (1999). Dendrogeomorphological analysis of a slope near Lago, Calabria (Italy). *Geomorphology* **30**, 165–174.

Fell, R. (1994). Landslide risk assessment and acceptable risk. *Canadian Geotechnical Journal* **31**, 261–272.

Fell, R. and Hartford, D. (1997). Landslide risk management. In *Landslide Risk Assessment* (eds D. Cruden and R. Fell). Balkema, Rotterdam, 51–108.

Fell, R., Finlay, P. and Mostyn, G. (1996a). Framework for assessing the probability of sliding of cut slopes. In *Landslides* (ed. K. Senneset). Balkema. Rotterdam **1**, 201–208.

Fell, R., Walker, B. F. and Finlay, P. J. (1996b). Estimating the probability of landsliding. *Proceedings of the 7th Australia/New Zealand Conference on Geomechanics, Adelaide.* Institution of Engineers Australia, Canberra, 304–311.

Finlay, P. and Fell, R. (1995). *A Study of Landslide Risk Assessment in Hong Kong.* Report for the Geotechnical Engineering Office, Hong Kong.

Finlay, P. J., Mostyn, G. R. and Fell, R. (1999). Landslides: prediction of travel distance and guidelines for vulnerability of persons. *Proceedings of the 8th Australia/New Zealand Conference on Geomechanics, Hobart.* Australian Geomechanics Society **1**, 105–113.

Fookes, P. G. (1997). Geology for engineers: the geological model, prediction and performance. *Quarterly Journal of Engineering Geology* **30**, 290–424.

Fookes, P. G., Baynes, F. J. and Hutchinson, J. N. (2000). Total geological history: a model approach to the anticipation, observation and understanding of site conditions. *GeoEng 2000, an International Conference on Geotechnical and Geological Engineering* **1**, 370–460.

Fort, D. S. and Clark, A. R. (2002). The monitoring of coastal landslides: a management tool. In *Instability – Planning and Management* (eds R. G. McInnes and J. Jakeways). Thomas Telford, London, 479–486.

Fort, D. S., Clark, A. R. and Savage, D. T. (2000). Instrumentation and monitoring of the coastal landslides at Lyme Regis, Dorset, UK. In *Landslides: In Research, Theory and Practice* (eds E. N. Bromhead, N. Dixon and M-L. Ibsen). Thomas Telford, London, 573–578.

Foster, H. D. (1980). *Disaster Planning.* Springer-Verlag, New York.

Francis, P. W. (1994). Large volcanic debris avalanches in the central Andes. In *Abstracts of the International Conference on Volcanic Instability on the Earth and Other Planets.* Geological Society, London.

Franklin J. A. (1984). Slope instrumentation and monitoring. In *Slope Instability* (eds D. Brunsden and D. B. Prior). John Wiley and Sons, Chichester, 143–170.

Franks, C. A. M. (1996). *Study of Rainfall Induced Landslides on Natural Slopes in the Vicinity of Tung Chung New Town, Lantau Island.* GEO Special Project Report SPR 4/96.

Franks, C. A. M. (1999). Characteristics of some rainfall-induced landslides on natural slopes, Lantau Island, Hong Kong. *Quarterly Journal of Engineering Geology* **32**, 247–260.

Fulton, A. R. G., Jones, D. K. C. and Lazzari, S. (1987). The role of geomorphology in post-disaster reconstruction: the case of Basilicata, Southern Italy. In *International Geomorphology 1986* (ed. V. Gardiner). John Wiley and Sons, Chichester, 241–262.

Fussell, J. B. (1973). A formal methodology for fault tree construction. *Nuclear Science Engineering* **52**, 421–432.

Fussell, J. B. (1976). Fault tree analysis: concepts and techniques. In *Proceedings of the NATO Advanced Study Institute on Generic Techniques in Systems Reliability Assessment* (eds E. Henley and J. Lynn). Noordhoff Publishing, Leyden, Holland, 133–162.

Gardner, T. (1754). *Historical notes on Dunwich, Blythburgh and Southwold.*

Garry, G. and Graszk, Ed. (1997). Plans de prévention des risques naturels prévisibles (PPR) – Guide général. *La Documentation Française*, Paris.

Gazzetta Ufficiale della Repubblica Italiana (1998). Misure urgenti per la prevenzione del rischio idrogeologico ed a favore delle zone col-pite da disastri franosi nella regione Campania. *Serie Generale*, Anno 139, n. 208, 7 September 1998 (in Italian), 53–74.

Gazzetta Ufficiale della Repubblica Italiana (1999). Atto di indirizzo e co-ordinamento per l'individuazione dei criteri relativi agli adempi-menti di cui all'art. 1, commi 1 e 2, del decreto-legge 11 giugno 1998, n. 180. *Serie Generale*, Anno 140, n. 3, 5 January 1999 (in Italian), 8–34.

Geomorphological Services Limited (1986). *Review of Research into Landsliding in Great Britain*. Reports to the Department of the Environment.

Geotechnical Engineering Office (GEO) (1996). *Report on the Fei Tsui Road Landslide of 13 August 1995*. Hong Kong Geotechnical Engineering Office.

Giddens, A. (1999). *Runaway World*. Profile Books, London.

Gilbert, E. S. (1981). On discrimination using qualitative variables. *Journal of the American Statistics Association* **63**, 1399–1412.

Glade, T. (1996). The temporal and spatial occurrence of landslide-triggering rainstorms in New Zealand. In *Heidelberger Geogr. Arb.* (eds R. Mausbacher and A. Schulte). Beitrage zur Physiogeographie – Festschrift fu Dietrich Barsch. **104**, 237–250.

Glade, T. (1997). *The Temporal and Spatial Occurrence of Rainstorm-triggered Landslides in New Zealand*. PhD thesis, Department of Geography, Victoria University of Wellington.

Glade, T. (1998). Establishing the frequency and magnitude of landslide-triggering rainstorm events in New Zealand. *Environmental Geology* **35**, 160–174.

Glade, T. and Crozier, M. J. (1996). Towards a national landslide information base for New Zealand. *New Zealand Geographer* **52**, 29–40.

Godt, J. W. and Savage, W. Z. (1999). El Niño 1997–98: direct costs of damaging landslides in the San Francisco Bay region. In *Landslides* (eds J. S. Griffiths, M. R. Stokes and R. G. Thomas). Balkema, Rotterdam, 47–55.

Gostling, Rev. W. (1756). Letter to *Gentleman's Magazine* **26**, 160.

Government of Hong Kong (1972a). *Interim Report of the Commission of Inquiry into the Rainstorm Disasters, 1972*. Hong Kong Government Printer.

Government of Hong Kong (1972b). *Final Report of the Commission of Inquiry into the Rainstorm Disasters, 1972*. Hong Kong Government Printer.

Graham, J. (1984). Methods of stability analysis. In *Slope Instability* (eds D. Brunsden and D. B. Prior). John Wiley and Sons, Chichester, 171–216.

Graszk, E. and Toulemont, M. (1996). Plans de prévention des risques naturels et expropriation pour risques majeurs. Les mesures de prévention des risques naturels de la loi du 2 février 1995. *Bull. Labo. P. and Ch.* **206**, 85–94.

Greenbaum, D. (1995). *Project Summary Report: Rapid Methods of Landslide Hazard Mapping*. Technical Report WC/95/30. British Geological Survey, Keyworth.

Gretener, P. E. (1967). The significance of the rare event in geology. *The American Association of Petroleum Geologists Bulletin* **51**, 11, 2197–2206.

Griffiths, J. S. (ed.) (2001). *Land Surface Evaluation for Engineering Practice*. Geological Society, London, Engineering Group Special Publication, 18.

Griffiths, J. S. and Marsh, A. H. (1986). The role of geomorphological and geological techniques in a preliminary site investigation. In *Site Investigation Practice: Assessing BS 5930* (ed. A. B. Hawkins). Geological Society, Engineering Geology Special Publication No. 2, 261–267.

Griffiths, J. S., Brunsden, D., Lee, E. M. and Jones, D. K. C. (1995). Geomorphological investigations for the Channel Tunnel Terminal and Portal. *The Geographical Journal* **161**, 3, 275–284.

Guadagno, F. M. (1999). The landslides of 5th May 1998 in Campania, Southern Italy: natural disasters or also man-induced phenomena? *Journal of Nepal Geological Society* **22**, 463–470.

Gumbel, E. J. (1941). The return period of flood flows. *Annals of Mathematical Statistics* **12**(2), 163–190.

Gummer, J. (2000). *Country Living*, 57–60, March, 2000.

Gutenberg, B. and Richter, F. (1949). *Seismicity of the Earth and Associated Phenomena*, Princeton University Press, Princeton, NJ.

Hadfield, P. (2001). Slip sliding away. *New Scientist* **169**, 2281, 20.

Haefeli, R. (1965). Creep and progressive failure in snow, rock and ice. In *Proceedings of the Sixth International Conference on Soil Mechanics and Foundation Engineering*, University of Toronto Press, Toronto **3**, 134–148.

Hahn, G. J. and Shapiro, S. S. (1967). *Statistical Methods in Engineering*. John Wiley and Sons, New York.

Sir William Halcrow and Partners (1986). *Rhondda Landslip Potential Assessment*. Department of the Environment and Welsh Office.

Sir William Halcrow and Partners (1988). *Rhondda Landslip Potential Assessment: Inventory*. Department of the Environment and Welsh Office.

Halcrow Asia Partnership (1999). *Stage 2 Detailed Site-specific QRA Report*. Technical report to Geotechnical Engineering Office, Hong Kong Government.

Halcrow Group Ltd (2001). *Preparing for the Impacts of Climate Change*. Report to SCOPAC.

Halcrow Group Ltd (2003). *Coastal Instability Risk: Ventnor Undercliff, Isle of Wight*. Report to the Isle of Wight Council.

Hall, A. P. and Griffiths, J. S. (1999). A possible failure mechanism for the AD 1575 'Wonder Landslide'. *East Midlands Geographer*, 21–22, 92–105.

Hall, J. W., Davis, J. P. and Blockley, D. I. (1997). Towards uncertainty management for coastal defence systems. *Proceedings of the 32nd MAFF Conference of River and Coastal Engineers*, MAFF, London, H.3.1–H.3.10.

Hall, J. W., Lee, E. M. and Meadowcroft, I. C. (2000). Risk-based assessment of coastal cliff recession. *Proceedings of the ICE: Water and Maritime Engineering* **142**, 127–139.

Hall, J. W., Meadowcroft, I. C., Lee, E. M. and van Gelder, P. H. A. J. M. (2002). Stochastic simulation of episodic soft coastal cliff recession. *Coastal Engineering* **46**, 159–174.

Hallet, B. (1987). On geomorphic patterns with a focus on stone circles viewed as a free-convection phenomenon. In *Irreversible Phenomena and Dynamic Systems Analysis in Geosciences* (eds C. Nicolis and G. Nicolis). NATO ASI, 533–553.

Hampton, M. A., Lee, H. J. and Locat, J. (1996). Submarine landslides. *Reviews of Geophysics* **1**, 33–60.

Hand, D. (2000). *Report on the inquest into the deaths arising from the Thredbo landslide*. State of New South Wales, Australia, Attorney Generals Department – Office of the NSW Coroner.

Hansen, A. (1984). Landslide hazard analysis. In *Slope Instability* (eds D. Brunsden and D. B. Prior). John Wiley and Sons, Chichester, 523–602.

Harley, J. B. (1968). Error and revision in early Ordnance Survey maps. *Cartographic Journal* **5**, 115–124.

Harp, E. L. (1997). *Landslides and Landslide Hazards in Washington State Due to February 5–9 1996 Storm*. US Geological Survey Administrative Report.

Harp, E. L. and Jibson, R. L. (1995). *Inventory of Landslides Triggered by the 1994 Northridge, California Earthquake*. US Geological Survey Open File Report: 95–213.

Harp, E. L. and Jibson, R. L. (1996). Landslides triggered by the 1994 Northridge, California earthquake. *Seismological Society of America Bulletin* **86**, S319–S332.

Harp, E. L., Jibson, R. L. and Wieczorek, G. F. (1981). *Landslides from the February 4 1976 Guatemala Earthquake*. US Geological Survey Professional Paper 1204-A.

Harr, M. E. (1987). *Reliability-Based Design in Civil Engineering*. McGraw-Hill Book Co. New York.

Hawley, J. G. (1984). Slope instability in New Zealand. In *Natural Hazards in New Zealand* (eds I. G. Speden and M. J. Crozier). UNESCO, Wellington, 88–133.

Head, K. H. (1982). *Manual of Soil Laboratory Testing. Volume 2 Permeability, Shear Strength and Compressibility Tests*. Pentech Press, London.

Head, K. H. (1985). *Manual of Soil Laboratory Testing. Volume 3 Effective Stress Tests*. Pentech Press, London.

Health and Safety Executive (HSE) (1988). *The Tolerability of Risk from Nuclear Power Stations*. HMSO, London.

Health and Safety Executive (HSE) (1992). *The Tolerability of Risk from Nuclear Power Stations (revised)*. HMSO, London.

Hearn, G. J. (1995). Landslide and erosion hazard mapping at Ok Tedi copper mine, Papua New Guinea. *Quarterly Journal of Engineering Geology* **28**, 47–60.

Hearn, G. J. and Griffiths, J. S. (2001). Landslide hazard mapping and risk assessment. In *Land Surface Evaluation for Engineering Practice* (ed. J. S. Griffiths). Geological Society, London, Engineering Group Special Publication **18**, 43–52.

Heim, A. (1882). Der bergsturz von Elm. *Z. Deutsch Geol. Ges.* **34**, 74–115.

Heim, A. (1932). Bergsturz und Menschenleben. *Beiblatt zur Viierteljahrsschrift der Naturforschenden Gesellschaft in Zurich*, **77**, 1–217. Translated by N. Skermer (Landslides and Human Lives, BiTech Publishers, Vancouver, BC, 1989).

Hendron, A. J. Jr. and Patton, F. D. (1985). *The Vaiont Slide: a geotechnical analysis based on new geologic observations of the failure surface*. Waterways Experiment Station Technical Report, US Army Corps of Engineers, Vicksburg, Mississippi.

Herd, D. G. and the Comite de Estudios Vulcanologies (1986). The 1985 Ruiz volcano disaster. *Eos* **67**, 457–460.

Hewitt, K. (1997). *Regions of Risk: A Geographical Introduction to Disasters*. Addison Wesley Longman, London.

High Point Rendel (1999). *Lyme Regis Environmental Improvements Preliminary Studies: Phases II & III Cobb Gate to Harbour Interpretative Geotechnical Reports*. West Dorset District Council.

High Point Rendel (2000). *The Holbeck–Scalby Ness Coastal Defence Strategy*. Report to Scarborough Borough Council.

High Point Rendel (2003). *Whitby Coastal Defence Strategy*. Report to Scarborough Borough Council.

Highways Agency (1997). *Design Manual for Roads and Bridges. Volume 13. Economic Assessment of Road Schemes, Section 2*. Highways Economic Note No. 2. Highways Agency.

Ho, K., Leroi, E. and Roberds, B. (2000). Quantitative risk assessment: application, myths and future direction. In *Proceedings of the Geo-Eng Conference, Melbourne, Australia*, Publication 1, 269–312.

Holling, C. S. (1979). Myths of ecological stability. In *Studies in Crisis Management* (eds G. Smart and W. Stanbury). Butterworth, Montreal.

Holling, C. S. (1986). The resilience of terrestrial ecosystems. In *Sustainable Development of the Biosphere* (eds W. Clark and R. Munn). Cambridge University Press, Cambridge.

Holz, R. D. and Schuster, R. L. (1996). Stabilization of soil slopes. In *Landslides: Investigation and Mitigation* (eds A. K. Turner and R. L. Schuster). Transportation Research Board, Special Report 247, National Research Council, National Academy Press, Washington DC, 439–473.

Hood, C. and Jones, D. K. C. (eds) (1996). *Accident and Design: Contemporary Debates in Risk Management*. UCL Press, London.

Hooke, J. M. and Kain, R. J. P. (1982). *Historical Change in the Physical Environment: a Guide to Sources and Techniques*. Butterworths, Sevenoaks.

Hooke, J. M. and Redmond, C. E. (1989). Use of cartographic sources for analysing river channel change with examples from Britain. In *Historical Change of Large Alluvial Rivers: Western Europe* (eds G. E. Petts, H. Moller and A. L. Roux). John Wiley and Sons, Chichester, 79–94.

Horowitz, J. K. and Carson, R. T. (1993). Baseline risk and preference for reductions in risk-to-life. *Risk Analysis* **13**, 457–462.

Hsü, K. J. (1975). On sturzstroms – catastrophic debris stream generated by rockfalls. *Bulletin of the Geological Society of America* **86**, 129–140.

Hsü, K. J. (1978). Albert Heim: observations on landslides and relevance to modern interpretations. In *Rockslides and Avalanches – 1, Natural Phenomena* (ed. B. Voight). Elsevier, New York, 71–93.

Huder, J. (1976). *Creep in Bundner Schist*. Norwegian Geotechnical Institute (Laurits Bjerrum Memorial Volume), 125–153.

Hughes, A., Hewlett, H., Samuels, P. G., Morris, M., Sayers, P., Moffat, I., Harding, A. and Tedd, P. (2000). *Risk Management for UK Reservoirs*. Construction Industry Research and Information Association (CIRIA) C542. London.

Hulme, M., Barrow, E. M., Jenkins, G. J., New, M., Osborn, T. J. and Viner, D. (1998). *Climate Change Scenarios for the UK Climate Impacts Programme*. UKCIP Technical Report. Norwich: Climatic Research Unit.

Hungr, O. (1997). Some methods of landslide hazard intensity mapping. In *Landslide Risk Assessment* (eds D. Cruden and R. Fell). Balkema, Rotterdam, 215–226.

Hungr, O. and McClung, D. M. (1987). An equation for calculating snow avalanche runup against barriers. In *Avalanche Formation, Movements and Effects*, IAHS Publication 162, 605–611.

Hungr, O., Morgan, G. C. and Kellerhals, R. (1984). Quantitative analysis of debris torrents for design of remedial measures. *Canadian Geotechnical Journal* **21**, 663–677.

425

Hungr, O., Evans, S. G. and Hazzard, J. (1999). Magnitude and frequency of rock falls along the main transportation corridors of southwestern British Columbia. *Canadian Geotechnical Journal* **36**, 224–238.

Hungr, O., Evans, S. G., Bovis, M. J. and Hutchinson, J. N. (2001). A review of the classification of landslides of flow type. *Environmental and Engineering Geoscience* **3**, 221–238.

Hutchinson, J. N. (1962). *Report on Visit to Landslide at Lyme Regis, Dorset.* Building Research Station Note c890.

Hutchinson, J. N. (1977). Assessment of the effectiveness of corrective measures in relation to geological conditions and types of movement. *Bulletin of the International Association of Engineering Geology* **16**, 131–155.

Hutchinson J. N. (1982). *Methods of Locating Slip Surfaces in Landslides.* British Geomorphological Research Group, Technical Bulletin, 30.

Hutchinson, J. N. (1987). Mechanisms producing large displacements in landslides on pre-existing shears. *Memoir of the Geological Society of China* **9**, 175–200.

Hutchinson, J. N. (1988). General report: Morphological and geotechnical parameters of landslides in relation to geology and hydrogeology. In *Landslides* (ed. C. Bonnard). Balkema, Rotterdam, 3–35.

Hutchinson, J. N. (1991a). The landslides forming the South Wight Undercliff. In *Slope Stability Engineering: Development and Applications* (ed. R. J. Chandler). Thomas Telford, London, 157–168.

Hutchinson, J. N. (1991b). Periglacial and slope processes. In *Quaternary Engineering Geology* (eds A. Forster, M. G. Culshaw, J. C. Cripps, J. A. Little and C. F. Moon). Geological Society Engineering Geology Special Publication 7, 283–334.

Hutchinson, J. N. (1992). Landslide hazard assessment. In *Landslides* (ed. D. H. Bell). Balkema, Rotterdam, 3, 1805–1841.

Hutchinson, J. N. (1995). The assessment of sub-aerial landslide hazard. In *Landslides Hazard Mitigation.* The Royal Academy of Engineering, London, 57–66.

Hutchinson, J. N. and Bromhead, E. N. (2002). Isle of Wight landslides. In *Instability – Planning and Management* (eds R. G. McInnes and J. Jakeways). Thomas Telford, London, 3–70.

Hutchinson, J. N. and Chandler, M. P. (1991). A preliminary landslide hazard zonation of the Undercliff of the Isle of Wight. In *Slope Stability Engineering: Development and Applications* (ed. R. J. Chandler). Thomas Telford, London, 197–205.

Hutchinson, J. N., Brunsden, D. and Lee, E. M. (1991). The geomorphology of the landslide complex at Ventnor, Isle of Wight. In *Slope Stability Engineering, Developments and Applications* (ed. R. J. Chandler). Thomas Telford, London, 213–218.

Ikeya, H. (1976). *Introduction to Sabo Works: the Preservation of Land against Sediment Disaster.* The Japan Sabo Association, Tokyo.

Institution of Chemical Engineers (IChemE) (1992). *Nomenclature for hazard and risk assessment in the process industries.* Rugby, IChem.

Institution of Civil Engineers (ICE) (1991). *Inadequate Site Investigation*. Report by the Ground Board of ICE, Thomas Telford, London.

Inter-Governmental Panel on Climate Change (IPCC) (1990). *Climate Change, the IPPC Scientific Assessment* (eds J. T. Houghton, G. J. Jenkins and J. J. Ephraums). Cambridge University Press, Cambridge.

Inter-Governmental Panel on Climate Change (IPCC) (1995a). *Climate Change 1995: The Science of Climate Change* (eds J. T. Houghton, L. G. Meira Filho, B. A. Callender, N. Harris, A. Kattenberg and K. Maskell). Cambridge University Press, Cambridge.

Inter-Governmental Panel on Climate Change (IPCC) (1995b). *Climate Change 1995: Impacts, Adaptations and Mitigation of Climate Change: Scientific-Technical Analysis* (eds R. T. Watson, M. C. Zinyowera and R. H. Moss). Cambridge University Press, Cambridge. A special report of IPCC Working Group II.

Inter-Governmental Panel on Climate Change (IPCC) (2001). *The IPCC third assessment report*. Summary for Policy Makers. Available at http://www.ipcc.ch/index.html.

International Association of Engineering Geology (IAEG) Commission on Landslides (1990). Suggested nomenclature for landslides. *Bulletin of the International Association of Engineering Geology* **41**, 13–16.

International Federation of the Red Cross and Red Crescent Societies (IFRCRCS) (1999). *World Disasters Report*. Edigroup, Chêne-Bourg, Switzerland.

International Pacific Salmon Fisheries Commission (1980). *Hell's Gate Fishways*. New Westminster, BC.

Ishihara, K. (1999). Liquefaction-induced landslide and debris flow in Tajikistan. In *Landslides of the World* (ed. K Sassa). Kyoto University Press, Kyoto, 224–226.

Jibson, R. W. (1992). The Mameyes, Puerto Rico, landslide disaster of October 7, 1985. In *Landslide/Landslide Mitigation* (eds J. A. Johnson and J. E. Slosson). Geological Society of America, Reviews in Engineering Geology, 9, 37–54.

Johnson, B. B. and Slovic, P. (1995). Presenting uncertainty in health risk assessment: initial studies of its effects on risk perception and trust. *Risk Analysis* **15**(4), 485–494.

Johnson, R. H. and Vaughan, R. D. (1983). The Alport Castles, Derbyshire: a South Pennine slope and its geomorphic history. *East Midlands Geographer* **8**, 79–88.

Jones, A. T. (1992). Comment on 'Catastrophic wave erosion on the southeastern coast of Australia: impact of the Lanai tsunami ca 105 ka?' *Geology* **20**, 1150–1151.

Jones, A. T. and Mader, C. (1996). Wave erosion on the southeastern coast of Australia: tsunami propagation and modelling. *Australian Journal of Earth Science* **43**, 479–483.

Jones, D. K. C. (1991). Environmental Hazards. In *Global Challenge and Change* (eds R. J. Bennett and R. C. Estall). Routledge, London, 27–56.

Jones, D. K. C. (1992). Landslide hazard assessment in the context of development. In *Geohazards: Natural and Man-made* (eds G. J. H. McCall, D. J. C. Laming and S. C. Scott). Chapman and Hall, London, 117–141.

Jones, D. K. C. (1993). Environmental hazards in the 1990s: problems, paradigms and prospects. *Geography* **79**, 339, 161–165.

Jones, D. K. C. (1995a). The relevance of landslide hazard to the International Decade for Natural Disaster Reduction. In *Landslides Hazard Mitigation*. The Royal Academy of Engineering, London, 19–33.

Jones, D. K. C. (1995b). Landslide hazard assessment. In *Landslides Hazard Mitigation*. The Royal Academy of Engineering, London, 96–113.

Jones, D. K. C. (1996). Anticipating the risks posed by natural perils. In *Accident and Design* (eds C. Hood and D. K. C. Jones). UCL Press, London, 14–30.

Jones, D. K. C. and Lee, E. M. (1994). *Landsliding in Great Britain*. HMSO, London.

Jones, D. K. C., Lee, E. M., Hearn, G. and Genc, S. (1989a). The Çatak landslide disaster, Trabzon Province, Turkey. *Terra Nova* **1**, 84–90.

Jones, D. K. C., Clark, A. R. and Lee, E. M. (1989b). The catastrophic landslide at Çatak, Turkey on 23 June 1988. *International Landslide News* **4**, 18–19.

Jones-Lee, M. W. (1989). *The Economics of Safety and Physical Risk*. Blackwell, Oxford.

Kaliser, B. and Fleming, R. W. (1986). The 1983 landslide dam at Thistle, Utah. In *Landslide Dams: Processes, Risk and Mitigation* (ed. R. L. Schuster). Geotechnical Special Publication No. 3, American Society of Civil Engineers, 59–83.

Kasperson, R. E. and Stallen, P. M. (eds) (1991). *Communicating Risk to the Public*. Kluwer Academic Press, Dordrecht.

Kates, R. W. (1978). *Risk Assessment of Environmental Hazard*. ICSU/SCOPE Report No. 8. John Wiley and Sons, Chichester.

Kauer, R., Fabbri, L., Giribone, R. and Heerings, J. (2002). Risk acceptance criteria and regulatory aspects. *Operation, Maintenance, Materials Issues* **1**(3), 1–11.

Keating, B. H. and McGuire, W. J. (2000). Island edifice failures and associated tsunami hazards. *Pure and Applied Geophysics* **157**, 899–955.

Keaton, J. R. and DeGraff, J. V. (1996). Surface observation and geologic mapping. In *Landslides: Investigation and Mitigation* (eds A. K. Turner and R. L. Schuster). Transportation Research Board, Special Report 247, National Research Council, National Academy Press, Washington DC, 178–230.

Keefer, D. K. (1984). Landslides caused by earthquakes. *Geological Society of America Bulletin* **95**, 406–421.

Khalili, N., Fell, R. and Tai, K. S. (1996). A simplified method for estimating failure induced deformation in embankments. In *Landslides* (ed. K. Senneset). Balkema, Rotterdam, 1263–1268.

Kiersch, G. A. (1964). Vaiont Reservoir disaster. *Civil Engineering* **34**, 3, 32–39.

King, J. P. (1996). *The Tsing Shan Debris flow*. GEO Special Report SPR 6/96.

King, J. P. (1997). *Natural Terrain Landslide Study: The Natural Terrain Landslide Inventory*. GEO Report No. 74.

King, J. P. (1999). *Natural Terrain Landslide Study: The Natural Terrain Landslide Inventory*. GEO Technical Note TN 1/97.

Kingdon-Ward, J. (1952). *My Hill So Strong*. Jonathan Cape, London.

Kingdon-Ward, J. (1955). Aftermath of the Great Assam Earthquake of 1950. *Geographical Journal* **121**, 290–303.

Knight, F. H. (1964). *Risk, Uncertainty and Profit.* Century Press, New York (originally published in 1921).

Knights, D. and Vurdubakis, T. (1993). Calculations of risk: towards an understanding of insurance as a moral and political technology. *Accounting, Organisations and Society* 18(7–8), 729.

Koirala, N. P. and Watkins, A. T. (1988). Bulk appraisal of slopes in Hong Kong. In *Landslides* (ed. C. Bonnard). Balkema, Rotterdam 2, 1181–1186.

Kong, W. K. (2002). Risk assessment of slopes. *Quarterly Journal of Engineering Geology and Hydrogeology* 35, 213–222.

Kovach, R. L. (1995). *Earth's Fury: An Introduction to Natural Hazards and Disasters.* Prentice-Hall, Englewood Cliffs, NJ.

Kuloshvili, S. I. and Maisuradze, G. M. (2000). Geological-geomorphological aspects of landsliding in Georgia (Central and Western Caucasus). In *Landslides: In Research, Theory and Practice* (eds E. N. Bromhead, N. Dixon and M-L. Ibsen). Thomas Telford, London, 861–866.

Kumamoto, H. and Henley, E. J. (1996). *Probabilistic Risk Assessment and Management for Engineers and Scientists.* Institute of Electrical and Electronics Engineers (IEEE) Press, New York.

Lacey, G. N. (1972). Observations on Aberfan. *Journal of Psychometric Research* 16, 257–260.

Lang, A., Moya, J., Corominas, J., Schrott, L. and Dikau, R. (1999). Classic and new dating methods for assessing the temporal occurrence of mass movements. *Geomorphology* 30, 33–52.

Lateltin, O. and Bonnard, C. (1999). Reactivation of the Falli-Holli landslide in the Prealps of Freiburg, Switzerland. In *Landslides of the World* (ed. K. Sassa). Kyoto University Press, Kyoto, 331–335.

Lee, E. M. (1992). Urban landslides: impact and management. In *The Coastal Landforms of West Dorset* (ed. R. Allison). Geologists Association Guide No. 47, 80–93.

Lee, E. M. (1995). Coastal cliff recession in Great Britain: the significance for sustainable coastal management. In *Directions in European Coastal Management* (eds M. G. Healy and J. P. Doody). Samara Publishing, Swansea, 185–194.

Lee, E. M. (1998). Problems associated with the prediction of cliff recession rates for coastal defence. In *Coastal Defence and Earth Science Conservation* (ed. J. M. Hooke). Geological Society Publishing, Bath, 46–57.

Lee, E. M. (1999). Coastal Planning and Management: The impact of the 1993 Holbeck Hall landslide, Scarborough. *East Midlands Geographer* 21, 78–91.

Lee, E. M. (2000). The management of coastal landslide risks in England: the implications of conservation legislation and commitments. In *Landslides: In Research, Theory and Practice* (eds E. N. Bromhead, N. Dixon and M-L. Ibsen). Thomas Telford, London, 893–898.

Lee, E. M. (2001). Geomorphological mapping. In *Land Surface Evaluation for Engineering Practice* (ed. J. S. Griffiths). Geological Society, London, Engineering Group Special Publication 18, 53–56.

Lee, E. M. (2002). A dynamic framework for the management of coastal erosion and flooding risks in England. In *Instability: Planning and Management* (eds R. G. McInnes and J. Jakeways). Thomas Telford, London, 713–720.

Lee, E. M. (2003). *Coastal Change and Cliff Instability: Development of a Framework for Risk Assessment and Management.* Unpublished PhD thesis, University of Newcastle upon Tyne.

Lee, E. M. and Brunsden, D. (2000). Coastal Landslides of Southern England: Mechanisms and Management. Post Conference Tour: Viii ISL Cardiff. Keynote Papers CD-ROM. Available from: e.bromhead@kingston.ac.uk.

Lee, E. M. and Clark, A. R. (2000). The use of archive records in landslide risk assessment: historical landslide events on the Scarborough coast, UK. In *Landslides: In Research, Theory and Practice* (eds E. N. Bromhead, N. Dixon and M-L. Ibsen). Thomas Telford, London, 904–910.

Lee, E. M. and Clark, A. R. (2002). *Investigation and Management of Soft Rock Cliffs.* Thomas Telford, London.

Lee, E. M. and Moore, R. (1991). *Coastal landslip potential assessment: Isle of Wight Undercliff, Ventnor.* Department of the Environment.

Lee, E. M., Moore, R., Brunsden, D. and Siddle, H. J. (1991a). The assessment of ground behaviour at Ventnor, Isle of Wight. In *Slope Stability Engineering: Developments and Applications* (ed. R. J. Chandler). Thomas Telford, London, 207–212.

Lee, E. M., Moore, R., Burt, N. and Brunsden, D. (1991b). Strategies for managing the landslide complex at Ventnor, Isle of Wight. In *Slope Stability Engineering: Developments and Applications* (ed. R. J. Chandler). Thomas Telford, London, 219–225.

Lee, E. M., Clark, A. R. and Guest, S. (1998a). An assessment of coastal landslide risk, Scarborough, UK. In *Engineering Geology: The View from the Pacific Rim* (eds D. Moore and O. Hungr). 1787–1794.

Lee, E. M., Moore, R. and McInnes, R. G. (1998b). Assessment of the probability of landslide reactivation: Isle of Wight Undercliff, UK. In *Engineering Geology: The View from the Pacific Rim* (eds D. Moore and O. Hungr). Balkema, Rotterdam, 1315–1321.

Lee, E. M., Brunsden, D. and Sellwood, M. (2000). Quantitative risk assessment of coastal landslide problems, Lyme Regis, UK. In *Landslides: In Research, Theory and Practice* (eds E. N. Bromhead, N. Dixon and M-L. Ibsen). Thomas Telford, London, 899–904.

Lee, E. M., Brunsden, D., Roberts, H., Jewell, S. and McInnes, R. (2001a). *Restoring Biodiversity to Soft Cliffs.* English Nature Report 398. Peterborough.

Lee, E. M., Hall, J. W. and Meadowcroft, I. C. (2001b). Coastal cliff recession: the use of probabilistic prediction methods. *Geomorphology* **40**, 253–269.

Lee, E. M., Meadowcroft, I. C., Hall, J. W. and Walkden, M. J. (2002). Coastal landslide activity: a probabilistic simulation model. *Bulletin of Engineering Geology and the Environment* **61**, 347–355.

Leone, F., Aste, J. P. and Leroi, E. (1996). Vulnerability assessment of elements exposed to mass movement: working towards a better risk perception. In *Landslides* (ed. K. Senneset). Balkema, Rotterdam **1**, 263–268.

Leroueil, S., Vaunat, J., Picarelli, L., Locat, J., Lee, H. and Faure, R. (1996). Geotechnical characterisation of slope movements. In *Landslides* (ed. K. Senneset). Balkema, Rotterdam **1**, 53–74.

Li, K. S. (1991). *Reliability Index for Probabilistic Slope Analysis.* Research Report No. R101, Department of Civil Engineering, University College, Australia Defence Force Academy, University of New South Wales.

Li, K. S. (1992). A point estimate method in calculating statistical moments. *Journal of Engineering Mechanics, ASCE* **118**, 1506–1511.

Li Tianchi, Schuster, R. L. and Wu Jishan (1986). Landslide dams in South-central China. In *Landslide Dams: Processes, Risk and Mitigation* (ed. R. L. Schuster). Geotechnical Special Publication No. 3, American Society of Civil Engineers, 146–162.

Lichtenstein, S., Slovic, P., Fischhoff, B., Laymen, M. and Combs, B. (1978). Judged frequency of lethal events. *Journal of Environmental Psychology: Human Learning and Memory* 4(6), 5512–581.

Linstone, H. A. and Truoff, M. (eds) (1975). *The Delphi Method; Techniques and Applications.* Addison-Wesley, Reading, MA.

Lipman, P. W. and Mullineaux, D. (eds) (1981). *The 1980 Eruptions of Mount St Helens.* US Geological Survey Professional Paper 1250.

Lipman, P. W., Normark, W. R., Moore, J. G., Wilson, J. B. and Gutmacher, S. E. (1988). The giant submarine Alika debris slide, Mauna Loa, Hawaii. *Journal of Geophysical Research* **93** B5, 4279–4299.

Lofstedt, R. E. and Frewer, L. (eds) (1998). *Risk and Modern Society.* Earthscan Publications Ltd, London.

McConnell, R. G. and Brock, R. W. (1904). Report on the Great Landslide at Frank, Alberta. In *Annual Report for 1903.* Department of the Interior, Ottawa, Canada.

McDonnell, B. A. (2002). Hazard identification and visitor risk assessment at the Giant's Causeway World Heritage Site, Ireland. In *Instability – Planning and Management* (eds R. G. McInnes and J. Jakeways). Thomas Telford, London, 527–534.

McGuire, W. J. (1995). Volcanic landslides and related phenomena. In *Landslides Hazard Mitigation.* The Royal Academy of Engineering, London, 83–95.

McGuire, W. J., Mason, I. and Kilburn, C. (2002). *Natural Hazards and Environmental Change.* Arnold, London.

McInnes, R. G. (2000). *Managing Ground Instability in Urban Areas: a Guide to Best Practice.* Cross Publishing, Isle of Wight.

McSaveney, M. J., Chinn, T. J. and Hancox, G. T. (1991). *Immediate Report – Mt Cook rock avalanche, 14 December 1991.* DSIR Geology and Geophysics, Lower Hutt.

McSaveney, M. J., Chinn, T. J. and Hancox, G. T. (1992). Mount Cook Rock Avalanche of 14 December 1991, New Zealand. *Landslide News* **6**, 32–34.

Mader, C. L. (2002). Modelling the 1958 Lituya Bay mega-tsunami, II. *Science of Tsunami Hazards* **20**, 5, 241–250.

Malone, A. W. (1998). Risk management and slope safety in Hong Kong. *Proceedings of Seminar on Slope Engineering in Hong Kong, Hong Kong.* Balkema, Rotterdam, 3–17.

431

Maquaire, O. (1994). Temporal aspects of the landslides located along the coast of Calvados (France). In *Temporal Occurrence and Forecasting of Landslides in the European Community* (eds R. Casale, R. Fantechi and J. C. Flageollet). Final Report Vol. 1, 211–234.

Maquaire, O. (1997). The frequency of landslides on the Normandy coast and their behaviour during the present climatic regime. In *Rapid Mass Movement as a Source of Climatic Evidence for the Holocene* (eds J. A. Matthews, D. Brunsden, B. Frenzel, B. Glaser and M. M. Weiss). Gustav Fischer Verlag, Stuttgart, 183–195.

Marin, A. (1992). *Costs and benefits of risk reduction. Appendix in Risk: Analysis, Perception and Management.* Report of a Royal Society Study Group, London, 192–201.

Mason, K. (1929). Indus floods and Shyock glaciers: the Himalayan Journal. *Records of the Himalayan Club, Calcutta* 1, 10–29.

Meadowcroft, I., Brampton, A. and Hall, J. (1997). Risk in an uncertain world – finding practical solutions. *Proceedings of the MAFF Conference of River and Coastal Engineers*, H.2.1–H.2.11. MAFF Publications, London.

Meadowcroft, I. C., Hall, J. W., Lee, E. M. and Milheiro-Olveira, P. (1999). *Coastal Cliff Recession: Development and Application of Prediction Methods.* HR Wallingford Report SR 549, Wallingford.

Merkhofer, M. W. and McNamee, P. (1982). *The SRI Probability Encoding Process: Experience and Insights.* Technical Report. SRI International, Menlo Park, CA.

Merriam, R. (1960). The Portuguese Bend landslide, Palos Verdes, California. *Journal of Geology* 68, 140–153.

Michael Leiba, M., Baynes, F. and Scott, G. (2000). Quantitative landslide risk assessment of Cairns, Australia. In *Landslides: In Research, Theory and Practice* (eds E. N. Bromhead, N. Dixon and M-L. Ibsen). Thomas Telford, London, 1059–1064.

Miller, D. J. (1960). *Giant Waves in Lituya Bay, Alaska.* Geological Survey Professional Paper 354-C, US Government Printing Office, Washington.

Miller, J. (1974). *Aberfan – a Disaster and its Aftermath.* Constable, London.

Ministry of Agriculture, Fisheries and Food (MAFF) (1993). *Project Appraisal Guidance Notes.* MAFF Publications, London.

Ministry of Agriculture, Fisheries and Food (MAFF) (1999a). *FCDPAG3 Flood and Coastal Defence Project Appraisal Guidance: Economic Appraisal.* MAFF Publications, London.

Ministry of Agriculture, Fisheries and Food (MAFF) (1999b). *High Level Targets for Flood and Coastal Defence and Elaboration of the Agency's Flood Defence Supervisory Duty.* MAFF, London.

Ministry of Agriculture, Fisheries and Food (MAFF) (2000). *FCDPAG4 Flood and Coastal Defence Project Appraisal Guidance: Approaches to Risk.* MAFF Publications, London.

Mitchell, R. C. and Carson, R. T. (1989). Using surveys to value public goods: the contingent valuation method. *Resources for the Future*, Washington DC.

Mompelat, P. (1994). *Unités cartographiques et évaluation de l'aléa mouvements de terrain en Guadeloupe (Antilles Françaises).* Thèse soutenue le 25 mars 1994 pour obtenir le titre de Docteur de l'Université Paris 6.

Montgomery, D. C. and Runger, G. C. (1994). *Applied Statistics and Probability for Engineers*. Wiley and Sons, New York.

Moon, A. T. (1997). Predicting low probability rapid landslides at Roxburgh Gorge, New Zealand. In *Landslide Risk Assessment* (eds D. Cruden and R. Fell). Balkema, Rotterdam, 227–284.

Moon, A. T., Olds, R. J., Wilson, R. A. and Burman, B. C. (1992). Debris flow zoning at Montrose, Victoria. In *Landslides* (ed. D. H. Bell). Balkema, Rotterdam 2, 1015–1022.

Moon, A. T., Robertson, M. and Davies, W. (1996). Quantifying rockfall risk using a probabilistic toppling failure model. In *Landslides* (ed. K. Senneset). Balkema, Rotterdam, 1311–1316.

Mooney, G. M. (1977). *The Valuation of Human Life*. Macmillan, London.

Moore, D. H. (1973). Evaluation of five discrimination procedures for binary variables. *Journal of the American Statistics Association* 68, 399.

Moore, J. G., Clague, D. A., Holcomb, R. T., Lipman, P. W., Normark, W. R. and Torresan, M. E. (1989). Prodigious submarine landslides on the Hawaiian Ridge. *Journal of Geophysical Research* 94 B13, 17465–17484.

Moore, J. G., Normark, W. R. and Holcomb, R. T. (1994). Giant Hawaiian landslides. *Annual Review of Earth and Planetary Sciences* 22, 199–144.

Moore, R., Lee, E. M. and Noton, N. (1991). The distribution, frequency and magnitude of landslide movements at Ventnor, Isle of Wight. In *Slope Stability Engineering: Developments and Applications* (ed. R. J. Chandler). Thomas Telford, London, 213–218.

Moore, R., Lee, E. M. and Clark, A. R. (1995). *The Undercliff of the Isle of Wight: a review of ground behaviour*. South Wight Borough Council.

Moore, R., Hencher, S. R. and Evans, N. C. (2001). An approach for area and site specific natural terrain hazard and risk assessment, Hong Kong. In *Geotechnical Engineering* (eds K. Ho and Li). Swets and Zeitlinger, Lisse, 155–160.

Moore, R., Lee, E. M. and Palmer, J. S. (2002). A sediment budget approach for estimating debris flow hazard and risk: Lantau, Hong Kong. In *Instability – Planning and Management* (eds R. G. McInnes and J. Jakeways). Thomas Telford, London, 347–354.

Morgan, G. C., Rawlings, G. E. and Sobkowicz, J. C. (1992). Evaluating total risk to communities from large debris flows. In *Geotechnique and Natural Hazards*, Bi Tech Publishers, Vancouver, 225–236.

Morgenstern, N. R. (1991). Limitations of stability analysis in geotechnical practice. *Geotechnia*, 61, 5–19.

Morgenstern, N. R. (1997). Towards landslide risk assessment in practice. In *Landslide Risk Assessment* (eds D. Cruden and R. Fell). Balkema, Rotterdam, 15–23.

Morgenstern, N. R. (2000). *Performance in Geotechnical Practice*. Inaugural Lumb Lecture, Transactions of the Hong Kong Institution of Engineers.

Mostyn, G. R. and Fell, R. (1997). Quantitative and semiquantitative estimation of the probability of landsliding. In *Landslide Risk Assessment* (eds D. Cruden and R. Fell). Balkema, Rotterdam, 297–315.

Mostyn, G. R. and Li, K. S. (1993). Probabilistic slope analysis – state-of-play. In *Probabilistic Methods in Geotechnical Engineering* (eds K. S. Li and S-C. R. Lo). Balkema, Rotterdam, 89–109.

Mueller, L. (1964). The rock slide in the Vaiont Valley. *Rock Mechanics and Engineering Geology* 2, 148–212.

Murray, J. G. (2000). The NZ landslide safety net. In *Landslides: In Research, Theory and Practice* (eds E. N. Bromhead, N. Dixon and M-L. Ibsen). Thomas Telford, London, 1075–1080.

Nash, D. (1987). A comparative review of limit equilibrium methods of stability analysis. In *Slope Stability* (eds M. G. Anderson and K. S. Richards). John Wiley and Sons, Chichester, 11–75.

National Research Council (1989). *Improving Risk Communication.* Committee on Risk Perception and Communication. National Academy Press, Washington DC.

Nguyen, V. U. and Chowdhury, R. N. (1984). Probabilistic study of spoil pile stability in strip coal mines – two techniques compared. *International Journal for Rock Mechanics, Mining Science and Geomechanics* 21, 303–312.

Nicoletti, P. G. and Sorriso-Valvo, M. (1991). Geomorphic controls of the shape and mobility of rock avalanches. *Geological Society of America Bulletin* 103, 1365–1373.

Nieto, A. S. and Schuster, R. L. (1999). Mass wasting and flooding induced by the 5 March 1987 Ecuador earthquakes. In *Landslides of the World* (ed. K. Sassa). Kyoto University Press, Kyoto, 220–223.

Noji, E. J. (1994). Earthquakes: casualty studies. In *Medicine in the International Decade for Natural Disaster Reduction (IDNDR)*. Royal Academy of Engineering, London, 18–20.

Olser, T. (1993). Injury severity scoring: perspectives in development and future directions. *American Journal of Surgery* 165(2A), 435–515.

O'Riordan, N. J. and Milloy, C. J. (1995). *Risk Assessment for Methane and Other Gases from the Ground.* Construction Industry Research and Information Association (CIRIA) Report 152. CIRIA, London.

Palmer, J. S. and Lee, E. M. (1999). *Tung Chung Natural Terrain Hazard and Risk Area Studies: Tung Chung East – Geomorphology and Engineering Geology Mapping.* Report to Halcrow Asia Partnership.

Palmer, J. S., Clark, A. R., Cliffe, D. and Eade, M. (2002). The management of risk on the chalk cliffs at Brighton, UK. In *Instability – Planning and Management* (eds R. G. McInnes and J. Jakeways). Thomas Telford, London, 355–362.

Palmer, L. (1977). Large landslides of the Columbia River Gorge, Oregon and Washington. In *Landslides, Reviews in Engineering Geology* (ed. D. R. Cruden). Vol. 3, Geological Society of America, Boulder, CO, 69–83.

Parfitt, J. P. (1992). Societal risk estimates from historical data for the United Kingdom and worldwide events. *AEA Technology*, May.

Parker, D. J., Green, C. H. and Thompson, P. M. (1987). *Urban Flood Protection Benefits: a Project Appraisal Guide.* Gower Press, London.

Pearce, D. W. *et al.* (1995). The social costs of climate changes: Greenhouse damage and the benefits of control. In *Climate Change 1995: Economic and Social Dimensions of Climate Change. Contribution of Working Group III to the Second Assessment Report of the IPCC* (eds J. P. Bruce, H. Lee and E. F. Haites). Cambridge University Press, Cambridge, 183–224.

Penning-Rowsell, E. C., Green, C. H., Thompson, P. M., Coker, A. M., Tunstall, S. M., Richards, C. and Parker, D. J. (1992). *The Economics of Coastal Management: a Manual of Benefits Assessment Techniques*. Belhaven Press, London.

Perry, A. H. (1981). *Environmental Hazards in the British Isles*. George Allen and Unwin, London.

Pethick, J. S. (1996). Coastal slope development: temporal and spatial periodicity in the Holderness cliff recession. In *Advances in Hillslope Processes* (eds M. G. Anderson and S. M. Brooks). John Wiley and Sons, Chichester **2**, 897–917.

Pethick, J. S. and Leggett, D. (1993). The geomorphology of the Anglian coast. In *Coastlines of the Southern North Sea* (eds R. Hillen and H. J. Vergagen). American Society of Civil Engineers, 52–56.

Petley, D. J. (1984). Ground investigation, sampling and testing for studies of slope stability. In *Slope Instability* (eds D. Brunsden and D. B. Prior). John Wiley and Sons, Chichester, 67–101.

Phipps, P. J. (2001). Terrain systems mapping. In *Land Surface Evaluation for Engineering Practice* (ed. J. S. Griffiths). Geological Society, London, Engineering Group Special Publication **18**, 59–61.

Phipps, P. J. and McGinnity, B. T. (2001). Classification and stability assessment for chalk cuttings: the Metropolitan Line case study. *Quarterly Journal of Engineering Geology and Hydrogeology* **34**, 4, 353–370.

Pierson, T. C. (1999). Rainfall-triggered lahars at Mt Pinatubo, Philippines, following the June 1991 eruption. In *Landslides of the World* (ed. K. Sassa). Kyoto University Press, Kyoto, 284–289.

Pierson, T. C., Janda, R. J., Umbal, J. V. and Daag, A. S. (1992). *Immediate and Long-term Hazards from Lahars and Excess Sedimentation in Rivers Draining Mt Pinatubo, Philippines*. US Geological Survey Water-Resources Investigations Report 92-4039.

Plafker, G. and Ericksen, G. E. (1978). Nevados Huascarán avalanches, Peru. In *Rockslides and Avalanches – 1 Natural Phenomena* (ed. B. Voight). Elsevier, Amsterdam, 277–314.

Plough, A. and Krimsky, S. (1987). The emergence of risk communication studies: social and political context. *Science, Technology and Human Values* **12**, 4–10.

Potter, H. R. (1978). *The Use of Historical Records for the Augmentation of Hydrological Data*. Institute of Hydrology Report No. 46. Wallingford.

Powell, G. (2002). Letters to the editor: Discussion 'Landslide Risk Management Concepts and Guidelines'. *Australian Geomechanics* **37**, 45–53.

Priest, S. D. and Brown, E. T. (1983). Probabilistic stability analysis of variable rock slopes. *Transactions of the Institution of Mining and Metallurgy Section A: Mining Industry* **92**, A1–A12.

435

Reeves, A., Ho, K. K. S. and Lo, D. O. K. (1999). Interim risk criteria for landslides and boulder falls from natural terrain. *Proceedings of the Seminar on Geotechnical Risk Management*, Geotechnical Division, Hong Kong Institution of Engineers, 127–136.

Regione Basilicata (1987). Gli Interventi per Senise, peril Consolidamento e il Transferimento di Insedimenti Abitati in Basilicata.

Reid, M. E. (1994). A pore pressure diffusion model for estimating landslide inducing rainfall. *Journal of Geology* 102, 709–717.

Rendel Geotechnics (1994). *Preliminary Study of the Coastline of the Urban Areas within Scarborough Borough: Scarborough Urban Area*. Report to Scarborough Borough Council.

Rezig, S. (1998). *Modélisation probabiliste de l'aléa Mouvements de terrain – Développement d'une éthode quantitative pour l'aide à l'expertise*. Thèse présentée le 30 octobre 1998 pour l'obtention du titre de docteur de l'Ecole Centrale de Paris. Spécialité: Mécanique des sols. Laboratoire d'accueil: Mécanique des sols, structures et matériaux.

Richter, C. F. (1958). *Elementary Seismology*. W. H. Freeman, San Francisco, CA.

Riggs, H. C. (1968). Frequency curves. In *Techniques of Water Resources Investigations of the United States Geological Survey*, Chapter A-2.

Roberds, W. L. (1990). Methods for developing defensible subjective probability assessments. *Transportation Research Record* 1288, 183–190.

Rosenbleuth, E. (1975). Point estimates for probability moments. *Proceedings of the National Academy of Science, USA* 72(10), 3812–3814.

Rosenbleuth, E. (1981). Two point estimates in probabilities. *Applied Mathematical Modelling* 5, 329–335.

Rozier, I. T. and Reeves, M. J. (1979). Ground movements at Runswick Bay, North Yorkshire. *Earth Surface Processes and Landforms* 4, 275–280.

Royal Society (1983). *Risk Assessment*. Royal Society, London.

Royal Society (1992). *Risk: Analysis, Perception and Management*. Report of a Royal Society Study Group. Royal Society, London.

RTM (1996). Isère: Inventaire des mouvements rocheux, Secteur de l'Y grenoblois, Service de Restauration des terrains en Montagne de l'Isère, Grenoble, France.

Sandilands, N. M., Noble, M. and Findlay, J. W. (1998). Risk assessment strategies for dam based hydro schemes. In *The Prospect for Reservoirs in the 21st Century*. Thomas Telford, London.

San Francisco Chronicle (1983). Highway 50 reopens and Tahoe rejoices. June 24, 2.

Sapir, D. G. and Misson, C. (1992). The development of a database on disasters. *Disasters* 16, 74–80.

Sapolsky, H. (1990). The Politics of Risk. *Daedalus* (Proceedings of the American Academy of Arts and Sciences) 119(4), 83–96.

Sassa, K. (1988). Geotechnical model for the motion of landslides. In *Landslides* (ed. C. Bonnard). Balkema, Rotterdam 1, 37–55.

Sassa, K. (1992). Landslide volume – apparent friction relationship in the case of rapid loading on alluvial deposits. *Landslide News* 6, 16–19.

Sassa, K. (1996). Prediction of earthquake induced landslides. In *Landslides* (ed. K. Senneset). Balkema, Rotterdam, 115–132.

Sassa, K. (ed.) (1999). *Landslides of the World*. Kyoto University Press, Kyoto.

Scheidegger, A. F. (1973). On the prediction of the reach and velocity of catastrophic landslides. *Rock Mechanics* 5, 231–236.

Schofield, J. (1787). *An historical and descriptive guide to Scarborough and its environs.* W. Blanchard, York.

Schrott, L. and Pasuto, A. (eds) (1999). Temporal stability and activity of landslides in Europe with respect to climate change (TESLEC). *Geomorphology*, Special Issue 30, Nos 1–2.

Schuster, R. L. (1983). Engineering aspects of the 1980 Mount St Helens eruptions. *Bulletin of the Association of Engineering Geologists* 20, 2, 125–143.

Schuster, R. L. (ed.) (1986). *Landslide Dams: Processes, Risk and Mitigation.* Geotechnical Special Publication No. 3, American Society of Civil Engineers.

Schuster, R. L. and Fleming, R. W. (1986). Economic losses and fatalities due to landslides. *Bulletin of the Association of Engineering Geologists* 23(1), 11–28.

Schuster, R. L. and Kockelman, W. J. (1996). Principles of landslide hazard reduction. In *Landslides: Investigation and Mitigation* (eds A. K. Turner and R. L. Schuster). Transportation Research Board, Special Report 247, National Research Council, National Academy Press, Washington DC, 91–105.

Schwarz, M. and Thompson, M. (1990). *Divided We Stand: Redefining Politics, Technology and Social Choice.* Harvester Wheatsheaf, Hemel Hempstead.

Selby, M. J. (1993). *Hillslope Materials and Processes.* Oxford University Press, Oxford.

Sellwood, M., Davis, G. M., Brunsden, D. and Moore, R. (2000). Ground models for the coastal landslides at Lyme Regis, Dorset, UK. In *Landslides: In Research, Theory and Practice* (eds E. N. Bromhead, N. Dixon and M-L. Ibsen). Thomas Telford, London, 1361–1366.

Shane, R. M. and Lynn, W. R. (1964). Mathematical model for flood risk evaluation. *ASCE Journal of the Hydraulics Division*, HY6 90, 1–20.

Siddle, H. J., Payne, H. J. and Flynn, M. J. (1987). Planning and development control in an area susceptible to landslides. In *Planning and Engineering Geology* (eds M. G. Culshaw, F. G. Bell, J. C. Cripps and M. O'Hara). Geological Society Engineering Geology Special Publication No. 4, 247–253.

Siddle, H. J., Jones, D. B. and Payne, H. R. (1991). Development of a methodology for landslip potential mapping in the Rhondda Valley. In *Slope Stability Engineering: Developments and Applications* (ed. R. J. Chandler). Thomas Telford, London, 137–142.

Sidle, R. C., Pearce, A. J. and O'Loughlin, C. L. (1985). *Hillslope Stability and Land Use.* American Geophysical Union, Washington DC.

Siebert, L. (1992). Threats from debris avalanches. *Nature* 356, 658–659.

Siegal, K. and Gibson, W. C. (1988). Barriers to the modification of sexual behaviour among heterosexuals at risk from acquired immune deficiency syndrome. *New York State Journal of Medicine* 14, 66–70.

Skempton, A. W. and Weeks, A. G. (1976). The Quaternary history of the Lower Greensand escarpment and Weald Clay vale near Sevenoaks, Kent. *Philosophical Transactions of the Royal Society, London* A283, 493–526.

Skempton, A. W., Leadbeater, A. D. and Chandler, R. J. (1989). The Mam Tor landslide, North Derbyshire. *Philosophical Transactions of the Royal Society, London* **A329**, 503–547.

Slovic, P. (2000). *The Perception of Risk*. Earthscan Publications Ltd, London.

Slovic, P., Fischhoff, B. and Lichtenstein, S. (1980). Facts and fears: understanding perceived risk. In *Societal Risk Assessment: How Safe is Safe Enough?* (eds R. Shwing and W. Albers). Plenum, New York, 181–214.

Smallwood, A. R. H., Morley, R. S., Hardingham, A. D., Ditchfield, C. and Castleman, J. (1997). Quantitative risk assessment of landslides: case histories from Hong Kong. In *Engineering Geology and the Environment* (eds P. G. Marinos, G. C. Koukis, G. C. Tsiambaos and G. C. Stournaras). Balkema, Rotterdam, 1055–1060.

Smith, K. (2001). *Environmental Hazards: Assessing Risk and Reducing Disaster. 3rd edition*. Routledge, London.

Soeters, R. and van Westen, C. J. (1996). Slope instability recognition, analysis and zonation. In *Landslides: Investigation and Mitigation* (eds A. K. Turner and R. L. Schuster). Transportation Research Board, Special Report 247, National Research Council, National Academy Press, Washington DC, 129–177.

Spiegelhalter, D. J. (1986). Uncertainty in expert systems. In *Artificial Intelligence and Statistics* (ed. W. A. Gale). Addison-Wesley, Reading, MA, 17–55.

Stalin Benitez, A. (1989). Landslides: extent and economic significance in Ecuador. In *Landslides: Extent and Economic Significance* (eds E. E. Brabb and B. L. Harrod). Balkema, Rotterdam, 123–126.

Starr, C. (1969). Social benefit versus technological risk. *Science* **165**, 1232–1238.

Stern, P. C. and Fineberg, H. V. (eds) (1996). *Understanding Risk: Informing Decisions in a Democratic Society*. National Academy Press, Washington DC.

Suleman, M. S., N'Jai, A., Green, C. H. and Penning-Rowsell, E. C. (1988). *Potential Flood Damage Data: a Major Update*. Flood Hazard Research Centre, Middlesex University.

Tatashidze, Z. K., Tsereteli, E. D. and Khazaradze, R. D. (2000). Principal hazard factors and mechanisms causing landslides (Georgia as an example). In *Landslides: in Research, Theory and Practice* (eds E. N. Bromhead, N. Dixon and M-L. Ibsen). Thomas Telford, London, 1449–1452.

Taype, V. (1979). Los desastres naturals como probleme de al defensa civil. *Bol Soc Geolog del Peru* **61**, 101–111.

Tavenas, F., Chagnon, J. Y. and LaRochelle, P. (1971). The Saint-Jean Vianney landslide: observations and eyewitness accounts. *Canadian Geotechnical Journal* **8**, 463–478.

Terlien, M. J. M. (1996). *Modelling Spatial and Temporal Variations in Rainfall-triggered Landslides*. International Institute for Aerospace and Earth Sciences (ITC) Publication 32, Enschede.

Terlien, M. J. M., De Louw, P. G. B., Van Asch, Th. W. J. and Hetterschijt, R. A. A. (1996). The assessment and modelling of hydrological failure conditions of landslides in the Puriscal area (Costa Rica) and the Manizalis Region (Colombia). In *Advances in Hillslope Processes* (eds M. G. Anderson and S. M. Brooks). John Wiley and Sons, Chichester, 837–855.

Terzaghi, K. (1950). Mechanisms of landslides. In *Application of Geology in Engineering Practice (Berkey Volume)* (ed. S. Paige). Geological Society of America, 83–123.

Thompson, M., Ellis, R. and Wildavsky, A. (1990). *Cultural Theory.* Westview, Boulder, CO.

Thomson, S. and Hayley, D. W. (1975). The Little Smokey Landslide. *Canadian Geotechnical Journal* **12**, 379–392.

Timmerman, P. (1981). *Vulnerability, Resilience and Collapse of Society.* Institute for Environmental Studies Environmental Monograph 1, Toronto.

Torrance, J. K. (1987). Quick clays. In *Slope Stability* (eds M. G. Anderson and K. S. Richards). John Wiley and Sons, Chichester, 447–474.

Turner, A. K. and Schuster, R. L. (eds) (1996). *Landslides: Investigation and Mitigation*, Transportation Research Board, Special Report 247, National Research Council, National Academy Press, Washington DC.

Tversky, A. and Kahneman, D. (1974). Judgements under uncertainty: heuristics and biases. *Science* **185**, 1124–1131.

UK Biodiversity Group (1999). Maritime cliff and slopes Habitat Action Plan. In *Action Plans, Volume V Maritime Habitats and Species*, 99–104.

University of Utah (1984). *Flooding and Landslides in Utah – an Economic Impact Analysis.* University of Utah Bureau of Economic and Business Research, Utah Department of Community and Economic Development and Utah Office of Planning and Budget, Salt Lake City, UT.

US Geological Survey (1989). Notes about the Armenian earthquake of 7 December 1988. *Earthquakes and Volcanoes* **21**, 68–78.

Valentin, H. (1954). Der landverlust in Holderness, Ostengland von 1852 bis 1952. *Die Erde* **6**, 296–315.

Van Asch, Th. W. J., Buma, J. and Van Beek, L. P. H. (1999). A view on some hydrological triggering systems in landslides. *Geomorphology* **30**, 25–32.

Van Dine, D. F., Jordan, P. and Boyer, D. C. (2002). An example of risk assessment from British Columbia, Canada. In *Instability – Planning and Management* (eds R. G. McInnes and J. Jakeways). Thomas Telford, London, 399–406.

Van Gassen, W. and Cruden, D. M. (1989). Momentum transfer and friction in the debris of rock avalanches. *Canadian Geotechnical Journal* **26**, 623–628.

Vanmarcke, E. H. (1977). Probabilistic modelling of soil profiles. *Journal of the Geotechnical Engineering Division, ASCE* **103**, 1227–1246.

Vanmarcke, E. H. (1980). Probabilistic stability analysis of earth slopes. *Engineering Geology* **16**, 29–50.

Varnes, D. J. (1978). Slope movement types and processes. In *Landslide: Analysis and Control* (eds R. L. Schuster and R. J. Krizek). Transportation Research Board, National Research Council, Washington DC, 11–33.

Varnes, D. J. (1984). *Landslide Hazard Zonation: a Review of Principles and Practice.* Engineering Geology Commission on Landslides and other Mass Movements on Slopes. UNESCO, Paris.

Varnes, D. J. and Savage, W. Z. (eds) (1996). *The Slumgullion Earth Flow: a Large-scale Natural Laboratory.* US Geological Survey Bulletin 2130.

Varnes, D. J., Smith, W. K., Savage, W. Z. and Poers, P. S. (1996). Deformation and control surveys, Slumgullion landslide. In *The Slumgullion Earth Flow: a Large-scale Natural Laboratory* (eds D. J. Varnes and W. Z. Savage). US Geological Survey Bulletin 2130, 43–50.

Vaughan, P. R. (1995). Possible actions to help developing countries mitigate hazards due to landslides. In *Landslides Hazard Mitigation*. The Royal Academy of Engineering, London, 114–122.

Voight, B. (ed.) (1978). *Rockslides and Avalanches – 1, Natural Phenomena*. Elsevier, Amsterdam.

Voight, B. (1990). The 1985 Nevado del Ruiz volcano catastrophe – anatomy and retrospection. *Journal of Volcanology and Geothermal Research* **42**, 151–188.

Voight, B. and Pariseau, W. G. (1978). Rockslides and avalanches: an introduction. In *Rockslides and Avalanches – 1, Natural Phenomena* (ed. B. Voight). Elsevier, Amsterdam, 1–67.

Vonder Linden, K. (1989). The Portuguese Bend landslide. *Engineering Geology* **27**, 301–373.

von Holstein, C. A. S. (1972). Probabilistic forecasting: an experiment related to the stock market. *Organisational Behaviour and Human Performance* **8**, 139–158.

Vrijing, J. K. and van Gelder, P. H. A. J. M. (1997). Societal risk and the concept of risk aversion. In *Advances in Safety and Reliability* **1**, 45–52.

Vulliet, L. (1986). *Modélisation des pentes naturelles en movement*. DSc thesis, École Polytechnique Fédérale de Lausanne.

Walkinshaw, J. (1992). Landslide correction costs on US State Highway systems. In *Transportation Research Record 1343*. TRB National Research Council, Washington DC, 36–41.

Weinberg, G. M. (1975). *An Introduction to General Systems Thinking*. John Wiley and Sons, New York.

Wesson, C. V. K. and Wesson, R. L. (1975). Odyssey to Tadzhik – an American family joins a Soviet Seismological Expedition. *US Geological Survey Earthquake Information Bulletin* **7**, 1, 8–16.

Whitman, R. V. (1984). Evaluating calculated risk in geotechnical engineering. *Journal of the American Society of Civil Engineers* **110**, 143–188.

Whitman, R. V. (1997). Acceptable risk and decision-making criteria. *Proceedings of the International Workshop on Risk-based Dam Safety Evaluation, Trondheim, Norway*.

Whittaker, M. (1984). *The Book of Scarborough Spaw*. Barracuda Press.

Wieczorek, G., Nishenko, S. P. and Varnes, D. J. (1995). Analysis of rock falls in the Yosemite Valley, California. In *35th US Symposium on Rock Mechanics* (eds J. J. Daemen and R. A. Schultz). Balkema, Rotterdam, 85–89.

Wilson, S. D. (1970). Observational data on ground movements related to slope instability. *Journal of the Soil Mechanics and Foundations Division*, ASCE **96**, SM4, 1521–1544.

Willows, R. I., Meadowcroft, I. C. and Fisher, J. (2000). *Climate Adaptation Risk and Uncertainty: Draft Decision Framework*. Environment Agency Report No. 21. London.

Wong, H. N. and Ho, K. K. S. (1996). Travel distance of landslide debris. In *Landslides* (ed. K. Senneset). Balkema, Rotterdam **1**, 417–423.

Wong, H. N. and Ho, K. (2000). Learning from slope failures in Hong Kong. Keynote Paper International Symposium of Landslides 2000, Cardiff, CD-ROM. Available from: e.bromhead@kingston.ac.uk.

Wong, H. N., Ho, K. K. S. and Chan, Y. C. (1997). Assessment of consequence of landslides. In *Landslide Risk Assessment* (eds D. Cruden and R. Fell). Balkema, Rotterdam, 111–149.

WP/WLI (International Geotechnical Societies' UNESCO Working Party for World Landslide Inventory) (1995). A suggested method of a landslide summary. *Bulletin of the International Association of Engineering Geology* **43**, 101–110.

Wu, T. H., Tang, W. H. and Einstein, H. H. (1996). Landslide hazard and risk assessment. In *Landslides Investigation and Mitigation* (eds A. K. Turner and R. L. Schuster). Transportation Research Board, Special Report 247, National Research Council, National Academy Press, Washington DC, 106–120.

Wyllie, D. C. and Norrish, N. I. (1996). Stabilization of rock slopes. In *Landslides: Investigation and Mitigation* (eds A. K. Turner and R. L. Schuster). Transportation Research Board, Special Report 247, National Research Council, National Academy Press, Washington DC, 474–506.

Yan, T. Z. (1988). Recent advances of quantitative prognoses of landslides in China. In *Landslides* (ed. C. Bonnard). Balkema, Rotterdam, 1263–1268.

Yin, K. L. and Yan, T. Z. (1988). Statistical prediction models for slope instability of metamorphosed rocks. In *Landslides* (ed. C. Bonnard). Balkema, Rotterdam, 1269–1272.

Yoshimatsu, H. (1999). A large rockfall along a coastal highway, in Japan. In *Landslides of the World* (ed. K. Sassa). Kyoto University Press, Kyoto, 203–204.

Yu, Y. F. and Mostyn, G. R. (1996). An extended Point Estimate Method for the determination of the probability of failure of a slope. In *Landslides* (ed. K. Senneset). Balkema, Rotterdam, 429–433.

Zaruba, Q. and Mencl, V. (1969). *Landslides and their Control*. Elsevier, Amsterdam.

Zêzere, J. L. (2000). Rainfall triggering of landslides in the Area North of Lisbon. In *Landslides in Research, Theory and Practice* (eds E. N. Bromhead, N. Dixon and M-L. Ibsen). Thomas Telford, London, 1629–1634.

Zêzere, J. L. (2002). Landslide susceptibility assessment considering landslide typology. A case study in the area north of Lisbon (Portugal). *Natural Hazards and Earth System Sciences* **2**, 73–82.

Zêzere, J. L., Ferreira, A. B. and Rodrigues, M. L. (1999a). Landslides in the North of Lisbon Region (Portugal): Conditioning and Triggering factors. *Physics and Chemistry of the Earth (Part A)* **24**, 10, 925–934.

Zêzere, J. L., Ferreira, A. B. and Rodrigues, M. L. (1999b). The role of conditioning and triggering factors in the occurrence of landslides: a case study in the area north of Lisbon (Portugal). *Geomorphology* **30**, 133–146.

Index

Page numbers in italics refer to charts and diagrams. Places mentioned in the index are in the United Kingdom unless otherwise stated.

Lightning Source UK Ltd.
Milton Keynes UK
19 March 2011

169541UK00001B/16/A